Into Tennessee and Failure

MERCER
UNIVERSITY

MERCER UNIVERSITY PRESS

Endowed by

TOM WATSON BROWN
and
THE WATSON-BROWN FOUNDATION, INC.

INTO TENNESSEE & FAILURE

JOHN BELL HOOD

STEPHEN DAVIS

MERCER UNIVERSITY PRESS
Macon, Georgia
2020

MUP/ H980

Published by Mercer University Press
1501 Mercer University Drive
Macon, Georgia 31207

25 24 23 22 21 20   9 8 7 6 5 4 3 2 1

Books published by Mercer University Press are printed on acid-free paper that meets the requirements of the American National Standard for Information Sciences—Permanence of Paper for Printed Library Materials.

Printed and bound in the United States.

This book is set in Adobe Caslon Pro (text) and Georgia (display).

Cover/jacket design by Burt&Burt.

ISBN   978-0-88146-720-8
CIP
Name: Davis, Stephen, 1948- author.
Title: Into Tennessee and failure : John Bell Hood / Stephen Davis.
Other titles: John Bell Hood
Macon, Georgia : Mercer University Press, [2019] | Includes
bibliographical references and index. |
Identifiers: LCCN 2020017057 | ISBN 9780881467673 (hardback)
Subjects: LCSH: Hood, John Bell, 1831-1879--Military leadership. |
Tennessee--History--Civil War, 1861-1865--Campaigns. |
Georgia--History--Civil War, 1861-1865--Campaigns. |
Alabama--History--Civil War, 1861-1865--Campaigns. | Confederate States of America. Army of Tennessee. | United States--History--Civil War, 1861-1865--Regimental histories.
Classification: LCC E476.8 .D38 2019 | DDC 973.7/3--dc23
LC record available at https://lccn.loc.gov/2020017057

Contents

Acknowledgments

Writing a book is like making a journey, and along the way for my previous *Texas Brigadier to the Fall of Atlanta* and this one, *Into Tennessee and Failure,* a lot of people helped me.

Sean Benjamin, Public Services Librarian at Tulane University's Howard-Tilton Memorial Library, provided me with a copy of Angela Winkler's *The Life and Character of Gen. John B. Hood* (Austin, c. 1885), a seldom-cited booklet. Sean also gave me access to the library's pristine copies of the *New Orleans Times* of May-June 1874, in which Hood's "Reply to Certain Statements Contained in Gen. Joseph E. Johnston's Book" appeared, a half-dozen years before these same articles resurfaced in the general's memoir, *Advance and Retreat.* Scholars, even Richard McMurry, have been unaware of Hood's articles in *Times.* I also thank Charles Nunez, president of the New Orleans Civil War Round Table, for his kind hospitality during my visit to the Crescent City.

I wish to thank Greg Biggs of Clarksville, Tenn., for sending me copy of several monographs in his possession, such as Houchens, "The Making of General John B. Hood: A Study in Command," as well as Governor Letcher's report on the cannon captured at Norfolk in April 1861. DeAnne Blanton at the National Archives in Washington confirmed the holdings of the Archives Reference Section's John Bell Hood Papers. I express appreciation also to Prof. Keith Bohannon of the University of West Georgia for transcripts of Gen. Stephen Lee's letters at Chapel Hill. Alinda Borell, Archivist at the Gilder Lehrman Institute of American History, helped me obtain the text of Hood's letter to James Chesnut, March 7, 1864. Its first page is pictured in Denkler, *Voices of the Civil War,* but otherwise the letter has not figured in the Hood literature.

Dr. David Bosshardt of Fort Oglethorpe, Georgia, first put me onto the location of the Hood division hospital at the Reed house near Chickamauga battlefield. Anthony DelRosario, Coordinator of the Research Help Desk, Howard-Tilton Memorial Library at Tulane, sent me articles in the *New Orleans Picayune* for January 8 and June 14, 1874. When I visited Mark Dunkelman at his home in Providence, he called my attention to Steve Newton's article on Resaca in *North & South.* Jamie Gillum of Franklin, Tenn. kindly transcribed the circular in the *Pulaski Citizen* of November 1866 when Hood asked his former officers to send him copy of their after-action reports from his campaigns.

Several years ago, Stephen M. Hood generously gave me photocopies of General Hood's several Telegrams Books, containing a number of missives not to be seen in the *Official Records*. In August 2013 "Sam" also provided me photocopies and typed transcriptions of Hood's personal papers that he had discovered in the Philadelphia attic of one of the general's great grandsons; since then these have now been published as *The Lost Papers* (2015).

Dr. Gordon L. Jones, Senior Military Historian at the Atlanta History Center, helped me find articles related to Hood in the postwar *Picayune*. Gordon also reminded me that the *Atlanta Intelligencer* of July 5, 1864, quipped about Joe Johnston being outflanked by Sherman all the way to the Gulf of Mexico. Decades ago Dennis Kelly, then at Kennesaw Mountain National Battlefield Park, gave me photocopy of the valuable war journal of W. L. Trask, an officer on Hardee's staff.

When I was researching the "Hood imbroglio" that worried Jeb Stuart after Gettysburg, John McClure, Reference Department Manager of the Virginia Historical Society, helped me with Stuart's letters to Flora. Two decades ago, Richard McMurry graciously gave me his transcription of Lt. Thomas Mackall's journal at William and Mary. The ever-helpful Helen Matthews, Atlanta History Center, verified Col. Allen Julian as author of the historical plaques placed in 1964 at Atlanta's Tanyard Branch Park. Thanks also to my friend Dr. Dent Purcell of Savannah for having given me copy of Pvt. Tress Smith's recollection of Schofield's retreat from Franklin.

Kathy Shoemaker, Research Services Associate Archivist at the Manuscripts, Archives, and Rare Book Library of Emory University, helped me find Hardee's letter to President Davis, June 22, 1864. My good friend Michael Shaffer of Kennesaw State University pointed me to Arthur Mason's little-known circular, dated July 7, 1864, instructing Johnston's infantry on how they were to fight in the unusual wood-and-dirt "Shoupades" on the north bank of the Chattahoochee.

We featured twenty-eight of Hal Jesperson's excellent maps in our first volume; here we are grateful to include five more samples of Hal's cartographic mastery.

I express my sincere gratitude for Robert D. Jenkins, Esq., for having read the manuscript text of *Texas Brigadier to the Fall of Atlanta: John Bell Hood*. And thank you, Sam Davis Elliott, Esq., for that breakfast in

Chattanooga after you read large sections of the present volume. You made a lot of good suggestions, and helped our narrative immensely.

It goes without saying that working with Dr. Marc A. Jolley, Director of Mercer University Press, has been an unmitigated delight. Marsha Luttrell, Publishing Assistant, has seamlessly ushered our manuscript through the editing and preparation processes. Mary Beth Kosowski, Marketing Director for the Press, has promoted our published work widely to both academic journals and popular magazines. Thanks, y'all!

And to my wonderful wife, Billie, I quote from the Beatles' "Michelle": *Je dirai les seulo mots que je sais (que) tu comprendaras, ma cherie: I love you, I love you, I love you!*

As I said, writing a book is like making a journey. Reading a book is journey, too. I hope yours with this one of mine proves worthwhile!

—Stephen Davis
Cumming, Georgia
February 2020

Preface

A few words about this work. First, this is essentially Part Two of a study of the Confederate generalship of John Bell Hood in 1864. Note that I'm not calling it a biography; Richard McMurry did that in 1982. Rather, I have sought to assess Hood's performance in the six months that he served as corps commander in the Army of Tennessee, February-July 1864, and the six months he served as commander of that army, July 1864-January 1865. In that effort I have kept a close eye on previous scholars' writing, especially Hood's three biographers, O'Connor (1949), Dyer (1950) and McMurry. To that extent my work has a bibliographical bent. I wish to thank Dr. Marc A. Jolley, Director of Mercer University Press, for having allowed me to take this approach, and also for having accepted my hefty manuscript, which we agreed should be split into the two volumes.

Texas Brigadier to the Fall of Atlanta: John Bell Hood (2019) offers a Prologue that follows Hood from youth in Kentucky to West Point and service in the old army, mainly in Texas. When war broke out, he volunteered for Confederate service and quickly distinguished himself for battlefield bravery. Promoted to colonel, then brigadier general, he took command of the Texas Brigade in Lee's army, further shining at Gaines' Mill, Second Manassas and Sharpsburg. He was made major general in October 1862, assuming divisional command. Hood was severely wounded at Gettysburg and again at Chickamauga, losing his right leg to amputation. When he returned to service it was in north Georgia as a lieutenant general and corps commander in Joseph E. Johnston's Army of Tennessee. There he performed well enough that when the Confederate government decided to relieve Johnston for his retreats back to Atlanta, President Davis named Hood to succeed him.

In his defense of the city, Hood fought three attacking battles, sent half his cavalry raiding into Sherman's rear and endured more than a month of bombardment. He ultimately failed to prevent the Federal capture of Atlanta, but I argue that he fully used every resource, met the government's expectations of him and probably held the city longer than Johnston would have. *Texas Brigadier* ends with Sherman's troops marching into Atlanta.

The title of my present work, *Into Tennessee and Failure,* signals my eyes-wide-open perspective on Hood's tragic Tennessee Campaign of November-

December 1864. When Hood's decision-making is defensible, I try to do so. For instance, I contend that when President Davis visited Hood at Palmetto, Georgia, in late September 1864, they agreed that Hood would not only lead the army into north Georgia, but that his ultimate objective would be Tennessee. I have done my best to marshal evidence on this point.

At the same time I am not blind to Hood's faults. After Davis appointed Gen. G. T. Beauregard to command the Military Division of the West, exercising at least nominal authority over him, Hood proved petulant, uncooperative, and at times downright insubordinate. I'm not sure historians have judged Hood sufficiently critically as to these flaws. Then, after the army headed into north Alabama, Hood's weaknesses as a logistician—which worried Beauregard so much—became painfully evident. He also seemed indecisive, shifting his army's crossing points of the Tennessee River more or less as circumstances dictated. It was not until November 21, after Forrest's cavalry had joined him, that the army got across.

For decades students have been writing about "the mystery of Spring Hill," so I have also given the events of November 29, 1864 considerable attention. Current consensus is that Major General Frank Cheatham bears chief blame for allowing John Schofield's troops to slip past the Confederates and join Thomas at Nashville. As for Hood's share of responsibility, I side with those who point to his evening retirement from the field after he had issued orders for a night attack.

My take on the bloody battle of Franklin has two prongs. First, Hood was determined to beat Schofield before he could retreat back to Thomas, and the most expeditious means of doing so was an all-out frontal attack. Second, in an extended comparison with Lee's decision for the charge of Pickett, Pettigrew and Trimble at Gettysburg, I try to show that Hood's attacking battle at Franklin stood at least as much chance of success to break the enemy line as Lee's.

One of the points I try to make with this campaign is that Hood in Tennessee faced fewer and fewer options. His decision to advance toward Nashville, dig in south of the city, and then hope for some kind of Fredericksburg-style defensive victory over George Thomas represented just about the best choice he had. Given his two-to-one numerical superiority and all the time he took to get things just right before he ordered his attack on December 15, Thomas' Federal victory at Nashville is not surprising.

I find fault with Hood not so much for his decision-making before Nashville, as his unflattering efforts afterward to hide the extent of the disaster from General Beauregard and the administration. After repeatedly requesting information on what he was hearing about the failed campaign, Beauregard became so angry that before he set out to inspect Hood's army personally in north Mississippi, he asked for and secured permission to relieve him of command.

Hood beat him, offering his resignation on January 13, 1865. Secretary Seddon snatched it up.

I am certainly not the first writer to judge that John Bell Hood was a victim of the "Peter Principle," the sociological precept that in an organization an individual tends to rise to his level of competence, but then be promoted to a sphere in which he proves *in*competent. To an extent, the fact of John Bell Hood's failed army generalship reflected as much the desperation of the Confederate government to do anything it could in 1864 to save the Southern nation's flagging fortunes, as it did its dwindling pool of human resources (talented generals) to achieve that end. Yet it cannot be denied that complicit in this sad story of Hood's pathetic record in Confederate history was his overweening ambition and desire for military promotion to the highest rank possible.

I once had a friend, Episcopal priest and psychiatrist, the Rev. Charles Fulghum, M.D. I still remember Charles' license plate here in Atlanta: HUBRIS. Father Fulghum, if only Bell Hood could have learned from you.

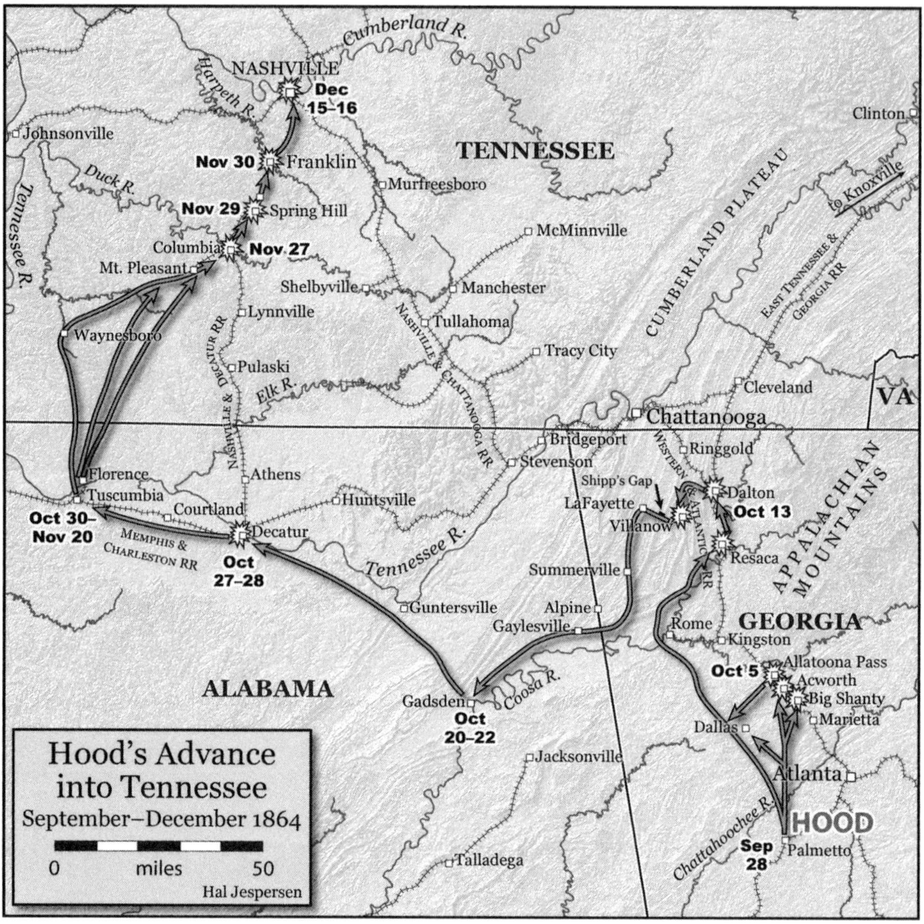

HOOD'S ADVANCE INTO TENNESSEE

September–December 1864

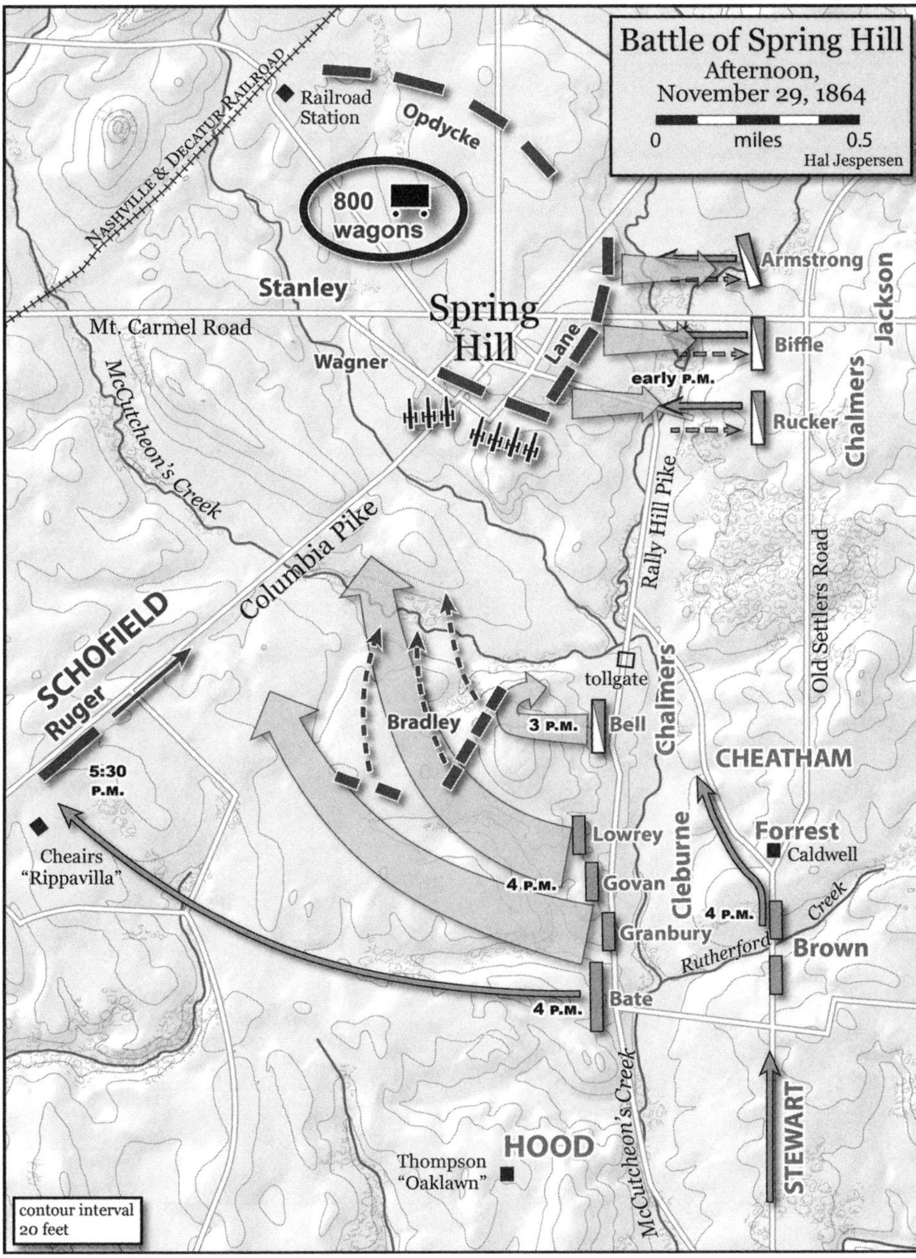

BATTLE OF SPRING HILL

November 29, 1864

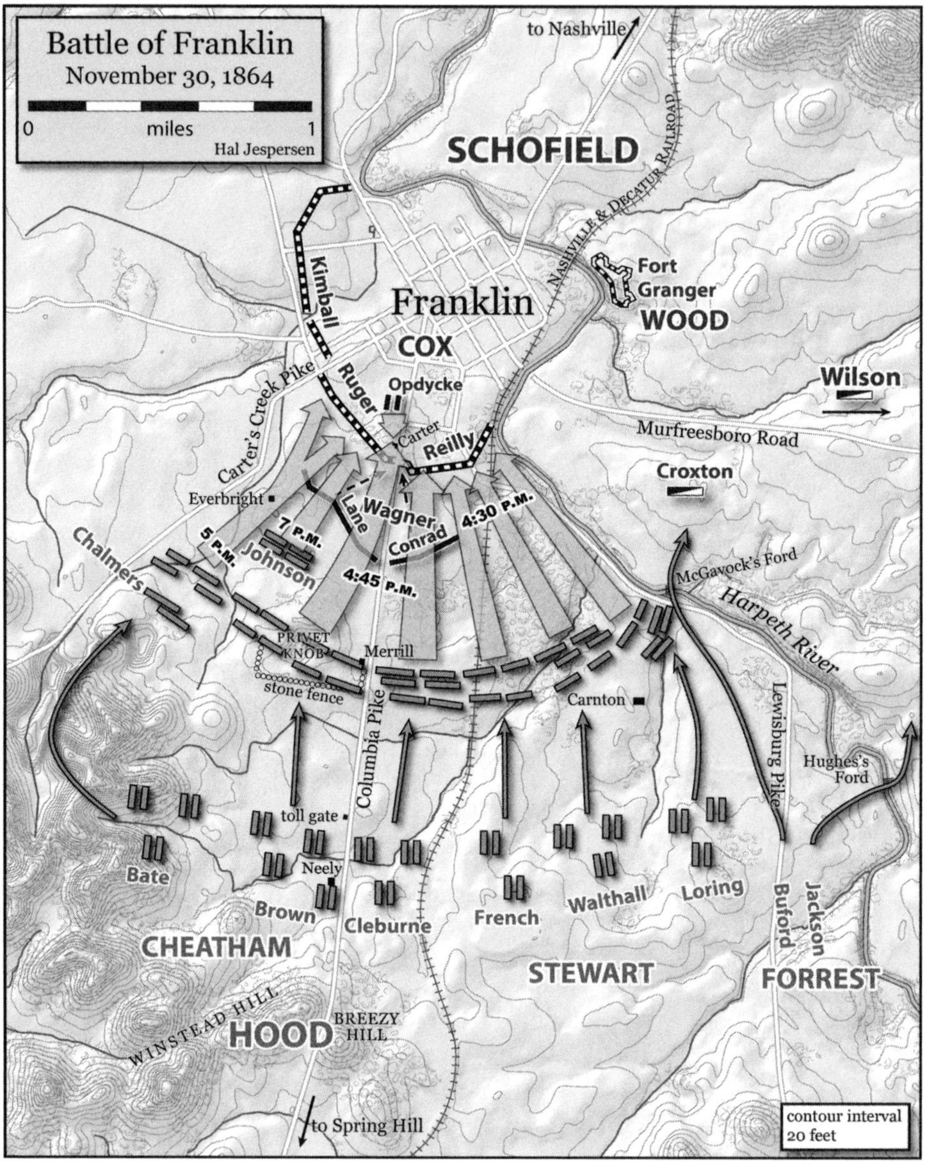

BATTLE OF FRANKLIN

November 30, 1864

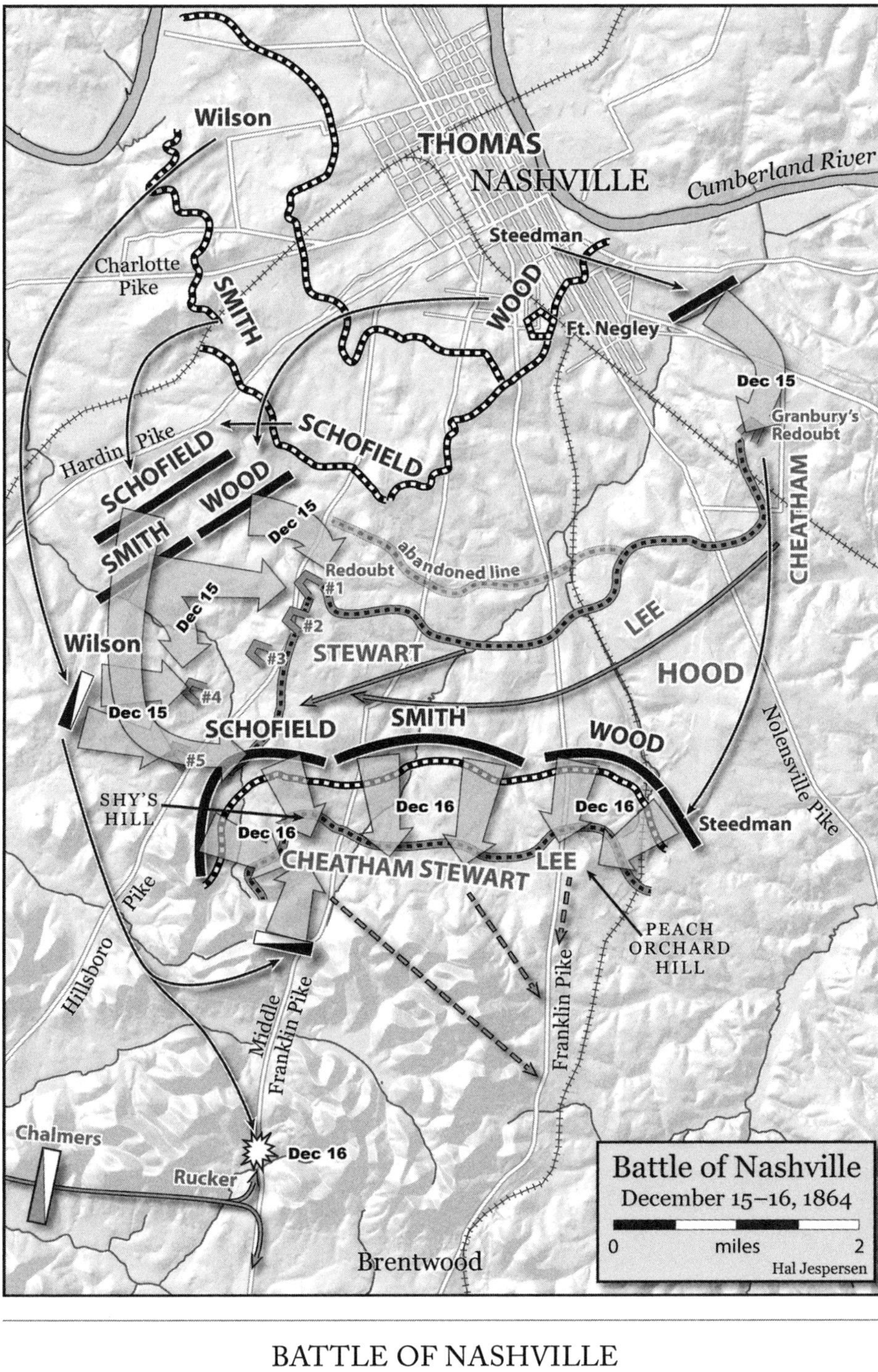

BATTLE OF NASHVILLE

December 15–16, 1864

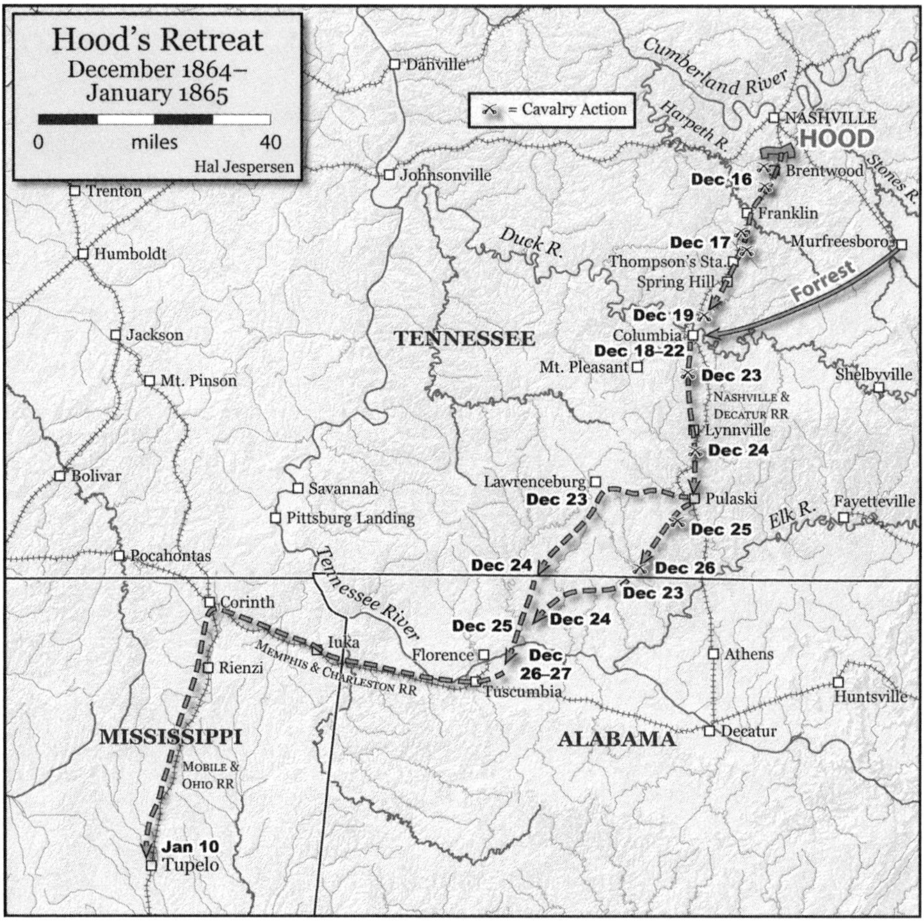

HOOD'S RETREAT

December 1864 – January 1865

Hood and the Tennessee Campaign

September 1864–January 1865

Chapter 1

September 1864

After Hardee's troops slipped away from Jonesboro on the night of September 1, they formed a new line to the south near Lovejoy's Station, six miles below Jonesboro. Hood, with Lee's and Stewart's Corps and the wagon train, headed there, marching all night September 1–2. Sherman ordered his infantry to feel Hardee's new line on September 2; two brigades advanced and were repulsed by "sweeping enfilading fire" of artillery, according to General Stanley. But then Sherman decided not to attack the Rebels further and ordered his six corps to begin withdrawing northward toward Atlanta. This allowed Hood's column (minus the militia, sent to Griffin), to reunite with Hardee's Corps. "I have often thought it strange," Hood later wrote, that Sherman failed to seize the advantage of "falling upon our main body on the march round to his rear." Hood himself arrived at Lovejoy's on the 3rd. The next day General Shoup could write, "Our army is all united at this point."[1]

---

[1] Shoup journal, entries of September 2–4, *OR*, ser. 1, vol. 38, pt. 3, 695 (henceforth, citations to *OR* refer to series 1 unless otherwise indicated); Stanley report, September 15, 1864, *OR*, vol. 38, pt. 1, 216; McMurry, *Atlanta 1864*, 176; Connelly, *Autumn of Glory*, 467 (on September 2, "Hood's army was spread over thirty miles between Atlanta, McDonough and Lovejoy's Station"); Shoup to Smith, September 3 (2:05 P.M.), *OR*, vol. 38, pt. 5, 1017; Hood, *A & R*, 208; Hood to Bragg, September 3, *OR*, vol. 38, pt. 5, 1017 (from Lovejoy's).

Otto Eisenschiml severely criticizes Sherman for letting Hardee escape September 2 to Lovejoy and doing nothing to prevent Hood and the other two corps from rejoining him there on September 3. "I fully believe that the battle of Lovejoy, which was not fought" he writes, "was more important than some crucial ones which were, for Hood's almost certain elimination or reduction to impotence would have shortened the war immeasurably" (Eisenschiml, "Sherman: Hero or War Criminal?" 29).

The fall of Atlanta was the subject of much conversation and correspondence across the Confederacy. General Hood's wire to Richmond prompted John B. Jones, the War Department clerk, to enter in his diary on the 4th, "Atlanta has fallen, and our army has retreated some thirty miles," calling the news "a stunning blow." "Our fondly cherished visions of peace have vanished like a mirage of the desert," he added. "This is a triumph for Johnston and his friends," observed Robert G. H. Kean, another War Department official. General Gorgas in Richmond recorded in his journal that day that the evacuation of Atlanta was "disastrous," even as he reasoned that "the mere acquisition of Atlanta by the enemy cannot be of much value to him now." A few days later he added that "the enemy have gained no great advantage." But to Gorgas the facts remained: First, Sherman's capture of Atlanta would embolden the war party in the North; "there will be little left for the peace party to stand on." And second, as he put it, "It is quite apparent that Hood besides being outnumbered is outgeneraled."

The Confederate Press Association issued a "Telegraphic" column on September 5 which sought to tamp down the impact of the loss of the city. "On the evening of the 30th ult. the enemy made a lodgment across the Flint river near Jonesboro," the short article stated. "We attacked them there on the evening of the 31st. ult. with two corps, but failed to

---

Errol Clauss is particularly critical of Sherman's allowing himself to be diverted from "the destruction of Hood's army" which, as he wrote in his *Memoirs*, "was the real object to be desired." Clauss sees in Sherman an indecision, a "vacillation and lack of drive" after Slocum's troops had entered Atlanta, as when he wrote late on the September 2 that he did not "care about punishing the enemy any farther." Sherman's "apparent inability to press for a final decision," Clauss concludes, was a key difference between him and Grant (Clauss, "Sherman's Failure at Atlanta," 323, 325, 327). Worse, Sherman seemed to blame George Thomas for his own inability. In a long letter to General Halleck, written on September 4, he acknowledged, "I ought to have reaped larger fruits of victory." But, he explained, "a part of my army is too slow"—echo of his earlier derision of how the Army of the Cumberland would halt at the sight of a plowed furrow (Sherman to Halleck, September 4, *OR*, vol. 38, pt. 5, 792).

Castel counts Sherman's failure to attempt to destroy Hardee's Corps at Lovejoy, or to intercept Stewart's and Lee's Corps as they marched from Atlanta, as one of Sherman's six command failures during the campaign. "Sherman, standing before the tottering wall of Confederate resistance in Georgia," he writes, "gave it a slight tap but then pulled back his hand" (*Decision in the West*, 541).

dislodge them. This made it necessary to abandon Atlanta, which was done on the night of the 1st inst." The P.A. column printed by the *Macon/Atlanta Intelligencer* on September 6 laconically filled in the particulars: "The Federals entered Atlanta in column by Peachtree road on Friday morning.... They took possession of the city very quietly and orderly, and citizens who remained were unmolested." In a later Press Association communiqué, the tone was even heartening: "While the fall of Atlanta is regretted, the army and people are not at all discouraged."

The news of Atlanta's fall brought comments from Southerners in the other major theater of the war. "Its loss is a heavy blow to us," wrote James Conner, a brigadier in Lee's army, "and at a time when everything looked bright and prosperous." He astutely observed, "it may and probably will have an effect on the election." The men in the trenches drew the same conclusion. "I am afraid that the fall of Atlanta will secure Lincoln's election," wrote a North Carolinian in Lee's army. General Lee mentioned it with his usual poise: "The fall of Atlanta is a blow to us," he wrote to his wife on September 5, "which is not very grievous & which I hope we will soon recover from." Edmund Ruffin, the fire-breathing Southern nationalist, viewed the event with unaccustomed equanimity. "Atlanta has been abandoned to the possession of the enemy," he wrote in his diary on September 5 from his home in Amelia County, Virginia. "Still, so far as yet made to the public our losses seem not to be important. Atlanta itself, under recent circumstances, was of little value." James Chesnut, Jr., who had served as a senior member of Jefferson Davis's staff, wrote the president on September 6 from Columbia, offering an optimistic opinion of the event:

> The recent and comparatively slight reverses in Georgia created, for a moment, great croaking and despondency among the weak kneed and faint hearted watchers of the hearth stone. But by the soldiers and by those who are constantly and earnestly engaged in maintaining the cause through good and through evil report, it was regarded with regret, it is true, and some anxiety but still more as a reason for increased effort—and manly resolution to do, and not to complain. It does appear to me that the real condition of the two armies is not much changed—except perhaps that the course of events may prove that Genl Sherman's difficulties may be increased.[2]

---

[2] Jones, *Rebel War Clerk*, 417; Kean, *Confederate Government*, 173; Escott, *After Secession*, 218 (quoting Jones); Gorgas, *The Civil War Diary of General Josi-*

Southern newspapers took the news of Atlanta's fall in their predictably diverse ways. "The loss of Atlanta is nothing," encouraged the *Richmond Examiner*, calling it "a trifling affair...puffed...out of all proportion" and harrumphing that it "may not be much more of a disaster than General Lee's evacuation of Hanover Junction." The editors added, "The prospects of the Confederacy were never brighter than at this moment, if she deserves success and is resolved to win it." Nevertheless, the *Examiner* voiced the obvious opinion that the fall of the city would affect the Northern presidential election. "The Democratick party would have been forever obliged to General Hood if he had managed to hold Atlanta for another fortnight," not to mention another two months. "In the nature of things reverses must come sometimes," philosophized the *Richmond Dispatch*. The *Richmond Sentinel* remarked, "All Yankeedoodledom is clapping hands, and huzzaing and flinging up caps, as though there was no longer a 'live rebel' in all America." Yet even while the North rejoiced over the news, it would demoralize the South only "if we are cowards and cravens.... It should only spur us on." Even if Atlanta's fall helped Lincoln to be re-elected, the *Fayetteville Observer* reasoned, so much the better, as a President McClellan would know how to prosecute the war against the South more efficiently.

On the other hand, the *Charleston Courier* admitted that "the news from Georgia...will, perhaps, depress and elongate the faces of some of our readers." Papers that had opposed Johnston's removal could not help but crow in the wake of Hood's defeat. "General Hood's egregious failures, it appears to us," declared the *Charleston Mercury*, "have proved, beyond question, that what was wanted by our army at Atlanta, was not a change of Generals, but reinforcements." Editors of the *Columbia South*

---

ah Gorgas, 139–40; "Telegraphic" (Richmond, September 4, and Lovejoy's, September 4), *Atlanta Intelligencer*, September 6; Risley, "The Confederate Press Association," 235; Andrews, *South Reports*, 462–63; Clauss, "Atlanta Campaign," 369 (Conner); Castel, *Decision in the West*, 546 ("Lincoln's re-election"); Dowdey and Manarin, *Lee Papers*, 851; Ruffin, *The Diary of Edmund Ruffin*, 3:555; Crist, et al., *Papers of Jefferson Davis*, 11:11–12 (Chesnut letter).

The famous diary entry, September 1, of General Chesnut's wife regarding the fall of Atlanta—"Atlanta gone. Well—that agony is over"—reinforces the assumption that in summer 1864 many Southerners had regarded the fall of Atlanta as a foregone conclusion. "Mr. Davis's mistake was to keep Johnston too long," she recorded under date of August 14. "Things are well-nigh hopeless" (Woodward, *Mary Chesnut's Civil War*, 636, 642).

Carolinian affirmed the obvious point that the conquest of Atlanta was a signal event "which the Republicans will strive to make useful in the approaching Presidential campaign." For this and other reasons, the *Mobile Advertiser & Register*'s correspondent in Montgomery wrote about the "sudden paralysis which fell upon the public mind" in the first capital of the Confederacy as news of the fall of Atlanta seeped out. Similarly, from Richmond, Peter W. Alexander surmised that neither Vicksburg not Gettysburg had brought about such a "painful impression" among the people as did the loss of Atlanta.

Editors of Georgia papers struggled to deal with the news. The *Intelligencer* in Macon used the occasion to issue another hortatory prediction.

> Whilst life lasts we have it in our hearts to shout defiance and hurl our hate at the accursed horde of Vandal foemen who haughtily confront us in their flush of victory and the pride of numbers. With the elasticity that peculiarly marks the character of Southern Soldiery, we will rapidly recover from the disasters of the past few months, and when once the tide of victory or success begins, we will push it on with a full flow of vengeance, that, we pray, may end in the destruction of the Northern host.

The *Columbus Times* tried to console its readers by reminding them that the cause had withstood the previous fall of New Orleans, Memphis, and other cities. "We are not conquered yet," it proclaimed. "Who ever heard of such a fuss being made over the fall of a twenty year old town," it asked, ignoring the fact that Columbus was not much older than that. On the other hand, the *Columbus Sun* admitted, "It is childish to say there is no cause for despondency." In similar style, the *Augusta Chronicle & Sentinel* acknowledged, "We have suffered a great disaster. We cannot conceal from ourselves the magnitude of the loss we have sustained in the fall of Atlanta." The *Savannah News* hopefully reminded everyone that General Hood's army was still intact and ready to continue the fight. The *Constitutionalist* of Augusta conceded, "The fall of Atlanta is a bitter pill to swallow," but added that the struggle must go on. "Our army is unwhipped and unconquerable," it declared; the struggle would continue, and the people must be "prepared to contest every foot of soil."

"Our good people have been thrown into consternation," the *Macon Telegraph* frankly observed on September 6. Then its editors offered a long view of the campaign and considered the issue of General Hood's culpability.

> This disastrous issue of the campaign in Georgia, opens wide the mouths of street corner critics.—Hood has been severely censured because he allowed himself to be flanked, and his connections cut, in a hopeless effort to hold on to Atlanta. But the enemy stole no march upon him, and as to flanking, there was the wide world opened to flank in, the moment the defensible country above the Chattahoochee had been abandoned. On the other hand, opposing critics are eloquent upon the faults of Johnston, and say the campaign was lost at the Allatoona Ridge. The great difficulty with both Generals, we suppose, has been that they could not multiply their forces by the figure 2. With twice as many men, no doubt they would have done well.

"But criticisms and regrets are now alike unavailing," the *Telegraph* concluded. "The enemy are coming down into the heart of Middle Georgia," and the city of Macon could be their next target. "It is difficult to exaggerate the critical nature of the military position, and all the energies [of] Government and people must be summoned to meet it at once."[3]

---

[3] Clauss, "Atlanta Campaign," 374–77 (*Examiner*, *Dispatch*, *Sentinel*, *Courier*, *South Carolinian*, *Intelligencer*, *Times*, *Sun*, *News*, *Constitutionalist*); Clauss, "The Battle of Jonesborough and the Fall of Atlanta," 128 (*Richmond Examiner* and "Hanover Junction"); D. E. Long, *The Jewel of Liberty*, 212 (*Examiner*, *Sentinel*, *Chronicle & Sentinel*); Miller, *Hood and Memory*, 136 (quoting *Charleston Mercury* of September 6); Escott, *After Secession*, 218 (*Examiner*); Moody, *Demon of the Lost Cause*, 28 (*Fayetteville Observer*); Andrews, *South Reports*, 463 (*Register*, Alexander); "Enemy's Advance in Georgia," *Macon Telegraph*, September 6; Castel, *Decision in the West*, 545 (quoting *Telegraph*).

Few Southern papers were as blunt in criticizing Johnston after his removal than (of all voices) the *New York Times*. "The rebels seem to have as much difficulty in finding a commander for their Western army as we had in finding one for the Army of the Potomac," began an editorial in the issue of July 23. Sidney Johnston and Beauregard had preceded Bragg as army commander; the latter was relieved because "he was forever on the retreat." No officer was thought able to surpass Bragg in that dubious regard, the *Times* opined, but Johnston had so proven himself.

> Since the beginning of May, JOHNSTON has retreated from twenty-nine impregnable positions; has evacuated a score of the mightiest mountain fastnesses on this continent, has forsaken three river lines of defence without fight, and has built gigantic earthworks all over Northern Georgia, only to fly from them when he heard the advancing tramp of our glorious army. He was only removed when he had

With the press in Georgia trying to keep the mood upbeat, how did the soldiers feel? Understandably many framed their opinions in terms of what their "Old Joe" would have done. Capt. Samuel Foster wrote of Hood, "He has virtually murdered 10,000 men around Atlanta trying to do that which Joe Johnston said could not be done," that is, to hold Atlanta with an outnumbered army. Dr. J. P. Cannon, surgeon for the 27th Alabama, wrote on September 4, "Although Gen. Johnston would probably have given up Atlanta, we believe he would have held it longer with much smaller loss and inflicted a heavier loss upon the enemy. Our army would now be stronger and in better condition." The doctor, probably like so many in the army, had mixed feelings about Hood. "We do not censure General Hood, for he is a brave and gallant officer and was placed in command for the express purpose of making a desperate struggle to hold the 'Gate City.'" Then Cannon commented that the government's decision to relieve General Johnston of command was "a fatal mistake." Hood's failure to hold Atlanta, after Johnston's removal, "intensified" some soldiers' disgust with the switch of commanders, wrote William P. Chambers. Captain Tom Key rued the loss of men and equipment in Hood's fight to save Atlanta: "I believe that if General Johnston had remained in command of this army there would have been 5,000 more men in it, and all these trains been saved." "Our troops are very much demoralized," wrote Lt. R. N. Colville to his father on September 7. "They do not have the confidence in Hood that they had in Johnston." Echoing was Missouri lieutenant George W. Warren, writing his brother on September 15. "There is great dissatisfaction in the army in regard to Gen Hood," he asserted. "It is observable to the humblest private that the army is not being handled as Gen. Johnston handled it. I believe that he could see just a month further ahead than Hood. At any rate we would be 15,000 men better off and an immense amount of property." J. M. Brannock, a surgeon with the army, concurred. "Sherman is more than a match for Hood, & has completely out-generaled him," he wrote his wife, Sarah, on September 12. "If President Davis could have been content to throw aside his personal prejudices & let Joe Johnston remain in command of this army we would still be in possession of Atlanta & Sherman with his disheartened troops falling back to-

---

reached our objective of Atlanta, and while he was preparing to retreat from that. ("Rebel Generals and Rebel Fortunes in Georgia," *New York Times*, July 23, 1864)

wards Chattanooga." Brig. Gen. Sul Ross also looked back ruefully. "When Johnston left this Army, it was the finest and best army I Ever heard of to the number," he wrote a friend on the 14th, "but now it is hacked and the men will not charge Breastworks, and is 20,000 men weaker now than when he left."[4]

"I hear many reports of desertions by wholesale, and of dissatisfaction and straggling," "Rover" reported from Hood's army, as early as September 2. Within the ranks, however, a number of Hood's soldiers seemed to shrug their shoulders over the loss of Atlanta. "Though beaten and outgeneraled," recorded Lieutenant William L. Trask in his journal, "we are still beligerent and rebellious. Most of our boys say: 'Well, we'll get 'em after a while.'" Col. Ellison Capers of the 24th South Carolina acknowledged the calamity of losing Atlanta but noted that his men appeared to accept the fact with greater equanimity than he expected. An Alabama soldier, James Hall, wrote home on September 4 that the army was "much less demoralized than you would suppose." Capt. S. H. Dent concluded about the loss of the city, "I do not think it renders our cause hopeless." Still, he complained that in the loss of Atlanta the army post office had been burned, "and no arrangements have been made for distributing the army mail." Captain Dent's remark leads one to recall McMurry's shrewd observation about soldiers' morale: "Some…seem to have been born pessimists. Many others were influenced by the events of the moment and their reactions depended on whether their suppers were adequate or they had received any mail that day."[5]

---

[4] Foster, *One of Cleburne's Command*, 129, 151 (diary entries of September 1, December 1); Miller, *Hood and Memory*, 134–35 (Cannon, Chambers); Cate, *Two Soldiers*, 129 (Key diary, entry of September 3); Daniel, *Soldiering*, 146 (Lieutenant Colville); Gottschalk, *In Deadly Earnest*, 403 (Warren); Bailey, *The Chessboard of War*, 12 (Brannock); Phillips, *Diehard Rebels*, 109 (Ross).

[5] "Rover," "The Evacuation of Atlanta," *Augusta Chronicle & Sentinel*, September 6; Trask diary, entry of September 1; Clauss, "Atlanta Campaign," 364 (Capers); Daniel, *Soldiering*, 90, 146 (Hall, Dent); McMurry, "Confederate Morale," 238–39.

The consequences of Atlanta's capture by Sherman continue to be debated by historians. The traditional argument states that the fall of Atlanta to Sherman's forces had immense political consequences, ensuring Abraham Lincoln's re-election. The idea is that if Robert E. Lee could hold Grant at bay before Richmond and Petersburg, and if Confederates could have held onto Atlanta, or defeated Sherman by early November 1864, the Democratic candidate, George

B. McClellan, could have won the presidency. Then, forced to honor his party's platform with its declaration that the war was a failure (the "Chicago platform," as some have called it), McClellan would have arranged an armistice or peace convention that would have recognized Southern secession and independence. Confederates themselves were much aware of the importance of the North's presidential election. As late as August 21, the Rebel war clerk John B. Jones recorded in his diary, "Lincoln and his party are now environed with dangers rushing upon them from every direction.... [E]verything depends upon the result of the Presidential election in the United States" (J. B. Jones, *A Rebel War Clerk's Diary* [2015], 2:243–44).

The hypothesis has been around for generations, fueled by the exultations of prominent Northerners after Sherman's victory. "We are going to win the Presidential election," joyfully predicted a New York journalist on September 6; "all divisions are going to be healed. I have never seen such a sudden lighting up of public mind as since the late victory at Atlanta" (Goodwin, *Team of Rivals*, 656). "Sherman and Farragut have knocked the bottom out of the Chicago platform," crowed Secretary of War Stanton, referring to the Union admiral's triumph at Mobile in August. Then came Sheridan's crushing victories over Jubal Early in the Valley. "Sheridan has knocked down gold and G. B. McClellan together," cheered George Templeton Strong, treasurer of the U. S. Sanitary Commission. "The former is below 200, and the latter is nowhere" (McPherson, *Battle Cry*, 775, 777–78). Larry E. Nelson is only one recent historian to accumulate further evidence showing how Confederates linked the fate of Atlanta, the election of a peace Democrat, and their hopes for Southern victory and independence. "Atlanta is, then, the great strategic point," an Augusta newspaper asserted just days before Sherman launched his campaign. "A crushing, decisive victory in Northwest Georgia will irretrievably crush the power of the enemy, and break down the war party of the North. A substantial victory now will lead to peace," the *Augusta Constitutionalist* declared on August 1. Two and a half months later, Sen. Ben Hill of Georgia still held out hope that Sherman could be defeated: if so, "Lincoln's power will be broken, and we shall speedily end the war on our own terms." But then Atlanta fell, and Southern hopes were dashed. John W. Graham, a North Carolinian serving in Lee's army, wrote home on September 5, "I am afraid that the fall of Atlanta will secure Lincoln's re-election and do much to prolong the war" (Nelson, *Bullets, Ballots, and Rhetoric*, 82, 119–20).

James M. McPherson helped set the stage for renewed discussion of the hypothesis by reminding us in 1992 of the importance of battlefield contingency. In his essay "American Victory, American Defeat," McPherson reviews the factors usually advanced as cause of Confederate defeat and finds them overstated. Northern superiority in numbers, logistics or leadership, Southern internal dissent or weakness of war-will: none of these, even applied together, meant that

Northern victory in the Civil War was inevitable. Rather, events on the battlefield, McPherson argues, determined the course and outcome of the war. Key turning points—the events of autumn 1862 and summer 1863—contributed mightily to the eventual Northern victory. So did Sherman's capture of Atlanta, coupled with Sheridan's victories in the Shenandoah Valley in September and October 1864. These latter events proved to be turning points as well, but they were scarcely inevitable. "Nor was Sherman's capture of Atlanta," McPherson argues, "any more inevitable than, say, McClellan's capture of Richmond in June 1862 had been." He concludes, "to understand why the South lost, in the end, we must turn from large generalizations that imply inevitability and study instead the contingency that hung over each military campaign, each battle, each election, each decision during the war" (McPherson, "American Victory, American Defeat," 40–42).

Albert Castel gave the traditional hypothesis renewed currency in *Decision in the West* (1992), in which he concludes that Sherman's capture of Atlanta sealed Lincoln's triumph at the polls. In late August 1864, a pessimistic Lincoln virtually predicted his electoral defeat while Jefferson Davis hoped for Confederate victory through political events: "all that the South needed to do to achieve it is to hold fast a while longer—just six more weeks—until the Northern elections get under way" (*Decision in the West*, 480). But then Atlanta fell, Lincoln was reelected, and the South's hopes for independence were dashed.

Castel reinforced the contention in "The Atlanta Campaign and the Presidential Election of 1864: How the South Almost Won by Not Losing," published in a collection of his articles in 1996. "There is, of course, no way of knowing how the election of 1864 would have gone had Hood had managed to keep Sherman out of Atlanta," he concludes, "until the voters of the North went to the polls or what would have occurred had McClellan become president."

> The historian can and should, however, point to two facts that are definite and indisputable: (1) That the fall of Atlanta, taking place when it did, turned what Republicans and Democrats alike had perceived as certain victory for McClellan in the upcoming presidential election into foregone defeat; and (2) by so doing, the fall of Atlanta destroyed the South's hope, its last realistic one, that it could gain its independence by depriving the North of its will to go on with the war. This is why the Atlanta Campaign was the final, decisive turning point of the Civil War: it assured Northern victory by defeating the South's attempt to win by not losing. (Castel, "The Atlanta Campaign and the Presidential Election of 1864," 29–30)

David E. Long considered the matter in his scholarly study of the 1864 presidential campaign. One aspect of the traditional hypothesis involves whether Union forces, after McClellan's electoral victory, could have launched operations against Confederate forces and crushed them before Lincoln left office in early March 1865. Long concludes, "the prospect of a forced offensive during the 1864–65 winter that would compel unconditional Confederate surrender was highly unlikely." On the key question—whether President McClellan would have repudiated the peace plank of the Democratic platform and continued to prosecute the war, Long shrugs, "there is conflicting evidence."

Long further points out the problems a new President McClellan would have had working with Congress. In the Senate, Republicans held a veto-proof majority, and would continue to, under Congressional procedure of the time, until December 1865. Had McClellan acted against its wishes, Long figures, "the Republican majority could have attempted impeachment, even if high crimes or misdemeanors had to be contrived" (D. E. Long, *The Jewel of Liberty*, 267–68).

Stephen W. Sears took a deep breath and traced McClellan's thinking on the Democrats' peace plank by reading the six drafts of his letter accepting the party's nomination. By the time he was finished, September 8, his letter, as Sears puts it, "accepted the Democratic presidential nomination on his own terms, and they never included, in thought or deed, the acceptance of peace at any price." A year after the election McClellan wrote a friend, "I can't tell what the secesh expected to be the result of my election—but if they expected to gain their Independence from me they would have been woefully mistaken" (Sears, "McClellan and the Peace Plank of 1864," 64).

In 1996 William C. Davis also argued strongly against the traditional hypothesis in "The Turning Point That Wasn't: The Confederates and the Election of 1864." "The notion of Lincoln's loss leading directly to Confederate independence is but a fable," he asserts. First, even as lame-duck president, Lincoln would have ordered Grant to press a winter campaign against Lee and Richmond. Grant would have brought Sherman and the bulk of his army to Virginia to help him. "Lee, his army, and Richmond as a result would inevitably have fallen before March 4," Davis reasons. And even if that had not happened by the new president's inauguration, McClellan's position all along was that peace could only come with the Southern states returning to the Union—a condition which Jefferson Davis would never agree to. As a result of all of these factors, "the conclusion that Confederate independence would or even might have been the result" of Lincoln's defeat in November 1864, Davis concludes, "becomes at best inconceivable, if not ridiculous" (*The Cause Lost*, 139–41).

The discussion continues. In his study of the 1864 election, John C. Waugh restates the traditional hypothesis: "if the Confederates on the battlefield could cause the fighting to go against Northern arms right up to election day, Lincoln might lose and the South might have peace on its terms." In their hope for the election of a peace Democrat, Waugh emphasizes Confederates' wishful thinking about the strength of the Northern peace faction and prevalence of war-weariness among the people of the North. "Whether it was true or not," the author concludes, "the South wanted to believe it" (Waugh, *Reelecting Lincoln*, 150–51).

Larry J. Daniel offered a rebuttal to Castel's argument in an article for *North & South* in 1998. "Atlanta, as a factor in the [Northern presidential] campaign, has been overplayed," he argues. Besides other battlefield victories (e.g., Farragut at Mobile Bay), Lincoln and the Republicans had strengths going into the election that a worried Lincoln overlooked. As proof, Daniel points out that in all but two cases, the states which Lincoln carried in 1864 were the same ones he carried in 1860 (in 1864 he also won Maryland and Missouri). "The capture of Atlanta, though significant, was thus far from being the turning point of the Civil War," Daniel concludes, adding that "there is much evidence that...Lincoln would have won anyway." Finally, as for the speculation around President George B. McClellan, to suggest that the former commander of the Army of the Potomac would turn his back on the sacrifices of thousands of Union soldiers and their families is to be naïve. "Hurl whatever insult you will at McClellan, call him an egotist, inept, or a moron," Daniel contends, "but at heart he was a patriot" (L. J. Daniel, "The South Almost Won by Not Losing," 47, 50–51).

Historians' discussion of the political impact of the fall of Atlanta is far from over. Richard McMurry in 2000 will thus not have the last word in his consideration of the issue. "In arguing that Lincoln had to have military success (or perceived success) in 1864 to win reelection, Castel was correct," he acknowledges. But McMurry finds a way also to agree with Daniel that Sherman's capture of Atlanta was not the military success that helped Lincoln win. Rather, it was Joe Johnston's failure to hold back Federal forces in early May at Snake Creek Gap. After Johnston showed he lacked the will to fight hard and Sherman showed he could maneuver around the Rebels' flanks, the conquest of north Georgia was the victory the Republicans needed for the election. "Absent a smashing Confederate military victory in Georgia after May 8, Lincoln's reelection was assured" (McMurry, *Atlanta 1864*, 207–208).

More recently, Steven Woodworth repeats the point that President McClellan would have faced a very pro-war Republican Congress. Even if President McClellan did not decide to collapse the war effort with an armistice bid, Woodworth suggests that "it's almost unimaginable that the Confederacy could have maintained its independence...through another two [years] under the staunchly Republican Congress northern voters had elected" (Woodworth, introduction to *The Tennessee Campaign of 1864*, 1).

McMurry recognizes that speculation about a McClellan presidency veers into counterfactualism, the currently popular realm of "what if?" history. The genre is not new. Joel Chandler Harris wrote a short story about Confederates kidnapping Abraham Lincoln; Winston Churchill wrote about Lee winning the

Soldiers of the Army of Tennessee would not have been encouraged had they known what their commanding general was saying about them to the authorities in Richmond. On September 4 Hood composed for General Bragg a review of the last days of the campaign, from August 26. As the enemy had pushed southward, Hood had shifted his strength to the left, leaving only enough troops in the city's fortifications to contend with the corps or more that Sherman had left at the Chattahoochee. On the 29th he had concluded that "two or three of his smaller corps" were marching on Jonesboro. Hood's statement is true concerning Howard's XV, XVI, and XVII, but it overlooks that two of the enemy's larger corps (IV and XIV) were advancing well north of Jonesboro. He barely mentions that "the road was cut by the enemy at Rough and Ready"—more "Jonesboropia." Hood then explained that he had sent Hardee's and Lee's Corps to drive the enemy back on August 31, but that the assault failed. "The whole affair proved entirely fruitless," he explained about the attacking battle he had ordered, "and my expectations, which I still believe entirely just, were terribly disappointed."

Here is what would have disappointed his officers and men:

> I can with justice blame no one with this failure. It seems the troops had been so long confined to trenches and had been taught to believe that intrenchments cannot be taken, so that they attacked without spirit and retired without proper effort.... I am officially informed that there is a tacit if not expressed determination among the men of

---

battle of Gettysburg. In our time the fate of Atlanta is the subject of conjecture in a few works. Robert Skimin's *Gray Victory* (1988) has Davis not firing Joe Johnston, who holds onto Atlanta past the Northern elections; McClellan is elected and an armistice is soon declared. Surely the wildest imaginative foray is Harry Turtledove's *The Guns of the South* (1992), in which time-travelers transport AK-47s back to the Confederacy, arming Lee's and Joe Johnston's armies. "General Johnston has held General Sherman at Rocky Face Ridge," reads a dispatch from Georgia, "with heavy losses on the Federal side, and then again at Resaca and Snake Creek Gap, when he tried to use his superior numbers to outflank us. Sherman's forces are now halted; prisoners report he dares not seek to outflank us again for fear of the casualties he would sustain from our rifles" (Harris, *On the Wing of Occasions Being the Authorised Version of Certain Curious Episodes of the Late Civil War*; Churchill, "If Lee Had Not Won the Battle of Gettysburg," 587–97; Skimin, *Gray Victory*; Turtledove, *The Guns of the South*, 146).

> this army, extending to officers as high in some instances as colonel, that they will not attack breast-works.

To pile on further, Hood wired Bragg, "our loss on the evening of the 31st was so small that it is evident that our effort was not a vigorous one." About the same time, he sent another wire from Lovejoy's: "To let you know what a disgraceful effort was made by our men in the engagement of August 31, I give you the wounded in the two corps: Hardee's corps, 539; Lee's 946; killed a very small number." Then, on September 6 came this zinger, addressed to the president, no less: "According to all human calculations we should have saved Atlanta had the officers and men of the army done what was expected of them."[6]

Hood had claimed that he could "blame no one with this failure," but he did precisely that. On September 13 he wrote President Davis that just as Hardee had to be blamed for the want of success at Peachtree Creek and on July 22, "our failure on August 31 I am now convinced was greatly owing to him." Hardee had earlier sought to be transferred from Hood's command and Hood had endorsed it, but the president had not approved. Now Hood tried again. "It is of the utmost importance that Hardee should be relieved at once," Hood implored Davis. "He commands the best troops of this army. I must have another commander." On the 13th Hood also telegraphed General Bragg. "Sent a dispatch to the President in cipher. Desire you to see it. You know something of the facts," he stated, referring evidently to the situation surrounding Hardee. Nevertheless, he closed, "Will do all in my power with resources in my hands."

A few days before, Hood had returned to the idea Hardee had himself favored, that of swapping his disgruntled corps commander with a department commander farther west. On September 8, Hood requested of Bragg that "Lieutenant-General Taylor be ordered to relieve General Hardee, bringing with him all the troops he can." Bragg's failure to reply prompted Hood to write the president directly, asking him on the 17th to "please appoint Cheatham to command Hardee's corps." (If Hardee were relieved, Frank Cheatham would replace him by seniority.) "The change will promote the efficiency of the army," he added. Again,

---

[6] Hood to Bragg, September 4, *OR*, vol. 52, pt. 2, 729–30; to Bragg, September 3, *OR*, vol. 38, pt. 5, 1016; to Bragg, September 5, *OR*, vol. 38, pt. 5, 1021; to Davis, September 6, *OR*, vol. 38, pt. 5, 1023.

Hood's request would not be acted upon by Davis—at least for the time being.[7]

Hood's censure of others did not stop with Hardee. In the army's reserve ordnance train at least 12,000 artillery shells, ten field pieces and nineteen caissons with limbers (not counting stores listed as in damaged condition) had been lost in the evacuation. Hood blamed his chief quartermaster, Col. M. B. McMicken. "Owing to the wanton neglect of the chief quartermaster of this army," he wrote to Bragg on September 4, "a large amount of ammunition and railroad stock had to be destroyed at Atlanta. He had more than ample time to remove the whole and had repeated instructions." Hood added vindictively, "I am reliably informed that he is too much addicted to drink of late to attend to his duties," before asking for a replacement officer. Hood's chief of ordnance, Lt. Col. J. M. Kennard, also came in for some of the blame. Soon McMicken, Kennard and even chief of staff Shoup asked for an investigation. Hood wrote Richmond that "Brigadier General Shoup is in no manner to blame for the loss of the cars, ordnance, &c., at Atlanta, Ga." The commanding general's endorsement did not assure Shoup; he resigned as chief of staff on September 14.[8]

---

[7] Hood to Davis and to Bragg, September 13, *OR*, vol. 39, pt. 2, 832; Hood to Bragg, September 8 (2:30 P.M.), *OR*, vol. 38, pt. 5, 1030; Hood to Davis, September 17, *OR*, vol. 39, pt. 2, 842; Hughes, *Hardee*, 247.

[8] W. D. Humphries, "Approximate statement of ordnance stores destroyed in the evacuation of Atlanta September 1, 1864," September 16, 1864, *OR*, vol. 38, pt. 3, 685–86; Hood to Bragg, September 4, *OR*, vol. 38, pt. 5, 1018; "Findings of the Court of Inquiry upon the loss of Confederate stores at Atlanta," *OR*, vol. 38, pt. 3, 991–92 (including Hood's endorsement of Shoup's innocence); General Field Orders No. 20, September 14, *OR*, vol. 39, pt. 2, 836 ("at his own request"); Bearss, "Francis Asbury Shoup," 5:151; McMurry, *Hood*, 153.

The court of inquiry which Shoup, McMicken, and Kennard requested did not issue judgment until February 1865. It concluded that five locomotives, eighty-one train cars, the reserve ordnance, plus small amounts of commissary goods, quartermaster stores and medical supplies, all had to be burned on the night of the evacuation. Shoup had issued orders on the 30th for commissary, quartermaster, and ordnance officers to pack their stores for removal from the city starting early on the 31st. Kennard was able to get his twenty-eight cars of ammunition out by noon that day (but as we have seen, the enemy's blocking of the railroad below Rough and Ready had forced the reserve ordnance train to chug back). Still, Lieutenant Colonel Kennard had done his duty and was exonerated by the court. McMicken got most of the blame for the material lost in the

Besides trying to blame his officers and men for the loss of Atlanta—"blaming everyone but himself," as Judith Hallock puts it—Hood also tried to cover up the extent of damage to the army, both in terms of morale and of numbers. He did his best to assure Richmond that the Army of Tennessee was holding together, after giving up the city it had fought so hard to defend. "I think that the officers and men of this army feel that every effort was made to hold Atlanta to the last," he wired Bragg on September 4; "I do not think the army is discouraged." Not everyone felt that way. Cavalry brigadier Alfred Iverson wrote President Davis on September 5, suggesting that General Lee be sent to the army in Georgia to restore its confidence. Nonetheless, a week later Hood offered the comment, "The army in good spirits."

A tougher matter involved the army's weakened numerical strength, which was undeniable. In a wire to General Bragg on September 11 Hood acknowledged that he had "about 26,000 infantry" (29,152 officers and men present in the three infantry corps). But he claimed that his losses in the fifty-four days since he took command only totaled "about 8,000." Under Johnston, on the other hand, in the seventy days of his campaign, Hood claimed that the army had lost fully 25,000 effectives. In other words, even if he had failed to keep Atlanta, Hood asserted that he had nonetheless lost only a third the number of troops Johnston had lost in falling back through north Georgia to the city's gates.[9]

---

evacuation. There were enough engines and cars available for the removal of his quartermaster supplies, the court concluded, but he had failed to issue the appropriate orders for railroad officials to act. That left Brigadier General Shoup. The court did not agree with Hood that his chief of staff was "in no manner to blame." Shoup had issued the correct orders, but starting on the morning of August 31 he had been too busy at headquarters—monitoring the intelligence on the enemy's movements coming in from the field—to see that they were carried out. To this extent the court found Shoup "censurable." The judgment, issued just a few months before the end of the war, probably did not matter to Shoup. About that time, he was ordered to report to Joe Johnston's forces assembling in the Carolinas against Sherman. He was paroled at Greensboro, North Carolina, on May 2 ("Findings of the Court of Inquiry"; Vandiver, "Hood as Logistician," 9 [Shoup ordered stores removed August 30]; Bearss, "Francis Asbury Shoup," 5:151).

[9] Hallock, *Braxton Bragg*, 201; Hood to Bragg, September 4, *OR*, vol. 38, pt. 5, 1018; Hood to Bragg, September 11, and abstract of return, September

Hood's assertion is so remarkable, given what we have seen of the army's casualties in its three big July battles alone, that it deserves scrutiny. First comes the Confederate casualty total for the campaign. As we have seen, the army medical returns maintained by Dr. Andrew J. Foard, medical director, were considered so accurate that both Hood and Johnston quoted them extensively in their respective memoirs. Foard relied upon the staffs of the several corps for his compilations, and he was apparently demanding. "'My Master,' the Doctor," Lieutenant Trask of Hardee's staff wrote on September 17, "informed me I could not start until all my returns and reports were made up—so since that time I have dived industriously into Medical returns, monthly reports, lists of casualties and so forth and last evening I got them all straight." It was remarkable after the hurried evacuation of Atlanta, with untold volume of army papers and reports likely left behind or burned, that Trask could specify Hood's loss in killed during August and early September as 482; as wounded he counted 3,223, for a total of 3,705. Then, even more remarkably, Trask recorded a precise counting of Johnston's and Hood's losses in all three categories for the entire campaign: 3,044 K + 18,962 W + 12,983 M = 34,979.

If Lieutenant Trask knew that the Army of Tennessee had lost nearly 35,000 men, then General Hood knew it too. (Historians, by the way, have accepted this number, including Richard McMurry, who has concluded that Confederate casualties for the Atlanta Campaign numbered "about 35,000," which "roughly equaled" Sherman's campaign loss of 34,514).

So how did Hood get the figure of "about 8,000" as his casualties for the campaign? By some meticulous subtraction and some shaky assumptions.

Hood led the way in his campaign report of February 1865. He started by positing Johnston's peak strength of 70,000, asserting that the army and militia of which he took command on July 18 numbered 47,750, deducting this latter from the former, and assuming the resulting figure was that of Johnston's casualties (22,750).

Elsewhere in his report Hood states that "my losses, including prisoners, during the entire campaign do not exceed ten thousand (10,000) men." But, by another display of meticulous math and questionable as-

---

20, *OR*, vol. 39, pt. 2, 828–29; Iverson to Davis, September 5, Crist, et al., *Papers of Jefferson Davis*, 11:10.

sumption, Hood also totals the army's strength on September 20 (40,403), adds in the 3,100 troops he had sent to other posts during the campaign, and subtracts this number (43,503) from that of his army when he took command (48,750), so as to allow him to finish with "a total loss of all arms of five thousand two hundred and forty seven (5247) men."

Whew! Even Hood could not repeat this mathematical feat a decade later when Johnston's memoir appeared in print and Hood rallied himself to refute its most objectionable statements. Hood wrote a long, detailed essay criticizing a number of Johnston's statements, which was published by the *New Orleans Times* in May 1874. He regretted having to do so, he wrote in his opening sentence, "but most unjust strictures passed upon me by Gen. Johnston, and which are derogatory to my character alike as a man and a soldier, compel me to speak in self-defense, or otherwise admit by silence the charges brought forth." A good portion of the article pertains to the Army of Tennessee's numerical strength and the casualties it sustained under Johnston's command. He argued that his predecessor's figure of 9,972 (Dr. Foard's total of killed and wounded) was far too low; that the army had lost 8,021 alone "in the last twenty miles of his retreat," based on returns of June 10 and July 10.

Then he called in the testimony of others.

> I sent for Major Kinloch Falconer, who was at that time one of my assistant adjutant generals, and called his attention to the outcry against me through the press, which charged that I had lost many more men during the siege of Atlanta than had Gen. Johnston during his campaign; and as he was the adjutant general of Gen. Johnston, I desired to know from him the entire loss from all causes during his retreat from Dalton to Atlanta. He at once replied that he could not give me the exact figures, as Gen. Johnston had with him all the books, records, etc., in Macon, but that we had lost twenty-five thousand men....

Hood even quoted sworn affidavits that had been mailed to him shortly after the end of the war by two staff officers, both of whom attested that the army had suffered 25,000 casualties during Johnston's tenure. Hood conducted further numerical calisthenics: the army's strength, reinforcements and Johnston's own statements of casualties in various stages of the campaign, all in order to conclude that Johnston had lost 22,750 officers and men. He then worked through two exercises to demonstrate that 1) his own losses were around 5,300, as he had stated in his report of February 1865; and 2) by yet another calculation he had lost 7,224 men during the forty-six days of his campaign command—all "against a loss of 25,000 sustained by Gen. Johnston whilst retreating sixty-six (66) days and abandoning one hundred (100) miles of territory."

For all of this, it can be said that Hood was more diligent in counting Johnston's casualties than he was in counting his own.[10]

---

[10] Trask diary, entry of September 17, 1864; J. P. Douglas, "Losses in the Atlanta Campaign," 273 (21,796 K & W + 12,983 C & M as 34,779 Confederate casualties in the campaign); McMurry, *Atlanta 1864*, 196–97; Hood report, February 15, 1865, *OR*, vol. 38, pt. 3, 634–37; "Gen. J. B. Hood's Reply to Certain Statements Contained in Gen. Joseph E. Johnston's Book," *New Orleans Times*, May 10, 1874.

Lieutenant Trask was not off the mark for Sherman's campaign loss, either, although he does not divulge how he got his numbers. Trask places the Federal casualties May–September as 4,423 killed; 22,822 wounded; and 4,442 missing for a total of 31,687. These are the figures Sherman eventually printed in his *Memoirs* (2:132). This raises the question of whether Trask edited his "journal" to include postwar data; I cannot offer comment.

J. W. A. Wright, formerly major in the 36th Alabama, was an early writer who grasped that Union and Confederate casualties in the Atlanta Campaign were roughly equal. In an article appearing in the *Philadelphia Weekly Times* in November 1884, Wright states that from May 6 to September 1, Sherman's strength dropped from 112,000 to 81,000 men while Johnston's/Hood's fell from 70,000 to 40,000 (Wright, "Hood's Nashville Campaign," 468).

Hood stuck tenaciously to his conclusion that Johnston had lost 25,000 effectives during his hundred-day retreat through north Georgia. Consequently, "the Army of Tennessee, when I assumed command...was not in condition for pitched battle," as he wrote a friend several years after the war (Hood to Sarah A. Dorsey, March 30, 1867, Hood, *Lost Papers*, 106).

A letter of Braxton Bragg to Hood after the war helps explain how Johnston's army could be said to have lost 25,000 men. "If my memory serves me," Bragg wrote, " Dr. F. [E. A. Flewellen] told me in 1874 that the numbers sick in hospital (He was Chief of Hospitals) increased during that campaign from about 2,500 to over 11,000. Add the half for this for ordinary sick (not wounded) and you have losses (with prisoners & deserters) of 25,000" (Bragg to Hood, August 24, 1875, Hood, *Lost Papers*, 70–71).

Richard McMurry states that "in his account of his own campaign Hood used a misleading method of determining losses in the battles around Atlanta...but his estimate of Johnston's casualties was accurate" (*Hood*, 201). He has more recently affirmed that if one adds all categories of Confederate casualties (deserters, prisoners, sick, et al.) for the Army of Tennessee during Johnston's

However he computed his forces' casualties, Hood needed reinforcements, and for the first time since assuming army command he called for them urgently. He telegraphed Bragg on September 3, warning that replacements were "absolutely necessary…to prevent this country from being overrun." Once this message reached the president, Davis could only answer that all available troops had already been sent. Even Hardee seemed alarmed, voicing his concerns directly to the president. "Unless this army is speedily and heavily re-enforced Georgia and Alabama will be overrun," he warned on September 4; "I see no other means to avert this calamity." (Hood in his memoirs admitted that he had suggested that Hardee write the president.) Davis wired back what he had told Hood: "No other resource remains. It is now requisite that…the means in hand be used with energy proportionate to the country's need." The commanding general was slow to understand that reinforcements could not come from that most improbable source, farther to the west. As late as September 8, he was still asking Maj. Gen. Frank Gardner in Louisiana if any of his troops had begun crossing the Mississippi to aid

---

phase of the Georgia Campaign, "Johnston's losses then total about 25,000" (foreword to Hood, *Lost Papers*, xvi).

Hood's article in the *New Orleans Times* provides the basis of chapter 4 (69–88) in *A & R*. I was tipped off to it by the letter of Mrs. Leonidas Polk to Hood of June 10 [1874], in which she wrote, "I have to thank you for a copy of the *Times* containing the second no. of your well discussed strictures on Gen. Johnston's book" (Hood, *Lost Papers*, 88). Confirmation came from Merrill, "Block-houses," 408 ("In a series of articles published in the *New Orleans Times* in the spring of 1874, General Hood reviews 'Johnston's Narrative'"). I found the articles during a visit to New Orleans and the Howard-Tilton Library of Tulane University in September 2014. (I thank Charles Nunez, president of the New Orleans Civil War Round Table, for his kindness and hospitality during my visit.) No previous author, not even McMurry, has written about Hood's writing for the *Times*, which carried several "Replies to Gen. Johnston." For instance, the column of May 31, 1874, forms the basis of chapter 5 in *A & R* (89–109).

Discovery of Hood's *Times* articles published in spring 1874 helps to explain why the six chapters of "Reply to General Johnston" in the middle of *A & R* read as if they were dropped into the middle of the book. It also adds comment upon Stephen M. Hood's complaint that Hood's "Reply" chapters in *A & R* read like "a polemic" (Hood, *Rise, Fall, and Resurrection*, xvii). If they *read* like a polemic, I want to say to Mr. Hood, it is because Hood *wrote* them as a polemic.

him. General Taylor had to write Hood from Mobile that day, "None; expect none." Absolutely the last straw was Hood's telegram to Bragg of the 10th: "Please ask General Lee to send me my old division." Richard McMurry is right in terming this melancholy missive "pathetic." It was worse than that; it was the self-pitying admission of failure by an officer who had climbed up the ranks so ambitiously.

In addressing his need for manpower, Hood was at least able to borrow on measures he had earlier enacted. Of the two thousand African American teamsters impressed during August, apparently a good number stayed with the army, even driving supply wagons into north Alabama when Hood launched his Tennessee Campaign.

There were other concerns Hood had to address after Atlanta's fall. Without a large supply depot serving his army's logistical needs, Hood had to wire subsistence officers for five days' supply of hard bread. Despite the army's disrupted supply line, commissary officers kept the men fed. B. C. Goodners, a Texan, entered into his diary on September 11, "We are in camp eating beef and corn bread once more but short rations." There were other necessities as well. Hood asked the president directly if he could send shoes and clothing for the men. To Bragg he had to send reminder that a good part of his army had not been paid for ten months. He followed this up a week later with a plea to the quartermaster general in Richmond, Brig. Gen. A. R. Lawton, that the army be paid "to prevent dissatisfaction and desertion" among the troops. Hood was now in a new field of operations, so he also had to ask Maj. Gen. Martin L. Smith, his chief engineer, for maps of the area southwest of Atlanta. In still another matter, the sick and wounded had to be sent farther to the rear, so Governor Brown was asked if he could provide railroad cars for the patients' transport.[11]

---

[11] Hood to Bragg, September 3 (1:45 P.M.), *OR*, vol. 38, pt. 5, 1016; Davis to Hood, September 5, *OR*, vol. 38, pt. 5, 1021; Hardee to Davis, September 4 (11:30 A.M.) and Davis to Hardee, September 5, *OR*, vol. 38, pt. 5, 1018, 1021; Hood, *A & R*, 245; Hood to Gardner, September 8 (2:30 P.M.) and Taylor to Hood, September 8, *OR*, vol. 38, pt. 5, 1031; Hood to Bragg, September 10, *OR*, vol. 39, pt. 2, 826; McMurry, *Hood*, 154; Mohr, *On the Threshold of Freedom*, 136 (black teamsters); Shoup to Maj. W. E. Moore, September 4, *OR*, vol. 38, pt. 5, 1019; Daniel, *Soldiering*, 62 (Goodner on rations); Hood to Davis, September 6, *OR*, vol. 38, pt. 5, 1023; Hood to Bragg, September 7, *OR*, vol. 38, pt. 5, 1027; Hood to Lawton, September 13, *OR*, vol. 39, pt. 2, 833; Hood

In the third week of September Hood received an infusion of veteran troops from a most unexpected source: the enemy. The fall of Atlanta put the large Confederate prisoner-of-war camp at Andersonville at risk much more than when Johnston and Hood had earlier suggested removal of its inmates. On September 5, Adjutant General Cooper telegraphed General Winder to transfer his POWs to a new prison that was being built in southeast Georgia near Millen, and to another at Florence, South Carolina. Until the Millen facility, named Camp Lawton, was completed, Winder was instructed to send the captives to Charleston and Savannah. The movement began within a day or two as trains arrived to carry off the Yankees, many of whom believed their transport was a happy omen that they were about to be exchanged. Winder and railroad authorities worked so quickly that within a week the prison population of Camp Sumter was reduced to fewer than 16,000.

Hood did his part to reduce the number of POWs needing to be moved by proposing to Sherman on September 8 that they arrange an exchange of prisoners. Sherman quickly agreed but capped the number of men at 1,800 or 2,000, which was the number of captured Rebels he had in Atlanta or Chattanooga. On the 9th he added another condition: he would only accept Union soldiers who belonged to regiments in his army and who could immediately resume their place in his ranks. This meant that he would not take any men captured in Virginia, Mississippi, or elsewhere. The two commanders arranged a truce and set the transfer point at Rough and Ready, between their armies' lines. Eventually, during September 19–30, 1,128 Federal and 1,332 Confederate officers and men were exchanged. (Among them was Major General Stoneman, who had been held at Charleston after his capture at Sunshine Church.) Not among them were 132 Union soldiers who had been captured by Forrest in Mississippi and taken to Rough and Ready; true to Sherman's orders, Union Lt. Col. Willard Warner, overseeing the exchange, had to turn them back to the Rebels and continued confinement. The least Sherman could do was to promise to send a supply of combs, scissors, and other goods for the Northern men still confined at Andersonville. Hood immediately arranged for General Iverson at Griffin to receive "these sup-

---

to M. L. Smith, September 4, *OR*, vol. 38, pt. 5, 1019; Hood to Brown, September 5, *OR*, vol. 38, pt. 5, 1022.

plies for the prisoners" although the record is silent on whether Sherman kept his promise.[12]

The prisoner swap was handled more civilly by Hood and Sherman than was an issue of far greater consequence to the three thousand civilians still staying in Atlanta. On September 7 Sherman sent a message to Hood that he did not intend to feed the people and intended to order them to leave the city. Atlantans who wanted to emigrate south into Confederate lines could do so if Hood agreed to a truce and was willing to let them into his lines for train transport to Macon. Hood answered on the 9th that he had no choice but to comply, yet he could not refrain from calling Sherman's plan utterly detestable. "The unprecedented measure you propose transcends in studied and ingenious cruelty," Hood wrote, "all acts ever brought to my attention in the dark history of war." That same day, a commissary officer in Hood's army, Maj. Isaac Scherck, wrote a friend, "Genl Sherman *the fiend in human form*, issues an order for all Men, Women, & Children, White and Black to leave Atlanta…Lucky are you that you removed your family when you did."

With Hood's letter of September 9 there began a week-long epistolary jousting between the two generals that included eventually five different missives and showed the bitterness, if not anger, that both men could express when their armies' conflict on the battlefield devolved into a personal war of words. Before they were done with their letter-writing, Hood and Sherman had argued over details of war history, appealed to God Almighty, and vied to see who could get the last word in (it was Sherman, even after Hood wrote on September 12, "I close this correspondence with you").

Despite their hot rhetoric, Hood and Sherman allowed a ten-day truce to take place, September 12–21. At Rough and Ready (fourteen miles south of Atlanta by the M. & W. Railroad), Lieutenant Colonel Warner and Confederate Maj. William Clare supervised the civilians' passage through the lines. Maj. A. P. Mason of Hood's staff made sure field officers knew that the truce pertained "only to the roads leading to the vicinity of Rough and Ready Station"; enemy sorties at other points

---

[12] Marvel, *Andersonville*, 198–202; Derden, *The World's Largest Prison*, 70; Speer, *Portals to Hell*, 273 (on Florence); S. Davis, *What the Yankees Did to Us*, 296–97, 299–300; Sherman to Hood, September 29, *OR*, ser. 2, vol. 7, 891; Hood to Iverson, September 27, Book IV Field Dispatches & Orders September 11, 1864–January 14, 1865, Hood Papers, National Archives.

were to be resisted as usual. Red Jackson's pickets oversaw the incoming traffic. The Federals provided wagons and ambulances for the expellees (and railway cars for the well-to-do). They were allowed to take some possessions with them. Union Lt. Col. Charles Morse kept a list, meticulously naming adults, number of children, servants, livestock, small conveyances, and bags or boxes of stuff. In Morse's "Book of Exodus" were eventually registered 1,651 individuals who carried 8,842 packages of belongings. Hood sent as many wagons as he could—a Confederate officer estimated some 400 mule teams had been assembled the night before the truce began—to receive them and to take them from Rough and Ready to Lovejoy's Station, where they boarded cars to travel to Macon.

So Sherman got rid of the secesh from Atlanta. What shocked the hundreds of Northern sympathizers still living in the city was that Sherman wanted to get rid of them, too. They were given the same terms as those traveling south: they could register for northbound transport, pack some possessions, and board trains to Nashville, Louisville, or other points. Lieutenant Colonel Morse did not keep a "Book of Exodus" for those expelled to the north (so far as we know). We therefore can only guess that the "secret Yankees" heading there roughly equaled in number Atlanta's southbound citizens. The *Atlanta/Macon Intelligencer* reported that "about one half of the population elected to go to Tennessee, and the rest were coming, or preparing to come, South." By this estimation some 1,000–1,500 Atlantans boarded trains to Chattanooga and beyond.

Not every civilian was kicked out. Perhaps fifty families were permitted to stay in the city and retain possession of their homes and property. These fortunate few fell into several categories: a) well-known Northern sympathizers prominent in the city for their position or wealth; b) those connected by family or friendship to the prominent, well-known Northern sympathizers; and c) professionals with some special skill or unique position. Thus Dr. Noel D'Alvigny, Atlanta physician who had served as Confederate contract surgeon during the campaign, was retained as a medical officer by the XX Corps. Father Thomas O'Reilly was priest of the city's only Catholic church; he was also able to stay in the city during the occupation.

When the people left, their residences became vacant; officers moved into some of the finer ones. Brigadier General Geary had been eyeing Edward Rawson's attractive home, dubbed "The Terraces" for its distinctive landscaping. He was staying in a house on Pryor Street when the Rawsons were packing up for their departure; the day after they left town he moved into their house.[13]

All the while Hood had to guess his opponent's next move. In the days after the fall of Atlanta, he guessed wrong. He wired Bragg on September 3 his belief that Sherman would not fall back on Atlanta from Jonesboro, "but will continue offensive movements. All the lieutenant generals agree with me." Hood thought the enemy would march south along the Flint River, so he ordered Iverson with his cavalry to watch in that direction. If the enemy advanced as he thought they would, Hood even envisioned launching an attacking blow against their flank or rear. It was evident that John B. Hood's fighting blood was still up. "It is necessary for me to give battle should the enemy continue to advance," he informed Bragg. Even as he sought reinforcements, Hood assured Richmond that he was making "every effort" to win back absentees to the army.[14]

With Sherman content to "retreat" (Hood's word) back toward Atlanta, the position of the Army of Tennessee remained unchanged near Lovejoy's. On September 6 Confederates advanced their lines and reoccupied Jonesboro. That day, since he was communicating now with the president directly rather than through the war secretary, Hood asked Davis if he or General Bragg "would visit the army." It was apparent that the commanding general wanted to secure the administration's approval of an ambitious plan. Even though he informed the president that "the army is much in need of a little rest," Hood spoke of assuming the offensive, despite his forces' weakened condition and the uncertainty of railroad networks to provide logistical support. After Camp Sumter was depopulated and his army was no longer needed to prevent a raid on the

---

[13] S. Davis, *What the Yankees Did to Us*, 295 (on c. 3,000 residents remaining after Confederate evacuation), 296–306, 310–20, 324–28; Rosen, *The Jewish Confederates* 307 (Major Scherck); Shoup journal, entry of September 12, *OR*, vol. 39, pt. 1, 804; Mason to J. T. Morgan, September 13, *OR*, vol. 39, pt. 2, 834 (truce); [Hood] to Jackson, September 13, Book IV, Field Dispatches, Hood Papers, National Archives (on civilian traffic); Anderson Humphreys and Curt Guenthes, *Semmes America* (Memphis: Humphreys, Ink, 1989) 357–58 ("Wagons, some 400 four- and six-mule teams").

[14] Hood to Bragg, September 3 (6:10 P.M.), *OR*, vol. 38, pt. 5, 1017; [Hood] to Iverson, September 4, *OR*, vol. 38, pt. 5, 1020; Hood to Bragg, September 4 (1 P.M.), *OR*, vol. 38, pt. 5, 1018; Hood to Bragg, September 5 (10:45 A.M.), *OR*, vol. 38, pt. 5, 1021.

huge prisoner-of-war camp near Americus, Hood proposed the following to Davis on September 6: "I think we should, as soon as practicable, place our army upon the communications of the enemy, drawing our supplies from the West Point and Montgomery Railroad. Looking to this, I shall at once proceed to strongly fortify Macon. Please do not fail to give me advice, at all times. It is my desire to do the best for you and my country. May God be with you and us." Hood was apparently anticipating another enemy cavalry raid against Macon, with its arsenal and other facilities. On the 7th, Shoup ordered Major General Smith, the chief engineer, to proceed there and oversee the strengthening of its defenses. Because Augusta and Columbus would also be vulnerable to raids if Hood's army marched out of Georgia, Smith was asked to do the same for those cities as well.

Even without the president's support, Hood broached his idea to Davis's chief adviser. "I wish to move by the left flank as soon as possible, resting my left on the Chattahoochee River, so as to interrupt the communications of Sherman," he wired Bragg on September 11. Hood reminded the administration that he had about 26,000 infantry and asked what reinforcements he might anticipate in the days ahead. On the 11th, he also asked Richard Taylor if Forrest even at this late date could damage Sherman's rail communications.[15]

McMurry has commented on Hood's negligence of administrative issues in his command; General Lee had earlier criticized him for it. Thus it is not flattering to Hood to see that General Shoup was instructed to write generals Hardee, Lee, and Stewart on September 7, "General Hood desires you to use all diligence in reorganizing and recuperating your commands." To an extent, recuperation of the army was being addressed by its commanding general (reinforcements, food, clothing, shoes, back pay); but his deferral to corps commanders for "recuperation" of their commands reads like a passing-of-the-buck. At the same time,

---

[15] Hood to Bragg, September 6, *OR*, vol. 38, pt. 5, 1023; Shoup to Ferguson, September 4 (9:15 A.M.), *OR*, vol. 38, pt. 5, 1020 (position of infantry); Hood to Bragg, September 5 (7:30 P.M.), *OR*, vol. 38, pt. 5, 1021 ("no change in position to-day"); Shoup to Morgan, September 6 (7 P.M.), *OR*, vol. 38, pt. 5, 1025 (occupying Jonesboro); Hood to Davis, September 6, *OR*, vol. 38, pt. 5, 1023–24; Shoup to M. L. Smith, September 7, *OR*, vol. 38, pt. 5, 1029; Hood to Bragg, September 11, *OR*, vol. 39, pt. 2, 829; Hood to Taylor, September 11, *OR*, vol. 39, pt. 2, 831.

Hood was justified in asking his corps leaders to submit recommendations of officers to fill the command vacancies then existing in the army. For instance, A.A.G. Withers in Richmond on the 8th assigned Maj. Gen. Arnold Elzey as chief of artillery for Hood's army, replacing Col. Robert Beckham. On September 13 Hood reminded the administration of his request to promote Col. Charles Shelley to command the brigade of James Cantey, away and still sick. Shelley's promotion occurred four days later.

One position he did not fill was that of army chief of staff. On September 14 Brigadier General Shoup, having submitted his resignation, was relieved of duty. Shoup had asked to be relieved ostensibly because of the allegation that he had misperformed in the last days of Atlanta's occupation, when supplies and ordnance had to be gotten out. His last journal entry was dated September 7. Hood did not announce a replacement and Maj. Arthur P. Mason, assistant adjutant general, assumed most staff duties.[16]

Just as Hood had earlier guessed wrong about Sherman's intention, Sherman guessed wrong about Hood's. "I think he will move back to Macon and send some men to Richmond," he wrote General Howard on September 20. The day before Sherman had wired Grant, warning him, "a deserter just in says Stewart's corps is moving back to Macon with a view to going to Virginia."

---

[16] McMurry, *Hood*, 153; Chilton to Hood, November 14, 1862, *OR*, vol. 19, pt. 2, 718–19; Shoup to Hardee, Lee and Stewart, September 7, *OR*, vol. 38, pt. 5, 1028; Withers Special Orders 213, September 8, *OR*, vol. 38, pt. 5, 1031; W. C. Davis, "Arnold Elzey," 2:103; Hood to Cooper, September 13, *OR*, vol. 39, pt. 2, 833 (Shelley); Kinloch Falconer, General Field Orders No. 20, September 14, *OR*, vol. 39, pt. 2, 836.

With General Elzey's appointment, Colonel Beckham became chief of artillery for S. D. Lee's corps though I cannot find in the literature explanation of this apparent demotion. Larry Daniel links it to Robert E. Lee's shelving of Dick Ewell out of his army: he states that Elzey, in command of the Department of Richmond, was sent west and Ewell was ordered to succeed him (L. J. Daniel, *Cannoneers in Gray*, 169–70). The problem with this, though, is that Ewell took over the Richmond Department on June 13, 1864, three months earlier than Elzey's assignment to command with Hood's army, which took place on September 8 (S. Martin, *The Road to Glory*, 322; W. C. Davis, "Arnold Elzey," 2:103).

Hood was thinking of doing no such thing. Instead, he was planning to position his army for an offensive campaign, or at least one of marching around Sherman at Atlanta. On the 15th Hood wired Bragg that he sensed a returning confidence among his officers: "I think we will make a better fight the next time than we could have done any time since we left Resaca." In the third week of September the army numbered in all three arms 39,070 "effective total present"—and this was with Wheeler still away. (Hood ordered Wheeler back on the 13th.) Another return for September 20 showed an even more encouraging total of 47,431 present for duty, with 40,403 effectives present.

This respectable force he ordered to start marching westward from Lovejoy's Station on the morning of September 18. The objective was Palmetto, on the West Point Railroad, twenty-five miles southwest of Atlanta. Now that the line to Montgomery had been repaired, that railway could now be Hood's route of supply. This meant that the army's ammunition stores would have to come from Selma; Hood even suggested that another powder works be established at Cahaba in east Alabama. Logistical support for the Army of Tennessee was thus shifting away from Georgia. Hood accordingly ordered that rails from both the Macon and Augusta roads be taken up and stored for future use. "It would be better for us if there were no railroad from Atlanta to Augusta," he wrote Bragg.

Hood's plan to get the army moving made such sense that mid-level officers in his army anticipated it. "I have no idea, except from conjecture, what will be our next move," wrote Capt. Benjamin L. Posey, 36th Alabama, in a letter published by the *Mobile Advertiser & Register* on September 20. "I presume we will at least change our position by shifting 20 or 30 miles westward, and covering the approaches from Atlanta towards Alabama." Posey then imagined the unimaginable: "If we cannot fall back, and cover and quarter ourselves upon Alabama, then this army must dissolve and commence *guerrilla* warfare."[17]

[17] W. Bedford to T. E. G. Ransom, September 20, *OR*, vol. 39, pt. 2, 430 (including Sherman to Howard); Sherman to Grant, September 19, *OR*, vol. 39, pt. 2, 404; Hood to Bragg, September 15, *OR*, vol. 39, pt. 837; abstract for September 15–24, 1864, *OR*, vol. 39, pt. 2, 838; abstract for September 20, *OR*, vol. 39, pt. 2, 851; Hood to Wheeler, September 13, *OR*, vol. 39, pt. 2, 834; Circular, September 17, *OR*, vol. 39, pt. 2, 842; Hood to Wright, September 18, *OR*, vol. 39, pt. 2, 844 ("the army to-morrow will be in position on the West Point

Hood announced his idea for an advance in a telegram to Bragg of September 22. Unless Sherman moved south from Atlanta, and as soon as he could collect sufficient supplies, he intended to cross the army over the Chattahoochee, "and form lines of battle near Powder Springs." This move would threaten the railroad to Chattanooga, draw the enemy out, "and force him to drive me off." If, however, Sherman marched south of Atlanta, "I shall fall upon his rear." In proposing this operation, Hood claimed, "Sherman is weaker now than he will be in future, and I as strong as I can expect to be."

Richmond was not impressed. Indeed, President Davis at the time was hearing from prominent citizens clear expressions of concern about Hood's ability. (Possibly for this reason—that is, to explain and justify his course of conduct in the recent campaign—Hood asked his three corps commanders on September 15 to "forward as soon as practicable a report of the operation of your command around Atlanta since [I] assumed command.") War clerk J. B. Jones recorded in his diary on September 20 that he had seen a letter from R. W. Barnwell approving the removal of Johnston, but "he thinks Hood hardly competent to command the army." Soldiers in Hood's army were apparently whispering along the same lines. One, Marcus B. Ely of the 54th Georgia, wrote his wife, Martha, on September 12, "It is a current rumor that Gen Beauregard will take command of the army of Tennessee." Lieutenant Ely even heard that General Beauregard had already arrived to do so.

---

railroad"); Hood to Bragg, September 20, *OR*, vol. 39, pt. 2, 847; J. B. Hood and [Mason] to A. R. Wright, September 20, *OR*, vol. 39, pt. 2, 850; Hood to Bragg, September 21, *OR*, vol. 39, pt. 2, 860; "Ben Lane," "Letter from the Army of Tennessee," *Mobile Advertiser & Register*, September 20, 1864.

"Mind games" is an anachronism for America in the 1860s, but Sherman was playing them with Hood. On September 20, the Federal commander received word from Washington that Sheridan had beaten Jubal Early badly at Winchester. "The enemy were driven off twelve miles," read Secretary Stanton's telegram, "2,500 prisoners were captured, 9 stand of colors, 5 pieces of artillery were taken, and the rebel killed and wounded left in our hands. The rebel Generals Rodes and Gordon were killed, 4 other rebel generals wounded." Sherman circulated Stanton's text in special field orders to his officers and men. The next day Sherman wired the secretary that he had sent copies of the telegram to Hood with his compliments, "but I know it does not afford comfort to the enemy" (Dayton, Special Field Orders No. 76, September 20, and Sherman to Stanton, September 21 [10 P.M.], *OR*, vol. 39, pt. 2, 423, 431).

Such was not the case, but in truth Jefferson Davis had begun to consider bringing General Beauregard to Georgia to oversee Hood's operations and offer advice and counsel. Beauregard was serving at the time as commander of the Department of Southern Virginia and North Carolina, but with headquarters at Petersburg, he was clearly in General Lee's shadow and without real autonomy. Beauregard hence sought transfer from Virginia to another theater. What he wanted was to take charge of an army in the field, as he had done in 1861–1862. "My greatest desire has always been to command a good army in the field," he wrote a friend on August 30, 1864. "Will I never be gratified?" Lee knew of Beauregard's dissatisfaction (and perhaps wanted him away from Virginia, too). In early September he suggested that Beauregard travel to Wilmington on an inspection tour. North Carolina governor Zebulon Vance helpfully expressed interest in having Beauregard take charge of the Wilmington defenses.

The fall of Atlanta and the administration's uncertainty as to Hood's plans led to another possibility. As he had done with Joe Johnston in 1862–1863, perhaps Davis would create a super-department in the west and put Beauregard in charge of it. That way everyone would get something of what he wanted: Beauregard away from Virginia (Lee), a position of respectable if nebulous authority (Beauregard), and an experienced general to look over Hood's shoulder (Davis).

Some writers, such as T. Harry Williams, Thomas L. Connelly, and William C. Davis, claim that the president asked General Lee to find out if Beauregard would accept such an assignment, but the documentary evidence of their statements is not apparent. They assume that Davis asked Lee to talk with Beauregard because of a remarkable letter written by Lee to the president on September 19, 1864. It is in the *Official Records*, in plain view, but it's the subject of little historical comment. Not one of Hood's three principal biographers—O'Connor (1949), Dyer (1950), or McMurry (1982)—even mentions it (nor do I see it mentioned by Stephen M. Hood in his chapter "Robert E. Lee's Opinion of John Bell Hood" in *Rise, Fall, and Resurrection* [2013]), even though it was an obvious effort by Lee to persuade Davis to relieve John Bell Hood and replace him with Gustave Toutant Beauregard. "Mr. President," Lee wrote, "I have had a conversation with General Beauregard with reference to the army and operations in Georgia." Note that Lee did not begin, "Mr. President. As you have asked, I have had...." Thus Williams, Connelly and Davis are within their rights to assume that Davis requested Lee to talk with Beauregard, but it is equally plausible from this text to assume that Lee could have approached Beauregard entirely on his own. Lee continued,

> I have endeavored particularly to explain to him the necessity of the commander in Georgia developing the latent resources of the department, drawing to him all absentees from the army, concentrating its strength, restoring its confidence, and, in a word, creating the means with which he must operate against the enemy and the impracticability at present of giving him any extraneous aid.

This sums up the Confederates' situation in Georgia after Hood's defeat and the loss of Atlanta. Strengthening his forces for future operations was what Lee had done after Gettysburg with the Army of Northern Virginia. Lee's implication here, that General Hood lacked the administrative ability and energy to do the same for the Army of Tennessee, reminds one of his remark to Davis the previous July: "Hood is a bold fighter. I am doubtful as to other qualities necessary."

Then he sought to build up Beauregard. "Of all this he is fully sensible, and while strongly impressed with the responsibility of the station and fearful of not being equal to the present emergency, being anxious to do all in his power to serve the country, he says he will obey with alacrity any order of the War Department placing him in command of that army, and do his best to expel the enemy." With his phrase "placing him in command of that army," Lee wanted to make absolutely certain the president knew that he was not recommending Beauregard for a loose supervisory role over Hood, but that he was recommending Beauregard to *replace* Hood. Lee must have known that he was asking too much of the president. Jefferson Davis had too much pride to admit that he had made a mistake in replacing Joe Johnston with Hood. To now replace Hood—to change army commanders twice in two months, no less—would have been admitting to the whole country that he had blundered. Besides, as Robert Kean had written a few months before, when the government wrestled over who could succeed Joe Johnston, "the only solution is to send Beauregard, but the President thinks as ill of him as of Johnston." So if the president was not considering replacing Hood with Beauregard, one must conclude that *this was Robert E. Lee's own idea.*

Entering the realm of presidential decision-making was most uncharacteristic of Lee. That he did so here suggests his sense of urgency for the situation in Georgia (and for the cause in general), yet the polite and mannerly way in which Lee couched his recommendation of Beauregard *was* characteristic: "Should you deem, therefore, a change in the commander of the army in Georgia advantageous, and select General Beauregard for that position, I think you may feel assured that he understands the general condition of affairs, the difficulties with which they are surrounded, and the importance of exerting all his energies for their improvement."

Lee wanted to repeat that he was recommending Beauregard for actual *army* command, and he wished to convince Davis that the Louisianan knew the challenges of the situation in Georgia and felt capable of facing them. If Beauregard were sent west, Lee even suggested that he take with him his chief of staff and quartermaster, both of whom "are conversant with that army and country." Note that throughout his letter Lee had not named "that army," much less identified General Hood as the officer to be replaced. But his intent is unmistakable.

Reinforcing Lee's remarkable letter is a memorandum drafted and signed by General Beauregard himself on September 19. Probably at Lee's request he wrote out an affirmation of their conversation, employing just such language ("anxious to do all in my power," "obey with alacrity") as Lee used in his letter to the president. And like Lee, Beauregard never referred to the Army of Tennessee by name, stating only that he was ready to obey any War Department order "which may put me in command of that army."[18]

There is no evidence that Davis acknowledged or replied to Lee's letter, which arrived just as the president was preparing to leave for Georgia. Coincidentally, on September 20, John Jones, the War De-

---

[18] Hood to Bragg, September 22, *OR*, vol. 39, pt. 2, 862; Hood, *A & R*, 252–53; [Hood] to generals Hardee, Lee, and Stewart, September 15, Book IV, Field Dispatches, Hood Papers, National Archives (which I do not see in *OR*, vol. 39, pt. 2); J. B. Jones, *A Rebel War Clerk's Diary* (1866), 2:287–88; McCardle, *A Just and Holy Cause*, 132; T. H. Williams, *P. G. T. Beauregard*, 240; Connelly, *Autumn of Glory*, 471; W. C. Davis, *The Man and His Hour*, 566; Horn, *Army of Tennessee*, 373 (quoting Lee's statement, "change in the commander of the army in Georgia," but without comment upon it); Lee to Davis, September 19, *OR*, vol. 39, pt. 2, 846; Kean, *Confederate Government*, 166 (on Beauregard); Roman, *The Military Operations of General Beauregard*, 2:275 (Beauregard's untitled memorandum of September 19, given to Lee); S. Davis, "Would P. G. T. Lead the A. o. T.?," 36–41; McMurry, *Hood*, 157–58; Hay, *Hood's Tennessee Campaign*, 28.

partment clerk, entered into his diary after noting that General Beauregard was in Wilmington, "the whole country is calling for his appointment to the command of the army in Georgia." Jones was surely exaggerating about "the whole country," but the fact that Robert E. Lee *was* calling for such an appointment is undeniable. The chief executive, however, was unswayed. He may have already decided to appoint Beauregard to lead a new Military Division of the West, which would comprise Hood's Department of Tennessee and Georgia and Richard Taylor's Department of Alabama, Mississippi, and East Louisiana. Williams says it well: "This decision was essentially a political device, designed to silence the critics of Hood, satisfy the friends of Beauregard, and save face for the administration." But it did offer some real strategic and tactical benefit; Beauregard would stay in the southeast and be available for consultations with the generals and inspections of the troops. Moreover, as Forrest's cavalry operated under Taylor, Beauregard's supervision could provide some needed unity of command for the Army of Tennessee's coming operations.

While Lee advocated for Hood's removal, there is no evidence that Davis let Beauregard in on his idea for a super-department in the West. The Louisianan, at Wilmington, at some point received the president's order to travel to Charleston to investigate a command issue in the area. He arrived there on September 25. The general's travel provoked some gossip, and the journalistic wags fanned it. In a column titled "Richmond Gossip," the *Augusta Chronicle & Sentinel* stated, "It is deemed certain that Gen. Beauregard will go to Georgia." "This is an auspicious sign," the editors affirmed although they did not say why. "GEN. BEAUREGARD.—The *Charleston Mercury* announces the arrival in that city of this illustrious commander," read one newspaper announcement. "From some rather spicy remarks of the *Mercury*, we infer that Gen. B. will not immediately assume command of Hood's army—if ever." Others read the situation differently. The *Savannah Republican* commented, "The *Charleston Mercury* of yesterday congratulates its readers upon information it has received from 'a source deemed unquestionable,' that President Davis has tendered to Gen. Beauregard the command of the Army of Tennessee." "This result," it added tellingly, "has been brought about by the earnest intervention and counsel of General Lee." The editors of the *Republican* weighed in with their own thoughts. "We hail with delight this announcement," they declared, "because the appointment of this gallant chieftain will have the effect of inspiring confidence among the troops of that army." The *Charleston Mercury* was more equivocal about Beauregard and his future. The paper opined in a column of September 26,

> The President has gone to the army in Georgia to endeavor to arrange matters *without putting General* BEAUREGARD *in command*—that is, to reconcile, if possible, the army to General Hood's continuation in its command. If he succeeds, according to his estimation, General Beauregard will be returned to his command near Petersburg. If he fails, General Beauregard will be ordered to the command of the army in Georgia.

"Our deductions are drawn from the facts," the *Mercury* asserted, "and our readers can judge of their correctness as well as we can."[19]

Having asked Jefferson Davis on September 6 if he could personally visit the army to buck up morale and to consult with him, General Hood got his wish. The president had twice before visited the Army of Tennessee to resolve command problems—in December 1862 and October 1863, amidst discontent in the officer corps over Braxton Bragg's leadership. In that context, this visit did not augur well, either for Davis or for Hood. In another of those letters written by officers of the Army of Tennessee apparently over the heads of their commanders, Major General French wrote the president on September 14.

> MR. PRESIDENT: Several officers have asked me to write to you in regard to a feeling of depression more or less apparent in parts of this army, and I have declined doing so, but for your satisfaction it might

[19] Jones, *A Rebel War Clerk's Diary* (1866), 2:288; T. H. Williams, *Beauregard*, 241; Connelly and Jones, *Politics of Command*, 163–64; Connelly, *Autumn of Glory*, 292–93; Woodworth, *Davis and His Generals*, 292; McPherson, *Embattled Rebel*, 210; Beauregard to Maj. Gen. Samuel Jones, September 25, *OR*, vol. 35, pt. 2, 630; Roman, *Beauregard*, 2:275 (the general reached Charleston September 25); "Richmond Gossip," *Augusta Chronicle & Sentinel*, September 30; "GEN. BEAUREGARD," *Augusta Constitutionalist*, September 27; "GENERAL BEAUREGARD," *Charleston Mercury*, September 26, reprinted in *Atlanta/Macon Intelligencer*, September 30.

Apparently all the newspaper gossiping about the possibility of Beauregard coming to Georgia prompted one citizen of Augusta on September 29 to write President Davis, urging that the Creole be named to replace Hood, whom the writer termed "a supercilious braggart" (Crist, et al., *Papers of Jefferson Davis*, 11:74, quoting the letter of "A Georgian," otherwise unpublished in the Davis Papers at Duke).

> be well that you send one or two intelligent officers here to visit the different divisions and brigades to ascertain if that spirit of confidence so necessary for success has or has not impaired within the past month or two. They might further inquire into the cause if they find in this army any want of enthusiasm. I am sure you will pardon my writing to you thus when I tell you it is dictated by the purest of motives and in the spirit of friendship.

This was as damning to Hood as anything other army leaders had expressed to Davis about Braxton Bragg in the several years before (after Perryville or Chickamauga). By September 14, Davis had decided to personally visit the army, as Secretary of War James Seddon endorsed French's letter, "The President will see and judge for himself."

Steven Woodworth has considered Davis's travel to Hood's army in Georgia as addressing several issues. First was to discuss Hood's strategic plans. "Davis himself seems to have been not altogether easy about the uncertain qualities of Hood's aggressive brand of generalship," he writes. Personal observations could help the president better evaluate Hood's ideas for the next campaign and weigh the need of appointing Beauregard as theater commander over him. Davis could also address the problem of General Hardee's requests for transfer of command, which evidently would not go away. Third, he could assess for himself the validity of General French's letter regarding the state of the army's morale. Finally, in essentially a whistle-stop train tour of what remained of the Confederate States, he could stage public appearances and give speeches to the people at large, assuring them that their cause was not lost.[20]

Davis left Richmond on the afternoon of September 20, accompanied by his aides Brig. Gen. Custis Lee and former Texas governor Francis Lubbock. At stops in Danville, Greensboro, Salisbury, Charlotte, and Augusta, crowds gathered, women especially, at the railroad stations. Davis spoke from his train car with "fierceness and determination," according to one observer. At Greensboro he told his audience that he was glad to see them "so cheerful and hopeful," a newspaper reported. At Salisbury he urged every soldier away from the ranks to return to his post; he enjoined the ladies "to resolve to marry no man who had shirked

---

[20] Hood to Davis, September 6, *OR*, vol. 38, pt. 5, 1023; Connelly, *Autumn of Glory*, 470–72; French to Davis, September 14, *OR*, vol. 39, pt. 2, 836; Seddon endorsement, Crist, et al., *Papers of Jefferson Davis*, 11:36; Woodworth, *Davis and His Generals*, 290–91.

the responsible duties of a soldier." "Now was the crisis of our fate," a reporter quoted him in Charlotte; "every man who could be spared should rush to the field."

When he reached Georgia, Davis became even more fervent. "It is no time now for despair, or despondency," he declared in Augusta on September 22. "We are struggling...to preserve the heritage bequeathed to us by our fathers—the right to govern ourselves—the right to be freemen," and unless every man did his duty, the Confederacy could fail and "constitutional government—political freedom, will fall with it." Yet Davis sounded hopeful notes, too. "He exhorted the people," the Augusta papers reported, "to be firm and faithful, and all would yet be well."[21]

These twenty-minute perorations may have been off the cuff, but when Davis arrived in Macon, where he arrived early in the morning of September 23, he apparently had a prepared speech. Citizens had scheduled a meeting at the Baptist church at 11 A.M. to discuss means of aiding the refugees recently arrived from Atlanta, but they changed their agenda so that the president could deliver an address. In his remarks at times Davis sounded optimistic:

> Sherman cannot keep up his long line of communication, and retreat sooner or later he must. And when that day comes, the fate that befell the army of the French Empire and its retreat from Moscow will be reacted. Our cavalry and our people will harrass [*sic*] and destroy his army as did the Cossacks that of Napoleon, and the Yankee General, like him will escape with only a bodyguard.

He grew somber as well. He regretted the loss of Atlanta. In a remark that came in for widespread criticism, Davis also claimed that "two-thirds of our men are absent." These soldiers included the sick and wounded, or course, "but most of them [are] absent without leave." "If one-half the men now absent without leave will return to duty," he added, "we can defeat the enemy."

In the course of his remarks, Davis could not resist a poke at Joe Johnston. The president commented that when Johnston was retreating through north Georgia, "I even heard that I had sent Bragg with pon-

---

[21] Crist, et al., *Papers of Jefferson Davis*, 11:58–60 (quoting Raleigh *Confederate* of September 24; Salisbury *Carolina Watchman*, September 22; *Charlotte Carolina Times*; *Augusta Constitutionalist* and *Chronicle & Sentinel*, both of September 23); Wooster, *Texas and Texans in the Civil War*, 97, 165 (on Lubbock); Cooper, *Davis, American*, 489.

toons to cross into Cuba." Then he added, "But we must be charitable." After the army had retreated across the Chattahoochee, "I then put a man in command who I knew would strike an honest and manly blow for the city." Joseph E. Johnston was living in Macon at the time, but his biographers do not record if he was in the church audience to hear Davis. One way or another, he got word of the president's remarks. Later, in a letter to Mansfield Lovell, he sneered at Hood's being labeled "the Striker of Manly Blows."

Davis continued, "Let us with one arm and one effort endeavor to crush Sherman. I am going to the army to confer with our Generals. The end must be the defeat of our enemy." "Let no one despond," he concluded. Ultimately one day the war would be over "and our independence won." "When the enemy is driven beyond Chattanooga," Davis predicted that the people of Macon "can join in the general rejoicing."[22]

The *Macon Telegraph* secured copy of the text Davis had read in his speech and printed it on the 24th. Within three days Sherman had a copy of the paper, for on the night of September 26 he wired Grant that he knew Davis was en route to see Hood and that "superhuman efforts will be made to break my road." A few days before, Sherman had written Howard, "I have no doubt Hood has resolved to throw himself on our

---

[22] Crist, et al., *Papers of Jefferson Davis*, 11:60–63; Ballard, "Breakdown in Macon," 31–33; Govan and Livingood, *Different Valor*, 328–29; Symonds, *Johnston*, 334.

Davis has been criticized for having "candidly discussed subjects that should have been taboo in a public forum" (Ballard, "Breakdown in Macon," 31). Chief among them was the president's admission that "two-thirds of our men are absent." That was an alarming claim, but in 1864, war department records did indeed show that at least 100,000 men were listed as AWOL (Woodworth, *Davis and His Generals*, 292–93). For the nine big departments enumerated by Steven H. Newton in his study of Confederate manpower in 1864, one can go through the *Official Records* and indeed tote up well over 100,000 soldiers counted as absent in returns for spring 1864 (Newton, *Lost for the Cause*, *passim*; Northern Virginia, April 20, 1864, *OR*, vol. 33, 1298; West Virginia, April 1864, *OR*, vol. 33, 1334; East Tennessee, March 31, 1864, *OR*, vol. 32, pt. 3, 721; Richmond, April 20, 1864, *OR*, vol. 33, 1299; North Carolina, February 1864, *OR*, vol. 33, 1201; South Carolina, Georgia, and Florida, April 1864, *OR*, vol. 35, pt. 2, 457; Tennessee, April 1864, *OR*, vol. 32, pt. 3, 865; Alabama, Mississippi, and East Louisiana, February 29, 1864, *OR*, vol. 32, pt. 2, 819; parts of the Trans Mississippi, March–April 1864, *OR*, vol. 34, pt. 2, 1034 and pt. 3, 800).

flanks to prevent our accumulating stores, &c., here, trusting to our not advancing into Georgia." Sherman was aware that the Rebels had shifted from Lovejoy's to Palmetto. "Hood seems to be moving, as it were, to the Alabama line," he informed Halleck on the 25th, but he was not worried. "I can whip his infantry," he added, "but his cavalry is to be feared." Indeed, a report that Forrest's cavalry had ridden into Athens in north Alabama (on the Tennessee & Alabama Railroad to Nashville) led Sherman to send an infantry division to Chattanooga and another to Rome. He worried most about that devil Forrest. "I know my line of communication will be attacked," Sherman predicted to Halleck on the morning of the 26th. "Forrest is now lieutenant-general and in command of all the cavalry." "His cavalry will travel a hundred miles in less time than ours will ten," he noted cynically.[23]

---

[23] Crist, et al., *Papers of Jefferson Davis*, 11:63; Sherman to Grant, September 26 (10 P.M.), *OR*, vol. 39, pt. 2, 479; Sherman to Howard, September 24, *OR*, vol. 39, pt. 2, 463 (Corse's division to Rome); Sherman to Stanton, September 21 (11:30 P.M.), *OR*, vol. 39, pt. 2, 431 ("Hood…has shifted…to the West Point road about Palmetto Station"); Sherman to Halleck, September 25 (6:30 P.M. [Newton's division to Chattanooga]), September 26 (10 A.M.), and September 29 (8:30 P.M.), *OR*, vol. 39, pt. 2, 464, 479 (with error as to Forrest's promotion, which did not occur till February 1865), 517.

The *Macon Telegraph* transcription of Davis's address in Macon ends with his statement that he hoped the city would care for Atlanta's "refugees and exiles." When the *Richmond Enquirer* reprinted Davis's text on September 29, the article ends with "his remarks were often interrupted by applause"—a deliberately pro-Davis addendum (Crist, et al., *Papers of Jefferson Davis*, 11:63 [*Telegraph* text]; Rowland, *Jefferson Davis Constitutionalist*, 6:344 [*Enquirer* text]).

The *Telegraph* may have done the administration another favor, though. A day or so after it printed the president's remarks, the *Telegraph* had a correction. As carried by the *Augusta Constitutionalist* on the 28th, it read, "GEN. HARDEE. In our report yesterday of the President's speech, we unintentionally omitted to mention the high and deserved compliment paid by the Executive to Gen. Hardee. 'Whoever may have failed,' said the President, 'that noble Georgian, Gen. Hardee, has done all that could be expected of a soldier or asked of a patriot.' Well said, and true to the letter.—*Macon Confederate Telegraph*."

The *Telegraph* editors' mea culpa ("unintentionally omitted") may have been a genuine typesetting error. It is at least plausible, however, that friends of General Hardee may have wanted Davis's remarks in Macon to have included a commendation of Hardee. It is also possible that Davis himself, reading his transcript in the press, might have realized that he had missed an opportunity to

Reading Davis's text in the Macon paper only confirmed Sherman's apprehensions. "Jeff. Davis was certainly at Macon on the 23d," he wired Grant on the 27th, "for he made a very significant speech." Sherman thought it so significant that he had its text telegraphed along his railroad lines all the way to Louisville as a way of warning officials that Rebel raiders could soon be on their way. "I take it for granted that Forrest will cut our road," he predicted to Stanton on the 29th, "but I think we can prevent his making a serious lodgment."

Guarding the long rail line back to Nashville would take a large number of troops, and Sherman asked if reserves from north of the Ohio could be sent forward. He even sent General Thomas back to organize troops in Tennessee; Slow Trot left Atlanta with a division on the 29th. At the same time, he worried that further detachments from his army would be ordered, and that he would not have sufficient force for his next campaign, which he was already planning. Sherman went so far as to write the president to express his concern. "Great efforts are being made to re-enforce Hood's army and to break up my railroads," he asserted, asking that there be "a good reserve at Nashville" to counter enemy raids. "It would have a bad effect if I were to be forced to send back any material part of my army to guard roads," he explained to Lincoln, "so as to weaken me to an extent that I could not act offensively."[24]

General Sherman's take on the phrase "act offensively" was a peculiar one, for, as he began planning his next campaign, he was not thinking of moving against Hood's army. Indeed, he was not planning to target any armed Rebel force at all; instead he aimed to march his army against the civilian populace of Georgia, leading it from Atlanta to Savannah. "I can sweep the whole State of Georgia," he bragged to Grant on September 10, as if leading sixty thousand troops against a helpless citizenry constituted some bold military feat. As he had done earlier in the year,

---

pay public tribute to the general who at the time was privately seeking to be transferred away from Hood's army. Regardless of the reason, the Macon paper (and others like the *Constitutionalist*) printed the laudatory paragraph. Neither Rowland nor Crist, who print the president's Macon speech text, make note of the *Telegraph*'s subsequent addition of the Hardee paragraph.

[24] Sherman to Grant, September 27 (10 A.M.), *OR*, vol. 39, pt. 2, 488; Sherman to Halleck, September 29 (8:30 P.M.), *OR*, vol. 39, pt. 2, 517; Sherman to Grant, September 26 (10 P.M.), *OR*, vol. 39, pt. 2, 479; Welcher, *The Union Army*, 2:583; Sherman to Lincoln, September 28 (11:30 A.M.), *OR*, vol. 39, pt. 2, 501.

marching across Mississippi from Vicksburg to Meridian, Sherman planned to subsist his troops from the peoples' farms and plantations, to wreck railroads, burn factories and shops, destroy cotton, and "make Georgia howl." (Come to think of it, "act offensively" was not a poor choice of terms.) "Why would it not do for me to leave Tennessee to the force which Thomas has and the reserves soon to come to Nashville, and for me to destroy Atlanta, and then march across Georgia to Savannah and Charleston, breaking roads and doing irreparable damage?" he asked Grant on October 1. Halleck liked the idea of taking the war to the Rebel civilians. "Your mode of conducting war is just the thing we now want," he wrote Sherman on September 26. "We have tried the kid-glove policy long enough." Yet Grant and Lincoln were not quick to give their approval. Grant, for instance, answered Sherman's telegram of September 10 by having Lt. Col. Horace Porter deliver a letter of reply and listen more to Sherman's ideas before bringing back a report. As it was, Grant, confessed, "What you are to do with the forces at your command I do not see." Clearly Cump would continue to have to lobby for his Georgia campaign.[25]

As Sherman discussed plans, so did his Confederate counterpart. The president arrived at Palmetto Station around 3 or 3:30 P.M. on the 25th. From Macon he had to take a circuitous rail route, as the Yankees held the railroad close to East Point. Davis probably traveled to Columbus, Opelika, and northeast from there toward Palmetto. (This would explain the two days spent traveling from Macon September 23 to his arrival at Palmetto on the afternoon of the 25th.) On the 24th Hood wired Brig. Gen. Robert Tyler, commanding at West Point along the president's route, "Please inform me by telegraph at what time President Davis will leave West Point, and what officers are with him." Woodworth states that at Palmetto Station Davis was greeted by a special

[25] S. Davis, *What the Yankees Did to Us*, 360–61; Sherman to Grant, October 1 (1 P.M.), *OR*, vol. 39, pt. 3, 3; Halleck to Sherman, September 26 (1 P.M.), *OR*, vol. 39, pt. 2, 480; Grant to Sherman, September 12, *OR*, vol. 39, pt. 2, 364.

Sherman replied to Grant's letter (delivered by Porter) with a telegram which the acerbic Otto Eisenschiml has called "a masterpiece of evasiveness." As example, he promised to "keep Hood employed," though by then Hood had moved his army well beyond Sherman's reach (Eisenschiml, "Sherman," 30).

guard. From there the president and aides boarded an unceremonious boxcar for a short train ride to Hood's encampments.

The two leaders met and began essentially two days of talks. They left no minutes, but we may surmise that they discussed Hood's options—or lack of them. The Confederate army was too weak to move against Sherman in Atlanta's fortifications, or even to risk an open field battle. It could not stay idle at Palmetto; despondency would further sink in among the men, and desertions would increase. Hood had to do something at least to try to reclaim the initiative. Sherman had the luxury to rest and refit his troops, but if left alone would sooner or later move against the Confederate supply centers at Macon, Augusta, or Columbus. Hood therefore resolved to march into north Georgia and Sherman's rear, tear up his Western & Atlantic supply line, and see whether this could lead to the enemy giving up Atlanta and withdrawing back to Chattanooga. If the enemy did so, the Confederates could be offered improved odds for an eventual battle. Hood had announced this intent in his telegram to Davis of September 6. Four days later a grossly inaccurate message from Wheeler—that he had destroyed fifty miles of track on the two railroads running south of Nashville—offered a measure of hope that Sherman might indeed by defeated through logistics. Then on the 22nd Hood outlined to Bragg his idea of crossing the Chattahoochee, threatening the W. & A., and forcing Sherman "to drive me off."

Davis, leaving for Georgia, may not have seen the telegram, so the general would have had to lay out his plans. As William J. Cooper describes them, "Hood proposed to move his army into north Georgia, believing that this maneuver would draw Sherman away from Atlanta. If Sherman refused to oblige the Confederates and marched south from Atlanta, then Hood would be close on his heels." The president was receptive, Cooper writes, because "Davis knew he could not approve a retreat by Hood after he had removed Johnston for following such a policy. He also knew that retreat toward the Atlantic or the Gulf of Mexico could have but one result. Aggressiveness he felt essential, so he supported Hood rather than those who urged a more cautious and static autumn for the Army of Tennessee." Connelly's conclusion is a fair one: Hood's plan had its risks and questionable rewards, "but he had no choice." "If Hood were going to halt Sherman's drive at all," he adds, "this appeared to be his last chance." McMurry agrees: "Hood's plan to operate on

Sherman's communications was as sound as anything the Confederates could have devised."[26]

The question is whether or when Hood and Davis during their time together at Palmetto, September 25–27, began to entertain the possibility that the Confederate advance into north Georgia might be extended to a march into Tennessee. In his Macon address the president had only talked about Southern forces pouncing on Sherman's railroad and forcing his retreat. This and his "Moscow" allusion contained no hint of the possibility that the Army of Tennessee would march from its base southwest of Atlanta all the way into the Volunteer State.

The question involves, as usual, historiography: the study of what has been written. "Just who did originate the idea of Hood's campaign into Tennessee is difficult to discover," admits Horn. After the war Davis claimed that at Palmetto he and Hood discussed the Army of Tennessee's marching no farther than Gadsden after it had torn up the railroad in north Georgia. He contended that Hood "decided to continue his march into Tennessee" sometime in late October, and that he learned about this "change of programme" from Beauregard. In his memoir Hood merely states that in their meeting at Palmetto on September 25 the two leaders discussed "the particulars of the proposed campaign in the rear of Sherman."

Historiography can be conflicting. Take Woodworth's description of the president's arrival at Palmetto.

> It was raining, and the bedraggled soldiers standing in the oozing red Georgia clay must have looked a forlorn sight. Davis decided to go straight into his morale-raising activities and addressed extemporaneous remarks to the dripping soldiers. "Be of good cheer," he conclud-

---

[26] Shoup, "Journal of Brig. Gen. Francis A. Shoup, C. S. Army, Chief of Staff," entries of September 10 and 25, *OR*, vol. 39, pt. 1, 804–805; Hood, *A & R*, 253; Hood to Tyler, September 24, *OR*, vol. 39, pt. 2, 868; Woodworth, *Davis and His Generals*, 291; Crist, et al., *Papers of Jefferson Davis*, 11:68 n.1 ("short ride with his aides"); Connelly, *Autumn of Glory*, 476–78; Losson, *Tennessee's Forgotten Warriors*, 199 ("to remain at Palmetto would risk demoralization and a loss of fighting spirit"); McMurry, *Hood*, 158; Hood to Davis, September 6, *OR*, vol. 38, pt. 5, 1023–24; Hood to Bragg, September 22, *OR*, vol. 39, pt. 2, 862; Kean, *Confederate Government*, 175 ("The President went south on Wednesday. About the time he started, Hood telegraphed that he had adopted a very bold purpose"); W. J. Cooper, "A Reassessment of Jefferson Davis as a War Leader," 198.

> ed his brief address, "for within a short while your faces will be turned homeward and your feet pressing the soil of Tennessee."

Why does Woodworth claim that Davis forecast the army's advance into Tennessee on the afternoon of September 25, before he had even begun to discuss strategy with General Hood? His source for Davis's "pressing the soil of Tennessee" remark, as we see in his endnotes, is only Shelby Foote's famous but unannotated narrative history. It is likely that Foote, and hence Woodworth, misdated Davis's remark, for a Confederate veteran two decades later claimed that Davis announced the Tennessee plan not on the 25th, but the next day. On September 26 the president reviewed and addressed the troops. "Soon we commence our march into Kentucky and Tennessee," Sam R. Watkins recalled Davis as saying. "Be of good cheer, for within a short while your faces will be turned homeward, and your feet will press Tennessee soil." Admittedly, Watkins was writing years after the war; his reminiscence was published in 1882. Still, his statement would make chronological sense. Hood and Davis met on the afternoon of September 25. If, as we assume, they discussed a march into Tennessee, the president would be eager the next day to share the news with the troops most happy to hear it (Private Watkins served in the 1st Tennessee Infantry). As Watkins remembered it, Davis promised, "We will flank General Sherman out of Atlanta, tear up the railroad and cut off his supplies, and make Atlanta a perfect Moscow of defeat to the Federal army." Watkins embellished his recollection by adding that Robert Toombs, then a brigadier general in the Georgia State Militia, followed Davis, "telling us how quick we were going to whip the Yankees, and how they would skedaddle back across the Ohio river like a dog with a tin oyster can tied to his tail." Some years later, in observances held at Franklin, Tennessee, on the forty-fifth anniversary of the battle, Confederate veteran Sam Cunningham gave an address in which he also recalled the president's speech at Palmetto, especially the forecast that Hood's army would march into Tennessee. Governor Isham Harris, of the Volunteer State itself, and Maj. Gen. Howell Cobb (commander of Georgia state troops) followed; with their combined oratory, "much enthusiasm was aroused." "It was great news to the Tennessee soldiers especially," Cunningham added.

Connelly accepts Watkins's recollection of what the president told the Tennesseans, but only up to a point. "In his speech to the First Tennessee Regiment at Palmetto, Davis promised that 'we will flank Sherman out of Atlanta, tear up the railroad and cut off his supplies,'" Connelly writes, "'and make Atlanta a perfect Moscow of defeat.'" The historian did not quote Watkins's "Be of good cheer…your feet will press" statement because Connelly argues that Hood and Davis *did not* discuss Tennessee in their meeting of the 25th. Thomas R. Hay *does* quote this, even adding that after the president spoke of "Tennessee soil," General Hood followed, declaring "within a few days, I expect to give the command 'forward.'" Christopher Losson, General Cheatham's biographer, takes a middle course: Davis told the troops they would return to Tennessee, but the president intended this statement only as a crowd-pleaser. "The latter was probably just a notion designed to win the applause of the Tennesseans in Cheatham's command," he concludes, "and was not seriously considered as a part of Hood's forthcoming move."

The three biographers are divided on the question. O'Connor believes that Davis and Hood did indeed discuss a Tennessee campaign, but that later Davis denied it. Dyer states that the president made his "Be of good cheer…pressing Tennessee soil" promise on the evening of the 26th, after the troop review and following a serenade by the 20th Louisiana regimental band. But then he asks how this could be so, when "Hood's Tennessee campaign as it later developed had not been planned at this time." McMurry takes this point of view also, contending that Hood's decision to march into Tennessee was only conceived well after the president's visit, and thus was a "new plan" devised after September 25–27.[27]

There is other evidence that Davis and Hood did indeed discuss an offensive march into Tennessee. John B. Jones, the Richmond war clerk, afterward recorded in his diary that the idea for sending Hood north was that of "the government itself." Other evidence comes from generals Taylor and Bragg. After he left Palmetto on the evening of the 27th the president traveled to Montgomery. There in the early afternoon of September 29 he addressed the Alabama state legislature and met with some

[27] Horn, *Army of Tennessee*, 374; J. Davis, *Rise and Fall*, 2:569; Hood, *A & R*, 253; Woodworth, *Davis and His Generals*, 291, 367 n.209; S. R. Watkins, *Co. "Aytch" Maury Grays*, 237–39; Scaife and Bragg, *Joe Brown's Pets*, 257 (General Toombs); S. A. Cunningham, "Events Leading to the Battle," 17; Connelly, *Autumn of Glory*, 479; Hay, *Hood's Tennessee Campaign*, 23, 205 n.25 (citing *A & R*, 255, which does not contain Hood's quotation); Losson, *Tennessee's Forgotten Warriors*, 199; O'Connor, *Cavalier General*, 223; Dyer, *Gallant Hood*, 275, 358 n.16; McMurry, *Hood*, 162.

leading citizens. That night the president met Dick Taylor for a conversation of several hours. According to Taylor, writing a dozen years after the war, Davis began by expressing gratification at the apparently good morale of Hood's officers and men, whereupon General Taylor "warned him of the danger of listening to narrators who were more disposed to tell what was agreeable than what was true." The president agreed.

Then came the strategic talk, as remembered by General Taylor.

> Proceeding to discuss the suggested movement toward Nashville, I thought it a serious matter to undertake a campaign into Tennessee in the autumn, with troops so badly equipped as were ours for the approaching winter. Every mile the army marched north, it was removing farther from supplies, and no reenforcements were to be hoped for from any quarter. Besides, Sherman could control force enough to garrison Chattanooga and Nashville, and, if time were allowed him to accumulate supplies at Atlanta by his one line of rail, could abandon everything south of Chattanooga, and with fifty thousand men, in the absence of Hood's army, march where he liked.

In the face of this opposition, Davis could only ask Taylor what troops could be brought eastward from across the river. The answer was, none (as Taylor had earlier made clear to Hood). Taylor offered hopefully that maybe Hood could operate on Sherman's railroad line north from Atlanta. At some point the issue of command of the Army of Tennessee came up. "As Johnston had been so recently removed from command, I would not venture to recommend his return," Taylor recalled, "but believed that our chances would be increased by the assignment of Beauregard to the army."

Thus, according to Taylor in his reminiscences, he was discussing Hood's advance into Tennessee within forty-eight hours after Davis had left Palmetto. He could only have done so if Davis had indeed talked of such a proposal with Hood. Further evidence that this occurred comes from General Bragg. Davis must have telegraphed General Bragg, who was in Richmond, informing him of the possible raid by Hood's army into Tennessee, for the president's adviser was to-the-point opposing it in his telegraphed reply of September 29—the same day as the Davis-Taylor meeting. "Operations in Middle Tennessee by a force unable to seize and hold the country will not benefit us and will seriously distress our people," he argued. Bragg knew this firsthand, from the failure of his own raid into Kentucky in fall 1862. "Raids are injurious to our troops and unprofitable in results." (Here Bragg properly used the term *raids* instead of *invasions*. Confederate armies could *raid* into northern territory without the expectation of holding onto it. On the other hand, northern armies *invaded* southern territory and maintained military control over large parts of it—albeit many times in conflict with local guerrillas and other pro-Confederate irregulars.) As for the injurious consequences of raids, one had only to recall the outcome of Lee's two grand raids into the North to confirm Bragg's point. "Our true field of operations is south of Tennessee," he therefore concluded, where Confederate cavalry could operate against Sherman's vulnerable rail link to Chattanooga and beyond.

As we have seen, a number of historians, including Hood's biographers, overlook these pieces of evidence as to Hood's planning of his eventual Tennessee campaign. At least William J. Cooper notices them. "Both Bragg and Lieutenant General Richard Taylor, the commander of the Department of Alabama and Mississippi whom Davis visited when he left Hood's army, objected to an offensive campaign," Cooper writes. "They argued that Hood's lack of strength and Sherman's overwhelming force made remote the chance of Confederate success." Then he wryly observes, "Of course, adopting the defensive in no way guaranteed success either."

One further piece of evidence lies in the Stephen D. Lee Papers at Chapel Hill: Jefferson Davis's letter to Lee of July 28, 1873. In it Davis asks Lee's recollections "in regard to our consultations when I visited the army in its encampment near to Atlanta"—when the president visited the army at Palmetto—"and the plan of operations agreed on." "It has been currently reported that I was the author of the movement into Tennessee," he wrote, "by which Sherman was left without molestation to the sea." Davis wanted to point out that this was untrue, because the Tennessee proposal was a "plan adopted in your council"—meaning during Davis's conferences with Hood and his corps commanders. Afterward, returning to Richmond, Davis remembered, "I gave to Genl Beauregard who I met in Augusta a statement of the plan" and that "Genl Beauregard expressed his satisfaction." From these statements it is evident that Jefferson Davis wanted to shirk blame for having given Sherman an eventual free pass to march across Georgia and pass off some of the responsibility to S. D. Lee or even to Beauregard. The "movement into Tennessee," according to Davis, thus had several "authors," but the conclusion that must be drawn here is that the plan had obviously been talked about at Palmetto in late September 1864.[28]

On the morning of the 26th Davis, Hood, corps commanders, staff, maybe other dignitaries—General Cobb and Governor Harris were also at Palmetto then—passed before the troops. "President Davis is here, and is expected to review the troops today," wrote Col. Lovick P. Thomas of the 42nd Georgia; "we are arranging to be under arms at 8 one half o'clock this morning." Robert M. Magill, 39th Georgia (Cumming's Brigade, Stevenson's Division) entered in his diary that day, "Monday, 26th.—Jeff Davis to come round at 9 A.M. Everything has to be cleaned up." According to one press report, Davis "was received by the men with great applause and made them a speech." "Cheer after cheer went forth as he passed by accompanied by the music from each band," recorded Capt. William Dixon of the 1st Georgia. General Shoup was also positive. "The President and General Hood, with their respective staffs, rode out to the front to-day," he recorded on the 26th, "and were enthusiastically received by the troops." In truth, some Rebs seem to have hallooed vociferously indeed. "The President together with Genl's Hood, Hardee &c road past on splendid chargers, bowing and touching their hats," Lt. Marcus Ely of the 54th Georgia wrote his wife the next day, "and we in return flourished our old hats and yelled like a set of maniacs. Splendid bands were discoursing sweet music all along the line—colors were flying &c—taking all in all we had quite a gala day."

Others, however, noted a more sullen reaction by the men. Capt. Benjamin L. Posey of the 36th Alabama later wrote, "there was an absence of enthusiasm" among the troops. Things got uglier when malcon-

---

[28] J. B. Jones, *A Rebel War Clerk's Diary* (1866), 461 (entry of December 17, 1864); R. Taylor, *Destruction and Reconstruction*, 250–51; R. Taylor, "The Last Confederate Surrender," 155 ("the interview extended over many hours, and the military situation was freely discussed"); Bragg to Davis, September 29, *OR*, vol. 52, pt. 2, 748; Seitz, *Bragg*, 462 (quoting Bragg's telegram); Hattaway and Jones, *How the North Won*, 233 (Bragg and Lee as "raiders"); Cooper, "Reassessment," 198 n.44; Davis to Lee, July 28, 1873, Stephen D. Lee Papers, #2440, Southern Historical Collection.

Michael Parrish, Taylor's biographer, accepts the general's recollection that he and the president discussed the idea of Hood raiding into Tennessee. "They discussed at length the concept of Hood moving the army north into Tennessee to lure Sherman away from Atlanta," Parrish writes, noting Taylor's disagreement with the plan (*Richard Taylor*, 413).

tents called out for a return of their old army commander. Corporal Magill recorded, “pretty weak cheering. Some shouting Johnston. Give us Johnston again.” Davis and Hood could only ride on, doubtless chastened. “I regretted I should have been the cause of this uncourteous reception to His Excellency,” Hood admitted in his memoirs. He was right. “The troops do not like Hood,” Robert Patrick of the 4th Louisiana entered in his diary after the review. Yet it was not Hood’s doings alone that the men objected to; it was the president’s as well. According to a correspondent of the *Montgomery Mail*, the soldiers’ cool response to the president reflected their lingering resentment that he had fired Old Joe. “The feeling in the army for Gen. Johnston is yet as enthusiastic as it was when he bade the troops a sad farewell,” the reporter claimed. “The sullen glances which were cast at the President, while here, were marks of the displeasure entertained in his presence.... They wished, in their own blunt way, to give expression to the estimation in which they held Mr. D.” And they clearly did so, according to General Arthur Manigault’s postwar recollection. “No cheers saluted him, countenances were depressed and sullen,” he remembered. “‘Give us Johnston! Give us our old Commander!’ and other remarks of a similar nature” were shouted; “this happened in many brigades.” Officers tried to still the outbursts. “Arrests and threats of punishment alone prevented the cry from becoming loud and general,” Manigault recalled. “One could not help remarking the expression of President Davis’ countenance as he passed by,” he added.

> He looked thin, care-worn, and angry. A scornful expression rested on it. He scarcely deigned to lift his hat from his head as we saluted. General Hood, with the Corps commanders, their staffs, and a large escort, accompanied him. All looked uneasy and apprehensive. I never remembered taking part in an affair of the kind so cheerless and unsatisfactory as this one was, where everyone seemed anxious to have it over.

After the war, James H. McNeilly, chaplain of the 49th Tennessee, remembered an incident of the Palmetto review.

> Our colonel was an enthusiastic man, and several who knew his temperament warned him not to call for cheers for the President. They told him that our men would not do anything to insult him, but they would not cheer him. So they urged the colonel to remain quiet in his position while the President passed, and the men would present arms and salute. He promised to do as advised. The President passed the

> Mississippi division and was greeted with ringing cheers. Our colonel's enthusiasm got away with him. As the President came opposite us our colonel spurred his horse out of the line and, swinging his hat about his head, cried: "Three cheers for our President!" But there was no response. The men were as silent as the grave. I never pitied a man more than I did the crestfallen colonel as he got back into line. We all loved him, for he was a grand man and a lovable one, but the men felt that they had warned him. Mr. Davis passed on as if not noticing, saluting as he rode by.

Yet all was not dour during the review, as Capt. Samuel Foster of the 24th Texas, in Granbury's Brigade, recorded a funny incident.

> Now it so happened that F R Lubbock late Governor of Texas was on Davis' staff, and he naturally supposing that the Texas soldiers would be glad to see him, thought he would take this occasion to introduce himself and we would give him a grand cheer—He made a serious mistake and so spoiled the whole thing. He stoped in front of an Irish (Brigade) Regt. just on our right before he got to us. Thinking he had found us, rode square up about the centre pulled off his hat and says "I am Governor Lubbock of Texas" and just when he expected to hear a big cheer, an Irishman says, "An who the bloody H—l is govener Lubbock?" with that peculiar Irish brogue, that made the Governor wilt. He turned his horse and galloped on to catch up with the President and party and pass by us without even looking at us.[29]

---

[29] Horn, *Army of Tennessee*, 372; Huff, *Four Families*, 84 (Colonel Thomas); B. D. Phillips, *Personal Reminiscences of a Confederate Soldier Boy*, 56; "From Hood's Army. Visit of President Davis," *Augusta Chronicle & Sentinel*, September 30; Dixon, *The Blues in Gray*, 237; Shoup journal, September 26, *OR*, vol. 39, pt. 1, 805; McCardle, *A Just and Holy Cause*, 135; L. J. Daniel, *Soldiering*, 20 (soldiers' cheers as "pretty weak"); Ben Lane, "Letter from the Army of Tennessee," *Mobile Advertiser & Register*, October 6, 1864; Hood, *A & R*, 253; Patrick, *Reluctant Rebel*, 230; "The Feeling in the Army of Tennessee," *Augusta Chronicle & Sentinel*, October 8; Manigault, *Narrative*, 256; McNeilly, "A Great Game of Strategy," 382; Brinsfield, et al., *Faith in the Fight*, 237 (McNeilly as chaplain of the 49th Tennessee); Foster, *One of Cleburne's Command*, 133 (diary entry of September 21–26).

Given the sour reception he received from some of Hood's soldiers, President Davis must have pondered his decision two months earlier to replace Johnston. Richard McMurry makes the point: "he must have realized that the morale of that command had been severely damaged by his order on July 18 which

Following his review of the troops, Davis met privately with Hardee. The general's desire for transfer he already knew, but at Palmetto he heard the senior corps commander's further recitation of why he wanted it. According to a memorandum he wrote after their conversation, Hardee resented that during Bragg's visit to the army in mid-July "he never said a word to me on the subject of the campaign." Then Hood had been placed in command of the army "because he was regarded as, par excellence, the fighting man"; Hardee told the president that Hood had been the one who several times had urged Johnston to retreat. He also smarted at Hood's accusation that he had failed to perform well at Jonesboro, characterizing Hood's conduct toward him as "unjust, ungenerous and unmanly." Hardee recounted that "if he wished to know what the Army wished, I would tell him that it wanted a change of Commanders." "I told him Johnston had a wonderful hold on the affections of the Army, that the men were devoted to him." From the president's reply, Hardee gathered that Johnston would not be brought back. When asked about Beauregard, he said that that change in command would be acceptable. In addition to his memorandum of their conversation, General Hardee prepared something else for the president, as lagniappe: a casualty report further indicting Hood. "The losses under Johnston and Hood are separately given," Hardee wrote, "shewing the loss in killed, wounded and missing to be 1951 more under Hood than Johnston."[30]

Davis met with the army's two other corps commanders as well, about which the president was not secretive. In his reminiscence Hood remembered that "the President held a long conference the next day with Lieutenant Generals Lee and Stewart, in a house not far from my tent." According to one account, Stewart and Lee advised the president that a change in army leadership—relieving or transferring Hood—would "better serve to satisfy the wishes of the men, conciliate existing differences between officers, and inspire greater confidence throughout the army." General Stewart apparently expressed preference for Johnston or Beauregard over Hood. A few years after the war Stewart wrote Johnston, laying out his remembrances, including that of Davis's visit at Palmetto Station. "He asked me whom the army preferred as its commander," Stewart re-

---

transferred the command of the army from Johnston to Hood" ("Confederate Morale," 16).

[30] Crist, et al., *Papers of Jefferson Davis*, 11:67–68 ("Conversation with William J. Hardee"), 71; Hughes, *Hardee*, 247–48.

called. "My reply was, in substance, they prefer General Johnston; next to him, of those available for the command, they prefer General Beauregard." Johnston was so gratified by Stewart's letter that he quoted it extensively in his *Narrative.* Connelly claims that S. D. Lee also suggested Johnston or Beauregard as a replacement for Hood, but the documentation of that statement has not come to light. Herman Hattaway, Lee's biographer, shrewdly concludes, "Lee's opinion is not recorded."[31]

Having thus heard from the army's corps commanders, Davis nevertheless seemed upbeat on the evening of the 26th when, again at Hardee's headquarters, he was serenaded by the band of the 12th Louisiana. Hundreds of soldiers gathered to hear the president. "Being called upon," Shoup noted, "he delivered a short and spirited speech," as did Hood, Cobb, and Harris following him. In the audience was Alabama Captain Posey, who heard "some cheering news." "He told us...that we must turn on the enemy and drive him out of Georgia—He expressed the hope that our advance would not stop there, but that we sweep over Tennessee, and finally bear our banners to the Ohio." Posey's article, in the *Mobile Advertiser & Register*, is thus one more source affirming that Davis and Hood had discussed the raid into Tennessee. Another is the recollection of Tennessean Sumner A. Cunningham, who years later remembered that in this firelight peroration, Davis announced that "the army would go into Tennessee for victory."

During the evening, the president received a few callers. One of them was Brig. Gen. Hiram B. Granbury of Cleburne's Division. James P. Douglas accompanied Granbury and wrote his wife the next day that he had been introduced to the president and had heard him conversing; "he seems in good spirits."

Still, Davis had a lot on his mind when he held his last conversation with Hood on the afternoon of September 27. In his memoirs Hood remembered, "I told him I was aware of the outcry against me through the press, since the removal of Johnston, and, if he adjudged a change of commanders expedient, not to hesitate to relieve me entirely from duty

---

[31] Hood, *A & R*, 251; Roy, "General Hardee," 377; Horn, *Army of Tennessee*, 372, 476 n.12 (citing the *Augusta Chronicle* of September 26 for the "better serve to satisfy" statement although I fail to find an issue for that date in the *Chronicle & Sentinel* microfilm at Emory); Connelly, *Autumn of Glory*, 472; Johnston, *Narrative*, 368 (quoting Stewart to Johnston, February 11, 1868); Hattaway, *Stephen D. Lee*, 131.

with the Army of Tennessee." Hood would even accept a demotion to corps or division command in order to stay with the army. Davis assured Hood that he would not be relieved. It was probably in this context that the president informed Hood of his intent to create a new western military division and to put Beauregard in charge of it. Connelly believes that Davis indeed discussed this idea with Hood at Palmetto, "but did not reach a final decision until he had left the army."

The president left Palmetto at 6 P.M. on the 27th. The next day he wrote to Hood from Opelika. "I have anxiously reflected upon the subject of our closing conversation, and the proposition confidentially mentioned," he began. Davis stated that he would talk with General Beauregard and offer him the super-command of Hood's and Taylor's departments. Beauregard could coordinate between the two "without relieving either of you of the responsibilities and powers of your special commands," except when he was present with the army, at which time he could assume command. "Before final action," he closed, "there will be time for you to communicate with me, and I shall be glad to have your views."

Davis entrusted his letter to a staff officer. When he got it on the 30th, Hood voiced no objection. "I concur fully in your views," he answered, promising a fuller reply later. But there is no evidence that Hood wrote anything more. After having offered his resignation to the president, he must have been relieved that he could retain command of his army. Whispers about all of this seeped through the ranks. On September 28, Captain Posey wrote his Mobile newspaper, "Gen. Hood is to remain in command. The reason assigned is, that time is precious, and there is not time to get a successor."[32]

Just as Hood got his wish to stay with the Army of Tennessee, Hardee got his to leave it. Even before he reached Opelika, the president wired Hood, "Relieve Lieutenant-General Hardee from duty with the

[32] Shoup journal, September 26, *OR*, vol. 39, pt. 1, 805; Lane, "Letter," *Mobile Advertiser & Register*, October 6, 1864; S. A. Cunningham, "Events Leading to the Battle," 17; L. R. Douglas, *Douglas's Texas Battery*, 136 (Douglas to "my dear Sallie," September 27); Hood, *A & R*, 254; Horn, *Army of Tennessee*, 373; Connelly, *Autumn of Glory*, 472; Davis to Hood, September 28, *OR*, vol. 39, pt. 2, 880; Hood to Davis, September 30, *OR*, vol. 39, pt. 2, 885; Hay, *Hood's Tennessee Campaign*, 207 n.56 ("No record of such a reply was found. Apparently Hood decided to make the best of the situation and say nothing.").

Army of Tennessee, and direct him to proceed at once to Charleston, S.C., and assume command of the Department of South Carolina, Georgia, and Florida." Hood could not act quickly enough. That same day Major Mason issued the orders relieving Hardee and reassigning him on the coast. Frank Cheatham took charge of the corps by seniority of rank.

Fifteen years afterward, Davis recalled his meeting with Hardee at Palmetto in a letter to Col. Thomas B. Roy, who had served as Hardee's assistant adjutant general. "At this day, so remote from the event, I cannot claim to remember any conversation upon incidental points which may have occurred," Davis wrote. He continued,

> but I can say with certainty that General Hardee was not relieved because of any depreciation of his capacity, his zeal or fidelity. General Hardee had earnestly requested to be relieved; it had been the subject of correspondence between us before my visit to the army, and my objections to complying with his wish were entirely complimentary to him. My assent to his persistent request to be relieved was finally given because of irreconcilable difference between himself and the officer commanding-in-chief.

As Hardee prepared to leave the army, he issued a farewell to his troops, confessing "heartfelt regret" at having to leave "the comrades of so many campaigns." The evening before his departure the men of his corps gathered around the general's quarters, cheering loudly for "Old Reliable."[33]

The president made a roundabout return to Richmond, giving more speeches at Columbus on September 30, Augusta on October 3, and Columbia and Greensboro, both on October 5. At a few stops Davis made subtle allusion to an offensive into Tennessee. At Augusta he noted that even though the conflict had come to Georgia, "I trust…that Providence may soon take the war beyond her borders." "We must beat Sherman, we must march into Tennessee," he declared; "there we will draw from twenty thousand to thirty thousand to our standard." General Hardee, in Augusta en route to his new assignment, followed the president with a few remarks of his own. According to the newspapers, Hardee an-

---

[33] Davis to Hood, September 28, *OR*, vol. 39, pt. 2, 879; Special Orders No. 5, September 28, *OR*, vol. 39, pt. 2, 880–81; Losson, *Tennessee's Forgotten Warriors*, 195–96; Roy, "General Hardee," 377; Hardee's untitled circular, September 28, *OR*, vol. 39, pt. 2, 881; Hughes, *Hardee*, 248–49.

nounced that before he departed the Army of Tennessee, General Hood had told him that "he hoped to lay his claws upon the State road (in rear of Sherman), and, having once fixed them there, it was not his intention to let them loose their hold." At this pronouncement, claimed the press, "vociferous cheers" erupted.

Davis evidently liked the "claws upon the State road" imagery and used it himself the next day in Columbia. Speaking of Hood, he said, "His eye is now fixed upon a point far beyond that where he was assailed by the enemy. He hopes soon to have his hand on Sherman's line of communications, and to fix it where he can hold it." "I therefore hope," he added, "that within thirty days that army…will be in search of a crossing on the Tennessee river." At Greensboro a reporter recorded that Davis claimed that Hood "expected to select his own position on that railroad, and if once he laid his paws on Sherman's rear, he expected to hold them there, wait for him to attack and then whip him." Davis announced that Sherman "would then be driven out of Tennessee and Kentucky, even across the beautiful Ohio, by our advancing and conquering armies." He even sounded a bit jaunty at Columbia, where he was quoted as declaring, "I believe it is in the power of the men of the Confederacy to plant our banners on the banks of the Ohio, where we may say to the Yankee, 'be quiet, or we will teach you another lesson.'"

While in Columbia, the president visited his friends, the Chesnuts. In conversation with Mary, "Jeff Davis stood up for his general" (meaning Hood) evidently so warmly that Buck "said she would kiss him for that—and she did." "He said his confidence in Hood was increased by these later events," Mrs. Chesnut recorded in her diary, apparently referring to Hood's plan to launch offensive operations.[34]

---

[34] Crist, et al., *Papers of Jefferson Davis*, 11:76, 79–80, 82–88, 91–92; Rowland, *Jefferson Davis Constitutionalist*, 6:358, 361; "President Davis in Columbia," *Macon Telegraph*, October 8; Hettle, *The Confederate Homefront*, 179 (Davis at Columbia); Woodward, *Mary Chesnut's Civil War*, 651–53 (entry of October 7, 1864).

Fifteen years later Jefferson Davis seems not to have recalled his oratorical references to Hood advancing into Tennessee. In a letter to Lucius Northrop in September 1879, Davis gave this recollection of his planning meeting with Hood at Palmetto: "my advice was to move the army to some elligable point upon the Chattanooga & Atlanta R. Rd. and there entrench it…. This it was supposed would compel Sherman to move to attack him…. If…Hood should not find the spirit of his army such as to justify him in offering battle, then he

President Davis's oratory—a dozen speeches, September 21–October 5—may not have had the hoped-for effect of raising the citizens' war-spirit. Richard Barksdale Harwell has sniffed that they "probably did more to improve the President's morale than that of the people." Others, like Woodworth, have judged that "overall, Davis' efforts to lift Confederate spirits seemed to be successful." Example is to be seen at Augusta, where the president was interrupted at least three times by applause and cheers. "Vociferous hand-clapping made him pause for some time," writes Horn. William J. Cooper agrees that Davis's speech-making tour was effective, and astutely points out that Davis's Northern counterpart "never did anything remotely comparable" to his three wartime tours of the Confederate States.

At the same time, Davis has come under criticism for having publicly announced that Hood and his army would march north, tear up Sherman's railroad supply line, and advance into Tennessee. Woodworth terms these statements "some rather severe breaches of military secrecy." Then he checks himself: "Perhaps Davis believed the next move was so obvious that it would be plain to any intelligent observer." He adds, "Or the president may simply have decided that raising morale was so desperately necessary as to justify the loss of the advantage of strategic surprise."

"Strategic surprise" is a strong statement, and it is a misleading one. As we have seen, even before he read Davis's Macon speech in the *Telegraph*, Sherman had already predicted that Hood and Forrest would be attacking his lines of communication. After Hood began moving his ar-

---

was to retreat towards Gadsden…to make a determined stand, and fight a conclusive battle." Davis claimed to Northrop that his first knowledge of Hood's intent to march into Tennessee came from "a telegram of the 24th of November, which unfortunately has been lost" (Davis to Northrop, September 25, 1879, Rowland, *Jefferson Davis Constitutionalist*, 8:415–416). The ex-president's remembrance has been accepted by such students as Donald Stoker: "In an 1879 letter, Davis recalled that in September 1864 he had advised Hood to move his army to some point along the railroad between Chattanooga and Atlanta and dig in" (Stoker, *The Grand Design*, 380). On the other hand, John D. Fowler is on firmer ground in questioning Davis's postwar memories. In his memoir, Davis wrote, "I must say after the event, as I did before it, that I consider this movement into Tennessee ill-advised" (J. Davis, *Rise and Fall*, 2:570). Fowler counters that "Davis gave his tacit if not explicit approval to the invasion," to the point, as we have seen, of announcing it publicly in his whistle stop speeches en route back to the capital from Palmetto (Fowler, "The Finishing Stroke," 146).

my from Lovejoy's to Palmetto and Forrest raided Athens, Sherman inferred, as he states in his reminiscences, that the Rebels' "real purpose was to assume the offensive against our railroads." On September 25 Sherman ordered John Newton's division of the IV Corps to Chattanooga. The next day, when Davis was addressing the troops at Palmetto, he sent John Corse's division (XV Corps) to Rome. Nevertheless, Davis's speeches as reported in the newspapers reinforced Sherman's apprehensions. "The whole burden of his song," he wrote Grant a month after Davis had returned to Richmond, "consisted in the statement that Sherman's communications must be broken and his army destroyed." "He promised his Tennessee and Kentucky soldiers that their feet should soon tread their 'native soil,' etc., etc.," Sherman declares in his *Memoirs*. "He made no concealment of those vainglorious boasts, and thus gave us the full key to his future designs." "To be forewarned was to be forearmed," he concludes.

This was not forewarning so much as confirming. Besides, as Hood's advance developed, and the Army of Tennessee marched from north Georgia into north Alabama, Sherman would have divined Hood's purpose even without Jeff Davis's speeches. Yet the president's rhetorical flourish in his address at Macon and elsewhere—"the fate that befell the army of the French Empire and its retreat from Moscow will be reacted" by Sherman's army (after his railroads were wrecked)—provoked sarcastic quips among the Yankees. "Mr. Davis," General Grant is said to have remarked, "has not made it quite plain who is to furnish the snow for this Moscow retreat through Georgia and Tennessee."[35]

Sherman, as usual, overstates the case. In his *Memoirs* he claims that when Jeff Davis addressed Hood's soldiers at Palmetto on September 26, "one of our spies was there at the time, who came in the next night, and

---

[35] McPherson, *Embattled Rebel*, 215 ("Davis' public speeches during his trip to the Deep South had alluded to an invasion that would take Hood's army all the way to the Ohio River"); Taylor, *Destruction and Reconstruction*, 360 n.5; Woodworth, *Jefferson Davis and His Generals*, 293; Strode, *Tragic Hero*, 96–99; W. J. Cooper, *Jefferson Davis and the Civil War Era*, 70; Sherman, *Memoirs*, 2:140–41; Welcher, *The Union Army*, 2:583; Sherman to Grant, November 6, *OR*, vol. 39, pt. 3, 659; H. Porter, *Campaigning with Grant*, 313.

Bell I. Wiley, generally critical of the Confederate president, nevertheless acknowledges Davis's "'effectiveness as a public speaker," crediting his speeches as "outstanding for clarity and logic," though not commenting on their ability to sway the public mood (Wiley, "Jefferson Davis, an Appraisal," 11).

reported to me the substance of his speech to the soldiers." Yet just a few pages later he states, "There was great difficulty in obtaining correct information about Hood's movements from Palmetto Station. I could not get spies to penetrate his camps...."

Sherman embellished his Palmetto claim a decade later in *Battles and Leaders*: "[H]ere also Mr. Davis made his famous speech, which was duly reported to me in Atlanta, assuring his army that they would make my retreat more disastrous than was that of Napoleon from Moscow. Forewarned, I took immediate measures to thwart his plans."

"Forewarned" is also an exaggeration, for Sherman had already begun taking steps, as we have seen, to protect his railroads into Tennessee. On September 28 another division of the XIV Corps, James D. Morgan's, was ordered from Atlanta to Chattanooga. The next day it set out, accompanied by Major General Thomas, whom Sherman had ordered to take charge of organizing defense troops in Tennessee. Thomas arrived in Nashville on October 3 and would not return to Sherman's army group. He would face Hood in Tennessee.[36]

If so named, Hood's Tennessee Campaign began on September 29, 1864, with Hood, as he telegraphed Davis, "very hopeful of good results." That morning the army started marching toward the Chattahoochee, Red Jackson's cavalry in the lead. According to returns of September 20, it numbered 4,455 officers and 42,976 men present for duty. In the artillery, which was reorganized before General Elzey's assumption of command on September 25, were counted 124 guns in thirty-two batteries.

Late in the day the column of march reached the river at Pumpkin Town and Phillips' Ferry near Campbellton, where the engineers had laid a pontoon bridge. As night fell, soldiers held torches so the troops could pass over. Stewart's and Lee's infantry plus cavalry got across. The corps of Frank Cheatham, who had promptly taken over from Hardee, remained on the south bank while Alfred Iverson's cavalry stayed in the

---

[36] Sherman, *Memoirs*, 2:141, 144; Sherman, "Grand Strategy," 254; Welcher, *The Union Army*, 2:583.

Some writers, particularly those critical of Davis for having relieved Johnston, accept Sherman's exaggeration. One is Atlantan Bill Scaife: "the President divulged in great detail, the strategy and logistics of Hood's upcoming invasion of Tennessee." He then quotes Sherman's claim, even capitalizing "FOREWARNED" (Scaife, *Campaign for Atlanta* [1985], 116–17).

rear near Palmetto, watching for enemy movement. On the 30th the march resumed. Hood, Major Mason and the staff pitched their headquarters at such rustic boroughs as Pray's Church and Dark Corner. From the latter place on October 1 Hood sent word to Iverson that he was still worried Sherman would yet strike out from Atlanta, perhaps to the southwest. To Brig. Gen. Robert Tyler, commanding at West Point on the A. & W. P. line to Montgomery, Mason wired, "Should such a movement as this be made by the enemy on a large scale this army will fall upon his rear."[37]

The men set out on their march in better moods than they had felt for some time. "Men in high spirits," recorded General Cleburne on the 29th in a diary he began keeping. One Tennessean recalled that when the men realized they were heading back toward their native state, "their spirits rose visibly." The Missourians of Francis Cockrell's brigade, according to Lt. George Warren of the 3rd, were "in very good spirits and all hopeful that our expedition will turn out successfully." "Jeff Davis has been here to review us and we have all sorts of rumors," Alabaman Hiram Williams entered into his diary on the 29th; "I think we are bound for Tenn." "I have scarcely ever seen a greater change that has taken place in the Army since the visit of President Davis," Dr. J. M. Brannock wrote to Sarah on October 5; "Gen'l. Hood has published his plans to the whole army & by this course has gained the confidence of the men." He added, "I believe they will fight for him now as well as they would for Joe Johnston."

The men, however, were quickly without food, except what they had stowed in their haversacks. "We crossed the Chattahoochee River about twelve miles southwest of Atlanta," remembered one officer, "and what a soldier could not pack was left behind." Hood had issued orders for the movement of the army's supply trains, but they did not keep up with the infantry columns. On October 1 Louisianan Robert Patrick

---

[37] Hood to Davis, September 28, *OR*, vol. 39, pt. 2, 880; Butkovich, *The Battle of Allatoona Pass*, 35–36; abstract for September 20, 1864, *OR*, vol. 39, pt. 2, 850; Daniel, *Cannoneers in Gray*, 167–68; [Shoup] journal, entry of September 25, *OR*, vol. 39, pt. 1, 805; Scaife, *Campaign for Atlanta*, 137; Mason to Lee and Stewart, September 29 (8 A.M.), *OR*, vol. 39, pt. 2, 884; Hood, *A & R*, 255–56; Scaife, *Allatoona Pass*, 8; Mason to Iverson, October 1, *OR*, vol. 39, pt. 3, 780 ("General H. expects you to give him prompt and reliable information on all movements"); Mason to Tyler, October 1, *OR*, vol. 39, pt. 3, 780.

complained, "I do not know how men can march without something to eat"; they had bacon but no bread rations. As usual, the man in charge got the blame. "General Hood or somebody else at the head of affairs doesn't seem to understand his business," the lieutenant observed.[38]

[38] W. L. White, "The Long Lost Diary of Major General Patrick R. Cleburne," 10; Purdue and Purdue, *Pat Cleburne*, 388; Bishop, "Twenty-ninth Tennessee Infantry," 1:437; Williams, *War So Horrible*, 114; Bailey, *Chessboard*, 34 (Brannock); Shotwell, "Hood's Campaign through North Georgia," 311; Mason to Jackson, September 28, *OR*, vol. 39, pt. 2, 881; Patrick, *Reluctant Rebel*, 231.

Chapter 2

October 1864

On October 1, all waited for Cheatham's Corps to come up. That afternoon Hood called a conference of not just generals Stewart and Lee but of the division commanders as well. Major General French remembered that Hood asked what they thought Sherman would do. French answered that he would head either for Mobile or Augusta, then Charleston or Savannah. Hood wondered how Sherman would feed his troops on such long marches. To French's reply that the enemy would find all the subsistence they wanted, Hood answered, "Well, I have nothing to do with that, as the President has promised to attend to that matter." French noted, "Every officer present disagreed with me save Gen. S. D. Lee."[1]

All the generals thus knew that Hood's plan called for marching north, reaching the Western & Atlantic, tearing it up, breaking the enemy's flow of supplies from Chattanooga, and seeing what Sherman would do. Not just the generals, but the men knew, too. Hood asked his division leaders to address their men about the forthcoming campaign—"the position of affairs and what he intended to do and what the men might be called on to endure," as Cleburne recorded in his diary. The Irishman complied, delivering "one of the most stirring patriotic speeches I ever listened to," one Confederate from Florida remembered. "In closing his address that night he turned his face towards the skies and with all the fervency of his soul he exclaimed," according to the Floridian, "'If this cause that is so dear to my heart is doomed to fail, I pray heaven may let me fall with it, while my face is toward the enemy and my arm battling for that which I know to be right.'"

Others were not so taken in. "Gen Hood has issued a circular in which he says he intends to strike the Rail Road," Private Patrick glumly

---

[1] Falconer to Lee, October 1, *OR*, vol. 39, pt. 3, 779 ("the general desires to see you and your division commanders at 5 o'clock this evening"); French, *Two Wars*, 223.

recorded in his diary on October 2. "The men must be prepared to live upon parched corn and fresh beef without salt for a time, until he can accomplish his purposes." To Patrick, this was an ill omen. "I fear this campaign will prove a failure though I will continue to hope for the best."

As his army began its new movement, Hood tried to keep General Bragg informed of his progress. On the morning of October 2, he telegraphed that on that day his forces would advance to a front stretching from Powder Springs (fourteen miles north of Campbellton) on his right to Lost Mountain (twenty miles) on his left. "This will, I think force the enemy to move on me or to move south," he speculated. The possibility that Sherman might "move south" meant that the enemy might also head east for Augusta, site of the Confederacy's essential gunpowder factory. Should that happen, Hood called for ably commanded reinforcements to be sent there. Then he added to Bragg, "Could General Lee spare a division for that place in such an event?"

This last remark reflected John Bell Hood's lack of strategic insight: Robert E. Lee could not spare a division. In its returns of September 10, the Army of Northern Virginia counted 51,045 officers and men present for duty (but that was with Early's Second Corps, 12,000 officers and men, detached to the Valley); Beauregard had another 10,799. In contrast, Grant, with his armies of the Potomac and James semi-besieging Petersburg, had at the time a combined effective strength of 84,122. Lee was on the defensive, with his outnumbered troops manning an entrenched line, which at the time stretched for thirty miles. Hood should have been aware of this situation; Confederate newspapers carried the Press Association's columns with their daily reports from the Virginia front. For Hood even to think that Lee's forces could help defend an area of Georgia that the Army of Tennessee was effectively abandoning to the enemy demonstrated a remarkable absence of strategic vision.[2]

---

[2] W. L. White, "The Long Lost Diary of Major General Patrick R. Cleburne," 10-11; Purdue and Purdue, *Pat Cleburne*, 388; Butkovich, *Allatoona Pass*, 59; Patrick, *Reluctant Rebel*, 231 (diary, October 2); J. T. Dodge and H. H. Ruger, "Map Illustrating the Fourth Epoch of the Atlanta Campaign," in *OR Atlas*, vol. 3, plate LX, pt. 1; Hood to Bragg, October 2, *OR*, vol. 39, pt. 3, 782; abstract for Lee's and Beauregard's forces, September 10, 1864, *OR*, vol. 42, pt. 2, 1243–44; "abstract...the armies operating against Richmond," September 30, 1864, *OR*, vol. 42, pt. 1, 39; J. I. Robertson, Jr., *Civil War Virginia*, 165.

October 2 was a day of movement, as Hood had forecast for Bragg. Cheatham's and Lee's infantry marched to Flint Hill Church west of Powder Springs. Stewart's Corps camped at Moon, north of it. Frank Armstrong's cavalry brigade made it all the way to the W. & A. near Big Shanty, where the Confederates caught a Yankee train and burned it.

The infantry would follow the cavalry. On October 3, Hood ordered Stewart to lead his three divisions, (French's, William W. Loring's, and Edward C. Walthall's) to Big Shanty. Major Mason's orders included nothing of the detail as those issued by Sherman when it came to railroad-wrecking, but Stewart's troops were to tear up the W. & A. (The next day Mason conveyed Hood's wishes about filling up railroad cuts and felling trees on parallel wagon roads.) Stewart was to complete his work, then retire on the 4th back to Lost Mountain, where the rest of the army would be encamped. (Hood's HQ was at a residence four miles from there.) Meanwhile General Hood was mindful of the enemy's position. Worried about a Union infantry division rumored to be at Marietta, he instructed Frank Armstrong to keep watch.

Stewart's infantry made lodgments on the railroad at two places on October 3. W. S. Featherston's brigade of Loring's Division struck at Big Shanty and after a brief skirmish captured the hundred Federals guarding the station. Two miles to the north, Daniel Reynolds's brigade of Walthall's Division occupied Moon's Station, bagging the eighty-odd Yankees there. Loring's other two brigades, John Adams's and Thomas Scott's, approached Acworth, another several miles above Moon's, by evening of the 3rd. The next day they overwhelmed the small Federal garrison there and took 250 prisoners.[3]

French's Division came up and all set to work tearing up the track. "All night every one was hard at work destroying the railroad," General French recorded. The men found the work tiring but enjoyable. "To destroy is a soldier's joy," wrote Capt. Joseph Buyce, a Missourian. "Here was property belonging to our friends"—meaning the State of Georgia, which owned the W. & A.—"but this was no concern of ours." The men pried up the rails, set their tie-fires, heated the iron, and bent the rails around trees. They also flung logs, dirt, and rocks into a railway cut near

---

[3] Butkovich, *Allatoona Pass*, 60–64; Mason to Stewart, October 2 and 3, *OR*, vol. 39, pt. 3, 783, 785; Scaife, *Needless Effusion*, 9; [Shoup] journal, entries of October 3 and 4, *OR*, vol. 39, pt. 1, 806 ("Carley's house"); Walthall report, January 14, 1865, *OR*, vol. 39, pt. 1, 825.

Big Shanty. By noon of October 4, the Confederates had torn up and twisted the rails for some eight miles of track, French thought. Peter Stewart believed the damage was more extensive. "By 3 P.M. of the 4th," General Stewart later reported, "the railroad was effectually torn up, the ties burned, and rails bent for a distance of ten or twelve miles." This destruction, and the capture of the three stations with several hundred prisoners, was achieved with twelve or fifteen casualties, mostly wounded.[4]

Sherman was being kept informed of the Rebels' movements. Judson Kilpatrick's horsemen were following Hood's army, sending reports back to headquarters. "I think Hood has crossed the Chattahoochee with two corps to attack our road, and has left one corps on this side, near Campbellton," Sherman wrote Jeff. Davis, his XIV corps commander on the 2nd. "We should interpose." He began planning "to put our whole force in motion in pursuit" of the Rebels. "I propose to attack the force on the other side" of the Chattahoochee, he announced to Jacob Cox, in uncharacteristically (for Sherman) bellicose fashion. Meanwhile he was fiddling with the organization of his army group. On September 22 the XVI Corps (which had only two divisions) was dismantled; its units were assigned to the XV and XVII corps. Of his now six corps, several divisions had already been sent north with General Thomas to begin preparing for enemy raiders. On October 3 he issued orders for the rest of his forces. Slocum's XX Corps would occupy Atlanta and guard the Chattahoochee bridge. The IV, XIV, XV, XVII, and XXIII (under temporary command of Major General Cox while Schofield was away at Knoxville), packed with ten days' rations, would head out from Atlanta, cross the river, and concentrate near Smyrna to await further orders.[5]

---

[4] French, *Two Wars*, 225; Gottschalk, *In Deadly Earnest*, 414; Stewart report, January 20, 1865, *OR*, vol. 39, pt. 1, 812.

McMurry states that Stewart's men "destroyed ten miles of track" on October 4 (*Hood*, 159), but a more probable number is General French's eight (Butkovich, *Allatoona Pass*, 69), or even seven by a little arithmetic. The Western & Atlantic measured fifteen miles from Marietta to Acworth (Barnwell, *City Directory*, 107). Trains ran eight miles from Marietta to Big Shanty (Rottman, *The Great Locomotive Chase*, 30). Thus one comes to the figure of seven miles from Acworth to Big Shanty as the length of line wrecked by Stewart's infantry that day.

[5] Butkovich, *Allatoona Pass*, 62; Sherman to Davis, October 2, *OR*, vol. 39, pt. 3, 27; Sherman to Howard, October 2, *OR*, vol. 39, pt. 3, 29; Sherman to Cox, October 2, *OR*, vol. 39, pt. 3, 34; Welcher, *The Union Army*, 2:300; Special

General Hood was issuing orders of his own. On the morning of the 4th he sent instructions to Stewart that he should bring two of his divisions back to the army's concentration at Lost Mountain, but "your third division (say French's) shall move up the railroad and fill up the deep cut at Allatoona with logs, brush, rails, dirt, &c." The railroad cut at Allatoona was a logical target. Dug out from a tall ridge decades before the war, the pass was 175 feet deep and some sixty feet wide. If it could be obstructed, the single-track W. & A. running through it would be blocked. But for how long? One may speculate that Federal fatigue parties working hard could clear away rocks, logs, and debris at least as fast as it would take French's troops to shove them into the cut. (French later complained, "he [Hood] knew we had but a few spades.") Hood's order is thus questionable on this, as well as other grounds. As Brad Butkovich, the recent chronicler of the battle at Allatoona, points out, of Stewart's three divisions, French's was farthest from the pass (at least eight miles, ten via the railroad). Loring's troops (numerically stronger than French's) at Acworth were only four miles away. Worse for the Confederates was Hood's apparent assumption that Allatoona was guarded by an enemy garrison no larger than those bagged by Loring and Walthall on the third. Because of the Federals' large stock of supplies at Allatoona, the place was held by three and a half regiments with a battery—over a thousand officers and men. In his memoirs Hood claims that French's priority at Allatoona was to capture the enemy's supplies there, yet he acknowledges the "two or three regiments" of Union troops guarding them. Hood did not know that the enemy at Allatoona were well dug in, with a fort and a redoubt on the high ground on each side of the railroad (constructed by Sherman's own chief engineer, Capt. Orlando Poe, no less). General French was no friend of Hood (recall his earlier letter to the president). "It would appear," he pointedly wrote in the autobiography based on his wartime diary, "that the general in chief was not aware that the pass was fortified and garrisoned that I was sent to have filled up." Worst still, in his orders Hood added that after laboring to obstruct the railroad French's exhausted troops were to march another five miles north to the Etowah River and burn the railroad bridge over it before returning to join the rest of the army near Lost Mountain and New Hope Church, another seventeen miles of marching. As Butkovich

---

Field Orders No. 83, October 3, *OR*, vol. 39, pt. 3, 43; McDonough, *Schofield*, 98 ("attend to business" in Knoxville).

puts it, "that was a lot to expect from anybody." Hood magnified the issue with a second order sent to Stewart four hours after the first. This one emphasized the burning of the Etowah train bridge as "of the greatest importance." "From the best information we have now," Mason wrote for the commanding general, "he thinks the enemy cannot disturb us before to-morrow, and by that time your main body will be near the remainder of our army." To Hood's tall order was now added an impracticable timetable. Bill Scaife is almost whimsical in his assessment, calling Hood's expanded assignment for French "mission impossible." The concluding question is, if Hood thought the destruction of the Etowah River train bridge "would be a great advantage to the army and the country" (as the second order put it), why did he not order Loring's and Walthall's divisions to join in the work? (Had Hood found out about French's earlier letter to the president?)[6]

Sherman knew the Rebels were on the march. Before they cut the telegraph wire north of Marietta on October 3, he was able to get off messages to Lt. Col. John Tourtellotte, commanding at Allatoona, and Brig. Gen. John M. Corse at Rome. To the former he directed, "I want him delayed only long enough for me to reach his rear"—Sherman was signaling his presumed intent to attack Hood. On the fourth, his five infantry corps marched toward Smyrna. The commanding general himself arrived there that night. He ordered Corse to move his division to Cartersville, reinforcing the Union troops already there. Then he directed Corse, via signal flags (the wire was down), to move with his

---

[6] Butkovich, *Allatoona Pass*, 50–51, 55, 67–69; Mason to Stewart, October 4 (7:30 and 11:30 A.M.), *OR*, vol. 39, pt. 1, 814; French, *Two Wars*, 229, 270; Hood, *A & R*, 257; Scaife, *Needless Effusion*, 9–10.

French after the war claimed that when Stewart handed Hood's order, he said, "General Hood does not seem to be aware that the place is fortified, and now, French, here is a fine opportunity for you" ("Movement against Allatoona—Letter from General S. G. French" [French to D. W. Sanders, May 30, 1881], 404–405). French adds, "If General Hood knew that the Allatoona Pass was fortified and garrisoned and then sent troops there to fill up the cut with logs, rails, etc., and did not inform the commander that it was so fortified and garrisoned, then he committed an almost criminal act." Maj. D. W. Sanders of General French's staff took his boss' side. After the war he claimed of Hood's order, "one more unjustifiable for a general commanding an army cannot be found in the history of the late war than that given by Gen. Hood to French on this occasion" (D. W. Sanders, "Strange Order of Hood at Allatoona," 507).

troops to Allatoona. From their observation site atop Kennesaw Mountain, Union signalmen under command of Brig. Gen. William Vandever wigwagged to the Federals holding the redoubts on Allatoona ridge: "General Sherman says hold fast. We are coming."[7]

So were French and his division. They left Big Shanty at midafternoon of October 4, got to Acworth before sundown and rested while the men cooked their dinners. The march resumed around 11 P.M. Four hours later, French's infantry approached Allatoona, all in darkness. Local residents informed the Confederates of the Yankees' strength and their two redoubts. General French planned for a dawn attack, so he placed his artillery where he thought it would best bear. With the help of a guide, French tried to reconnoiter but decided to wait till daylight. Thereupon the Southerners first saw the challenges ahead of them: "Instead of two redoubts there were disclosed to us three redoubts on the west of the railroad cut," French wrote, "and a star fort on the east with outer works and approaches defended to a great distance by abatis, and nearer the works by stockades and other obstructions." Nevertheless, French deployed his three brigades for assault.[8]

General Corse arrived at Allatoona after midnight on October 4–5, bringing with him a thousand infantrymen as reinforcements. With Tourtellotte's garrison, the Federals now numbered about 2,000 (including six guns of the 12th Wisconsin Battery) against some 3,200 in French's Division. But helping the Federals was luck, always a factor in war, and here it was the weather. With the telegraph down, Sherman could only get messages to Corse by flags atop Kennesaw, but clouds on October 3 had prevented such communication. Fortunately for the Northerners, the weather cleared. "The orders bringing reinforcements to Allatoona were signaled over the heads of the enemy the morning of the 4th," wrote one Union veteran after the war. "If the cloudy weather of the day before had continued this could not have been done, and the reinforcements could not have reached us. Without it the garrison could not have resisted for five minutes."

---

[7] Butkovich, *Allatoona Pass*, 65, 69–71; Sherman to "Commanding Officer, Allatoona" and to Corse, October 3, *OR*, vol. 39, pt. 3, 53; Vandever to Corse, October 4, *OR*, vol. 39, pt. 3, 75; Howard report, December 28, 1864, *OR*, vol. 39, pt. 1, 730 ("the telegraph wires having been cut"); Vandever to "Commanding Officer, Allatoona," October 4 (6:30 P.M.), *OR*, vol. 39, pt. 3, 78.

[8] French, *Two Wars*, 230, 232; Butkovich, *Allatoona Pass*, 91–92.

The Confederates did not know the odds they were facing. One Southern officer told Sgt. William Chambers of the 46th Mississippi, that the taking of the Yankee position was merely a breakfast task. Chambers replied, "anybody was welcome to *my* share of *that* breakfast."

After some initial artillery firing, around 9 A.M. General French sent in a staff officer, Maj. David W. Sanders, with a demand for the enemy's capitulation. "I have the forces under my command in such positions that you are now surrounded," read the note, addressed to the "Commanding Officer U.S. Forces, Allatoona"—French did not know that Corse had arrived to take charge. "To avoid a needless effusion of blood," French stated, "I call on you to surrender your forces at once and unconditionally." He allowed five minutes for an answer.

General Corse could well have been expected to refuse to give up, but he did so in language quite saucy. "Your communication demanding surrender of my command I acknowledge receipt of," he wrote, "and respectfully reply that we are prepared for the 'needless effusion of blood' whenever it is agreeable to you."

Corse's reply, delayed somewhere, never reached Major Sanders. After a quarter hour, he "broke off the interview and returned," as French recorded. He later argued that Corse never wrote a reply, or at least that he never received it. Corse argued back that he did and pointed to his inclusion of the text of French's demand note in his own report, written a few weeks after the battle. Who was right? Butkovich shrugs, "we may never know for sure." The uncertainty allows Southern partisans to call Corse out for lying. After the war, Joseph M. Brown (son of the governor) wrote of Corse's reply that some Confederate veterans "believed that it was written after the battle, because it would read so pretty in history."[9]

The effusion of blood occurred, needless or not. The Confederates charged hard against the Yankees' western redoubt, named for Col. Richard Rowett and eventually overran it. The Federal survivors fled eastward to the Star Fort. The Southerners charged the fort fiercely—

---

[9] Butkovich, *Allatoona Pass*, 53, 55 (1,092 officers and men), 75, 83, 89–93; J. Q. Adams, "Hold the Fort!," 799; Chambers, *Blood & Sacrifice*, 175; French report, November 5, *OR*, vol. 39, pt. 1, 816; Corse report, October 27, *OR*, vol. 39, pt. 1, 763; French, *Two Wars*, 232; Peter Osterhaus report, December 26, *OR*, vol. 39, pt. 1, 741 (complimenting Corse's answer for its "laconic style"); "Autobiography of Maj. D. W. Sanders," 371; Scaife, *Needless Effusion*, 17, 20, 23, 24 (quoting Brown).

four assaults in all—but could not take the position. General Corse was knocked out by a minie ball to his face; Tourtellotte was also wounded. But the Federals held their fort and repulsed a lesser assault against the redoubt east of the railroad gap. French reluctantly gave up the attack around 1:30. His men were running short of ammo and he heard from Frank Armstrong that enemy infantry were marching up the railroad from the south (which proved untrue). French withdrew his troops from the field mid-afternoon. Brig. Gens. William H. Young and Francis M. Cockrell protested the decision to end the attack, but French was adamant.

Casualties attested to the severity of the 3½–hour battle (10 to 1:30). Corse reported 707 K W M (32 percent of 2,190 present). Some regiments were hit harder: the 7th Illinois lost 47 percent of its strength; the 39th Iowa 59 percent. French's loss came to 898 (40 percent of perhaps 2,250 engaged). Among the Southerners taken prisoner was General Young, who had fallen with a severe wound in the foot. (Young was taken by comrades to the rear and put on an ambulance, which took the wrong road and ran into Federal cavalry at New Hope Church.)

Also among the wounded captured was, surprisingly, a woman, fighting in the Confederate ranks like the stoutest of men. Lt. William Ludlow, a Federal engineer, recalled a visit after the battle to a hospital in which a surgeon told him there was a wounded woman among his patients. "See if you can pick her out," the doctor said, whereupon the two walked through the wards.

> We went through the hospital, and I saw no woman, but passing through again on the way back, the doctor stopped at a bed where a tanned and freckled young rebel, hands and face grimy with dirt and powder, lay resting on an elbow, smoking a corn-cob pipe. The doctor inquired, "How do you feel?" and the answer was, "Pretty well, but my leg hurts like the devil." As we turned, the doctor said, "That is the woman."

The formidable female belonged to Cockrell's Missouri Brigade. She was Mary Bowen, widow of Maj. Gen. John S. Bowen, who had died of dysentery in Mississippi in July 1863. After her husband's death, Mary "donned a Confederate uniform and picked up a musket, joining the ranks beside her two brothers and friends from Carondelet and St. Louis," as Philip Tucker explains; "at the field hospital located at the base of the Allatoona ridge, the Yankees were surprised to find the captured female Rebel in gray who mentioned that she had lost her husband and a brother in the war."

Thus ended the battle of Allatoona. The Confederates had not obstructed the railroad cut because the enemy still held fortified positions guarding it. They had also certainly not marched on to the Etowah to burn the railroad bridge across it, which Hood had unrealistically added to their list of tasks. And there was the question of the Northerners' immense commissary stores warehoused at Allatoona, said to be anywhere from a million to three million rations. French offered various excuses: he only learned of their presence when he was about to end the battle; he did not have enough wagons to carry away the supplies; the men were helping themselves to them anyway; before marching away he ordered the warehouses burned, but "the matches furnished would not ignite and no fire could be procured." French's excuses (bad matches?!) are not only lame but contradicted by the recollection of Sgt. Maj. Harvey Trimble of the 93rd Illinois. He remembered seeing some hundred Rebels rushing with flaming firebrands toward the warehouses, but Federal musket volleys killed and scattered them all. Colonel Tourtellotte concurred: his troops from the high ground could pick off the Rebels approaching the stockpiles. "Small parties of the enemy dashed across the dirt road," he remembered, "toward the storehouses, but they died on the way."

French had clearly failed. As one historian of the engagement has concluded, "the supplies at Allatoona, the railroad north, and the Etowah bridge, were all intact." French sought to salvage something positive from the expedition when he reported the capture of the enemy blockhouse at Allatoona Creek and the burning of the 200-foot long railroad bridge over it. Nonetheless after the war he vented his resentment against Hood for having issued him such "remarkable orders." All in all, as Scaife states, French "appears to have made the best of his 'mission impossible' to Allatoona," before leading his division back to rejoin the army on October 6.[10]

---

[10] Hicken, "'Hold the Fort,'" 22–23, 26–27; Scaife, *Needless Effusion*, 37–38, 41, 43, 47; Butkovich, *Allatoona Pass*, 125, 131, 134–35, 137, 144; D. Evans, "Hold the Fort," 24; Emery, *Hold the Fort*, 4–11; Corse report, October 27, *OR*, vol. 39, pt. 1, 766; French report, October 8, *OR*, vol. 39, pt. 1, 813; French, *Two Wars*, 232–24, 264, 276 ("nor had I any means to carry stores away with me"), 283 (Hood's "remarkable orders"); S. Bennett, "'The Storm Broke in All Its Fury,'" 40; Welsh, *Medical Histories of Confederate Generals*, 243 (Young);

Cump Sherman, who had signaled Corse, "Hold the place, I will help you," did nothing of the sort. Horn quips, "Sherman's much publicized assistance was no more than a cheerleader's at a football game." His orders for October 5 to the infantry corps assembled at Smyrna called for only slight movement. By nightfall of the 5th Sherman's advance had gotten no closer than eleven miles from Allatoona—and that was by direct line. In his campaign report Sherman claimed that he ordered Cox's XXIII Corps to "move rapidly from the base of Kennesaw…threatening the rear of the forces attacking Allatoona." A decade later, in his memoirs, Sherman stated his hope "to interpose this corps between Hood's main army at Dallas and the detachment then assailing Allatoona." Concerning these two assertions, Brad Butkovich politely judges, "unfortunately, neither was true." "Either Sherman's memory on the subject was faulty," he concludes, "or he had begun to embellish the facts by January 1865 to make it look like he did more to support the besieged Allatoona garrison than he actually did." Remembering Albert Castel's trenchant characterization of Sherman's memoirs—"prevaricating through Georgia"—I side with the latter of Brad's two options.

For all its malarkey, Sherman's promise to come to Corse's aid was made into a popular postwar hymn, "Hold the Fort," composed by Philip P. Bliss ("we meet to-day in freedom's cause and raise our voices high; we'll join our hands in union strong to battle or to die. Hold the fort, for we are coming, Union men, be strong; Side by side we battle onward, Victory will come"). Writing fifty years ago, Campbell Brown aptly characterized Bliss's song as "a sort of 'Praise the Lord and pass the ammunition' thing." Some were taken in by all this, even Union veterans. "General Sherman," wrote Sgt. William C. Johnson decades later, "signaled

---

Ludlow, "The Battle of Allatoona October 5th, 1864," 782 ("That is the woman"); Cross, *They Sleep Beneath the Mockingbird*, 143 (General Bowen); P. T. Tucker, *The Forgotten "Stonewall Jackson of the West,"* 322–23; Flint, "The Battle of Allatoona," 741 (rations); Gottschalk, *In Deadly Earnest*, 427 (quoting Tourtellotte); French report, November 5, *OR*, vol. 39, pt. 1, 817–18 (Allatoona Creek); F. E. Brown, "The Battle of Allatoona," 295.

Joseph M. Brown takes French's side in the matter of the Northern supplies stored at Allatoona. "General French was not aware that it was an army depot of supplies," he wrote in his monograph on the battle. "Had General French been made aware of the great number of rations in store there he would surely have had them burnt at the time when in his possession" (Brown, *The Battle of Allatoona. October 5, 1864*, 8).

the commander, 'Hold the fort, for I am coming,' and immediately pushed his troops, by a forced march, to the relief of that battalion of men, who did gallantly and bravely 'hold the fort' until the near approach of our troops."

Unfortunately, John Bell Hood also showed some carelessness of fact. As we have seen, his two sets of orders to Stewart for General French mentioned as his objectives only the obstruction of the Allatoona gap and the destruction of the Etowah bridge. Thus we must question the general's memory when he states in *Advance and Retreat* that part of his orders to French were "to gain possession of the supplies," the enemy stores at Allatoona. To be sure, Hood is correct that the journal begun by his chief of staff, General Shoup, mentions in its entry of October 4, "a large supply of provisions." But no information of such had been sent to French. With this in mind, Hood's statement in his recollections, "Major L. Perot, adjutant of Ector's Brigade, has informed me by letter that our troops were in possession of these stores during several hours, and could easily have destroyed them," appears to be something of a stretch.[11]

After French's Division rejoined his army, Hood marched northwestward via Dallas, Van Wert, and Cedartown to Cave Spring, near the Alabama border, where he pitched his headquarters on the afternoon of October 9. The day before, Joe Wheeler rejoined the army with what was

---

[11] Butkovich, *Allatoona Pass*, 70 ("Hold the place"), 116; Horn, *Army of Tennessee*, 376; Ludlow, "Allatoona," 72 ("11 miles distant in an air line"); Sherman report, January 1, 1865, *OR*, vol. 39, pt. 1, 581; Sherman, *Memoirs*, 2:147; Castel, "Prevaricating through Georgia," 97 ("Offense Number Three: Prevarication, fabrication, and falsification for the purpose of making it seem that he, Sherman, knew and did things that in fact he did not know or do"); Scaife, *Needless Effusion*, 47, 49; C. H. Brown, "To Rescue the Confederacy," 14; W. C. Johnson, "The March to the Sea," 805; Hood, *A & R*, 257 (Major Perot); [Shoup] journal, October 4, *OR*, vol. 39, pt. 1, 806.

Leopold Perot, formerly private in the 15th Texas Cavalry, wrote Hood in early 1876, asserting that at Allatoona, if General French had not ordered a cessation of the battle, "the whole of his [Sherman's] Army stores...could have been destroyed" (Hood, *Lost Papers* [Perot to Hood, February 8, 1876]). The statement led Hood to write, "if this assertion be correct, I presume Major General French forbade their destruction, in the conviction of his ability to successfully remove them for the use of the Confederate Army" (*A & R*, 257). The issue of Northern supplies at Allatoona during the battle of October 1864 is one which historians have been unable to resolve.

left of his mounted command. Having left two months earlier with half of the army's cavalry to tear up Sherman's railroad, Wheeler had ridden up the Western & Atlantic, to be sure, knocking it at places. But before Chattanooga, he headed northeast along the East Tennessee & Georgia Railroad toward Knoxville (which was not part of Sherman's logistical network). From Knoxville, the Confederates rode two hundred miles westward to strike the Nashville & Chattanooga line above Tullahoma. Then they headed back south. On September 2—the day Federal soldiers occupied Atlanta—Wheeler's troopers crossed the Tennessee River. They spent the next few weeks resting at Tuscumbia. There Bedford Forrest met with Wheeler and reported on September 20 what he learned to General Taylor. "His command is in a demoralized condition," Forrest claimed. It was reduced in strength to 2,000 men (probably fewer) and Wheeler himself was so dispirited that he talked of being relieved of command. Hood was not aware of this sad state. On September 15, Major Mason wrote Wheeler, "General Hood desires that you should move in this direction…tearing up the railroad as you come between Chattanooga and Atlanta." Needless to say, neither Wheeler nor his horsemen were in any shape to conduct such operations. The headquarters journal begun by General Shoup (and now kept up by staff officers, probably Mason) noted simply on the 8th, "Major-General Wheeler returned to-day with a portion of his command."[12]

Thus far Hood seems to have ignored General Beauregard. Davis had clearly stated his intent in his letter of September 28 ("to place him in command of the department, embracing your army and that of General R. Taylor"), and two days later Hood had quickly agreed to the president's proposal ("I concur fully in your views"). Beauregard and Davis conferred in Augusta on October 2 and confirmed the new arrangement. With all the public swirl about Beauregard possibly succeeding Hood, T. Harry Williams is doubtless correct in stating that the Creole "must have been surprised and also disappointed that he was not to have a field command." Nevertheless, Beauregard accepted Davis's offer of the super-department. The president wrote out his expectations in a memorandum,

[12] [Shoup] journal, entries of October 6–9, *OR*, vol. 39, pt. 1, 806; Hood, *A & R*, 259; Longacre, *Cavalry of the Heartland*, 295–99 (including good map of Wheeler's route); Dyer, *"Fightin' Joe" Wheeler*, 195–96; Forrest to Taylor, September 20, *OR*, vol. 39, pt. 2, 859; Mason to Wheeler, September 15, *OR*, vol. 39, pt. 2, 837.

copying Hood (but not Taylor). Beauregard was to set his headquarters within the military division; "your personal presence is expected wherever in your judgment the interests of your command render it expedient." When personally present with an army in the field he was to assume command of it. Davis also explained his discussions with Hood about making north Georgia a "Moscow of defeat." Orders from the adjutant and inspector general's office in Richmond announcing Beauregard's new appointment went out October 3.

In his memoirs, Hood quotes Davis's memo, pronouncing it "most satisfactory," yet he made no effort to communicate with Beauregard, much less congratulate him on his new role. Indeed, it was the latter who had to track down Hood for a meeting. Beauregard left Augusta on October 4 on a train route that would include other conferences in Milledgeville (Governor Brown) and Macon (General Cobb). On the 7th he wired the president from Opelika that he was headed for Hood's army. Riding by train to Newnan, he set out on horseback for rendezvous with Hood at Cave Spring, seventy miles away. Beauregard arrived there on the evening of October 9. The Shoup journal entries of October 9 and 10 do not even mention General Beauregard's arrival, nor does Hood in *Advance and Retreat.* But the commander of the Army of Tennessee was about to hand his departmental superior the deepest cut of all: an *announcement* that he had changed his campaign plan. The army had crossed the Chattahoochee and marched north, grabbing the initiative. It had struck the Western & Atlantic, torn up miles of track, and compelled Sherman with the bulk of his forces to set out from Atlanta in a loose sort of "pursuit." But Sherman's sloth did not portend a battle any time soon, much less a general engagement in which the odds favored the Confederates. (Sherman's strength with his five corps was said to be roughly 40,000, about the size of the Army of Tennessee.) Thus the objective Hood had announced on September 22, forcing Sherman to "drive me off" the railroad to Chattanooga, seemed unattainable, at least in the short term.

As a consequence, Hood hatched a variant of plan. So far, the cutting of the Chattanooga railway had been the Confederates' most effective tactical achievement. Following the traditional military maxim of reinforcing success, Hood decided to do more of it. "In truth, the effect of our operations so far surpassed my expectations," he wrote in *Advance and Retreat*, "that I was induced to somewhat change my original plan to draw Sherman to the Alabama line and then give battle." Instead, he aimed to go after the Western & Atlantic again. This he announced in a wire to Bragg on October 8. He would cross the Coosa below Rome, march eastward, strike the railroad and tear it up from Kingston to Tunnel Hill. This could force Sherman to march northward. If he did so, Hood would march into Alabama to the Tennessee River at Gadsden. He would then move his trains and surplus artillery to Jacksonville, Alabama, which would become the army's new base.[13]

Having announced his intentions to Bragg, Hood now presented them to Beauregard as a *fait accompli* when the latter arrived at Cave Spring on October 9 ("his arrival creates unusual enthusiasm among the troops," observed Cleburne). There was accordingly little for the Creole to say. Technically Beauregard had not even assumed command of the Military Division (that would happen a week later), and, even if he had the authority, he was reluctant to use it, so new was he to his role and to Hood's theater of operations. In a report of their meeting written a few days afterward, Beauregard informed General Cooper that Hood planned to strike the railroad between Kingston and Dalton. He does not mention the part, announced by Hood to Bragg, about marching to the Tennessee River at Gadsden. This leads Connelly to believe that "evidently Hood, either deliberately or not, misled him." Nonetheless, to Hood Beauregard urged two points of caution: 1) that he guard his crossing point at the Oostanaula in case Sherman brought superior numbers down upon him; and 2) that he not give battle "unless with positive advantage on our side of numbers and position."

It became apparent that Beauregard had other concerns about Hood's plan. According to Alfred Roman, Beauregard's "biographer," the general worried that "the movement had been rather hastily undertaken"; that the army was changing base (from Palmetto to Jacksonville) without careful preparation; that various details seemed not to have been thought through; "and that a great deal had been left to future determi-

---

[13] Davis to Hood, September 28, *OR*, vol. 39, pt. 2, 880; Hood to Davis, September 30, *OR*, vol. 39, pt. 2, 885; Roman, *Beauregard*, 2:276, 280–81; T. H. Williams, *Beauregard*, 241–42; Davis to Beauregard, October 2, *OR*, vol. 39, pt. 3, 782; Special Orders No. 234, Richmond, October 3, *OR*, vol. 39, pt. 3, 785 (Beauregard's assignment); Hood, *A & R*, 258–59; Beauregard to Davis, October 7, *OR*, vol. 39, pt. 3, 800; [Shoup] journal, entries of October 9–10, *OR*, vol. 39, pt. 1, 806–807; Hood to Bragg, October 8, *OR*, vol. 39, pt. 3, 804–805.

nation, and even to luck." In the words of Roman (which were probably Beauregard's own), "It was easy to discover in the details of the plan evidences of the fact that General Hood and Mr. Davis were not accustomed to command armies in the field, especially armies like ours, for the management of which much had to be foreseen, and much prepared or created." "Sadly impressed with what he had seen and heard," Beauregard nevertheless pledged to do what he could to make it all happen and proceeded to Jacksonville. At least he was giving Hood better treatment than he had received. To Hay, this first meeting was merely the beginning of "Hood's indifference and neglect, which were continuous" toward his nominal superior.[14]

Hood had chosen Jacksonville (forty-seven miles southwest of Rome) as a point for the accumulation of supplies. But the town itself had no railroad; the nearest depot was at Blue Mountain (near today's Anniston), twelve miles to the south. Blue Mountain was linked by track to Selma, the munitions center. Beauregard tried to help by directing that the rail line from Blue Mountain be extended to eliminate the dozen miles of wagon traffic (the construction was never begun).

But Hood was contemplating a more northerly base for his march into Alabama. On October 8 he asked Bragg if the Memphis and Charleston Railroad, the main west-east line, could be repaired from Corinth to Decatur. From the latter place, the Tennessee & Alabama runs north to Nashville. (Horn considers this request a signal from Hood that he was indeed intending to march into Tennessee.) Despite the hard work of Maj. George Whitfield, weeks into the campaign the rail line between Corinth and Decatur had been restored only to Cherokee, Alabama, thirty-six miles east of Corinth. Richard Goff, expert on Confederate supply, claims that even though "subsistence existed in abundance in central Alabama and eastern Mississippi," the long, battered route such stores would have to take—from Selma to Meridian to Tupelo and Corinth, then east to Decatur—was tortuous, ramshackle, and unreliable.

---

[14] Beauregard to Cooper, October 12, *OR*, vol. 39, pt. 1, 795–96; W. L. White, "The Long Lost Diary of Major General Patrick R. Cleburne," 14 (entry of October 9); T. H. Williams, *Beauregard*, 242–43 ("Hood *informed* Beauregard" [emphasis added]); General Orders No. 1, October 17, *OR*, vol. 39, pt. 3, 824–25; Connelly, *Autumn of Glory*, 481–82; Roman, *Beauregard*, 2:281; Hay, *Hood's Tennessee Campaign*, 209 n.64.

There is little evidence that Hood gave much thought to all this. (One can only imagine how the compulsively logistics-minded William T. Sherman would have directed his energies to improve such a lamentable situation.) At least Beauregard applied himself to help make things better. For instance, on the 13th from Jacksonville, he informed Hood of the lack of officers there to oversee activity and recommended that at least an engineer officer be sent to construct works to protect the place against raids. Hood readily complied. Beauregard also tried to set up a courier system to deliver messages between Hood and Taylor at Selma, as some of the telegraph stations in north Alabama were along railroad lines in enemy-held territory. Beauregard was doing so much behind the scenes to help Hood that John D. Fowler shrewdly observes, "Beauregard functioned more as Hood's chief of staff than his commanding officer."[15]

The Army of Tennessee embarked on the next stage of its campaign by crossing the Coosa River on October 10. The next day Hood's headquarters were at Armuchee, ten miles northwest of Rome, in the area of Colonel Frank Little's home, where Hood had begun his post-amputative recovery after Chickamauga. General Cleburne noted in his diary that the return of the Confederate army to the region thrilled the residents of the area. "The people are all overjoyed," he wrote; "the women look like some terrible incubus has been lifted from their hearts." Late on the 12th, Stewart's troops reached the railroad near Resaca and began wrecking the track northward. Resaca was held by an enemy garrison, and around 4 P.M. General Lee sent in a demand for surrender signed by Hood, but the Union commander, Col. Clark Wever, refused. Wever wrote back, "In my opinion I can hold this post; if you want it come and take it." Lee decided, "I did not deem it prudent to assault the works,

---

[15] McMurry, *Hood*, 160; Black, *Railroads of the Confederacy*, 264–65 (including good map, "Railroad Supply Lines Hood's Tennessee Campaign October–December 1864"); Roman, *Beauregard*, 2:281; Connelly, *Autumn of Glory*, 482; Hood to Bragg, October 8, *OR*, vol. 39, pt. 3, 805; Horn, *Army of Tennessee*, 378; Goff, *Confederate Supply*, 217; Roman, *Beauregard*, 2:600 (A. R. Chisolm to Hood, October 13, which I do not see in the *OR*); Hood to Beauregard, October 16, *OR*, vol. 39, pt. 3, 822 (Hood's reply); N. Carpenter, *A Slight Demonstration*, 32; Fowler, "Finishing Stroke," 145.

On Sherman's mastery of military logistics, see McCarley, "'He Was So Well Provided for that He Could Sweep the World for Gain," 105–15, and J. J. Cooke, "Feeding Sherman's Army," 97–114.

which were strong and well manned, believing that our loss would have been severe."

The Confederates tore up the railroad as they marched along it toward Dalton. General Cleburne watched as his men performed the work. "This was a far more exciting business than I had supposed," he entered into his diary on the 13th.

> I can hardly understand the effect on my mind of the scene that followed. The men would lift shear up one hundred yards of the track at once and then pitch the immense grating of wood and iron over on its back. The heavy fall would break rails and cross ties loose. The shouting of the long line of lifters, the ringing of pick and ax heads against the iron pins fastening the rails to the wood, the interminable line of fires as the cross-ties were given to the flames, the thousand columns of smoke, the scene so suddenly and completely changed, it bewildered me as I have been by the roar and sights of a great city and a few minutes before this road was still as death.

Once at Dalton on the 13th, the Confederates threatened the enemy fort there. The Federal commander, Col. Lewis Johnson of the 44th U.S. Colored Infantry, surrendered his garrison. Most of its 715 men were black, and upon their capitulation many Confederates shouted insults. Division commander Maj. Gen. William B. Bate was particularly vindictive against the white officers of the colored troops. The black soldiers were stripped of their shoes and ordered to work tearing up the railroad. One sergeant was shot when he refused. As was their practice, Confederates returned 600 of the black Federals either to their masters or sent them to Mississippi. Elsewhere, farther up the W. & A., smaller contingents surrendered at Tilton and Mill Creek Gap. The commander of a blockhouse garrison on the Oostanaula, ordered by Stewart to give up, answered, "Present my compliments to General Stewart, and tell him to come and take me." A well-directed cannonade within five minutes brought about the Federal officer's swift surrender. (On the 13th Hood reported that with these several captures his troops had bagged "about 1,000 prisoners and some stores.")

The Southerners continued their track-tearing to Tunnel Hill. Hood later claimed that the W. & A. from Resaca to Tunnel Hill had been wrecked, but General Stewart was more precise, stating that the damage began 1½ miles north of Resaca. Nevertheless, Hood's north Georgia railroad damage, October 12–13, was some of the most extensive conducted by Confederates during the war. The seven or eight miles of track cut between Big Shanty and Acworth on October 4 were repaired by the Northerners as of the 16th. But the Resaca-Tunnel Hill stretch stayed out of operation from October 12 to the 29th. One Federal officer gauged the damaged length to be twenty-four miles (which was about the distance of W. & A. track measured from Resaca to Tunnel Hill). Thus, all told, during October 4–13, Hood's soldiers knocked out over thirty miles of railroad line in north Georgia.[16]

Sherman meanwhile was continuing his march behind Hood's army, reaching Resaca on October 14 and Dalton the next day. By that time, having torn up the Western & Atlantic, Hood was marching away from it toward Gaylesville, Alabama, west of Rome. On the 15th he wired Secretary Seddon in Richmond a report on his progress. That day the Federals got close enough to skirmish with Lee's infantry, the rear guard, at Snake Creek Gap, but that was the extent of the serious battle which both Hood and Sherman had been writing about for weeks. Wheeler, whose cavalry was resisting the Federals' advance, gave estimates of the enemy's strength that led Hood to guess they might number as many as 65,000. At Cross Roads (a village nine miles south of Lafayette, near Gaylesville), Hood asked his corps commanders whether their troops could prevail in general battle. "The opinion was unanimous," he wrote in his reminiscence, "that although the Army had much improved in spirit, it was not in condition to risk battle against the numbers reported by General Wheeler."

---

[16] [Shoup] journal, entries of October 10–11, *OR*, vol. 39, pt. 1, 807; McMurry, *Hood*, 161; W. L. White, "The Long Lost Diary of Major General Patrick R. Cleburne," 16–17 (entries of October 12–13); Stewart report, January 20, 1865, *OR*, vol. 39, pt. 1, 812; Wever report, October 23, *OR*, vol. 39, pt. 1, 753; Lee report, January 30, 1865, *OR*, vol. 39, pt. 1, 810; Johnson report, October 17, *OR*, vol. 39, pt. 1, 720 (715 men surrendered); Bailey, *Chessboard*, 36–37 (treatment of the Dalton garrison's black soldiers); Losson, *Tennessee's Forgotten Warriors*, 200 (Dalton); Hay, *Hood's Tennessee Campaign*, 46; D. W. Sanders, "Hood's Tennessee Campaign. Chapter II," 200; Hood report, February 15, 1865, *OR*, vol. 39, pt. 1, 802; S. Davis, *What the Yankees Did to Us*, 348; Col. William W. Wright to Thomas, October 18, *OR*, vol. 39, pt. 3, 353; Barnwell, *City Directory*, 107 (23 miles from Resaca to Tunnel Hill); Ward, "Twisting Confederate Iron," 27 ("units of the Union's Construction Corps had one, ten-mile break repaired before Rebels had finished wrecking a second, twenty-five mile section").

"I remained two days at Cross Roads in serious thought and perplexity," Hood later recalled, as he offered his troops some rest. He had the luxury of this time (October 15–16) thanks to General Sherman's none-too-eager "pursuit." But the Confederates could not count on the enemy to remain lethargic. If they could not give battle, they also could not remain on the defensive; Sherman had too many men and resources to throw at Hood. Thus, as McMurry writes, "Hood soon arrived at the only logical solution...further maneuver." Hood records it in his memoir: "In this dilemma, I conceived the plan of marching into Tennessee."

Hood is being disingenuous. As we have seen, President Davis left Hood at Palmetto and within two days informed generals Bragg and Taylor of the possibility that Hood would march into the Volunteer State. Davis's several remarks in the week after his conference with Hood ("your feet pressing the soil of Tennessee") also provide evidence that the general and president had discussed this possibility during September 25–27. Yet we may conclude that it was at Cross Roads, October 15–16, when Hood *resolved conclusively* to lead his army into Tennessee. Word got out among the troops. "The spirit of our men is improving daily," Maj. Gen. Henry Clayton wrote his wife on October 15; "the Tennesseans are perfectly jubilant at the prospect of going again into their state."

Hood may have been less forthcoming to his superiors. After having made up his mind, October 15–16, he did not immediately inform Bragg, Seddon, or Davis, much less ask their opinion of his idea or seek their permission to implement it. (Perhaps he thought they already knew it.) At the very least, he avoided opportunities to directly communicate his plan to Richmond. All he told Seddon on the 19th was that he intended to march to Gadsden tomorrow, from whence "I shall move for the Tennessee River."[17]

Most importantly, Hood failed to keep General Beauregard, his superior, informed of his "new" intentions. (A general order and drum-thumping proclamation from the Military Division of the West, issued at

[17] Welcher, *The Union Army*, 2:584; Horn, *Army of Tennessee*, 378; Hood to Seddon, October 15, *OR*, vol. 39, pt. 3, 820; S. D. Lee, "From Palmetto, Ga., to Defeat at Nashville," 257 (Lee held Snake Creek Gap October 13–15); Hood, *A & R*, 262–64; McPherson, *Embattled Rebel*, 215 ("Davis's public speeches during his trip to the Deep South had alluded to an invasion that would take Hood's army all the way to the Ohio River"); Hood, *Rise, Fall, and Resurrection*, 74 (quoting Clayton); Hood to Seddon, October 19, *OR*, vol. 39, pt. 3, 831.

Jacksonville on the 17th, announced the Creole's formal assumption of command.) On October 20 Hood wired Richard Taylor, then at Mobile, *asking* for help in moving supplies to Tuscumbia in preparation for his march. (Worse, Hood lied to Taylor when he claimed "I have not seen General Beauregard"; he had seen the military division commander just ten days before.) Hood seems to have refused to recognize that General Beauregard, if brought in as a partner to his operations, could have *ordered* Taylor to cooperate in them. By this kind of conduct it is apparent that John Bell Hood, about to launch the most important campaign of his career, was taking about every step he could to keep those in authority above him, and who could lend assistance, in the dark. Connelly pounces on this deliberate deception and disrespect: "Hood planned to move into Tennessee without even informing Beauregard." "Inexcusable neglect" is McMurry's polite phrase."[18]

Thus, it was a dutiful but oblivious G. T. Beauregard who set out from Jacksonville on October 19 for another round of talks with Hood. Thirty miles out, heading toward the army at Blue Pond (he thought), Beauregard was surprised to learn from Wheeler that Hood had moved to Gadsden, twenty-seven miles to the west. "Retracing my steps," the Creole rode there, arriving late morning of the 21st.

How the two officers greeted one another at Gadsden is not a matter of record. Nor do we know whether the military division commander discussed protocols of communication with his subordinate. Months later, in his campaign report, Beauregard stated only that Hood "had commenced his movement toward Middle Tennessee without advising me," and that upon arrival at Gadsden "he informed me that he was then en route to Middle Tennessee." In his report, Hood merely states that after loading up the supply of shoes and other stores he had gotten at Gadsden, he intended to march north, crossing the Tennessee at or near Guntersville, then marching northeast (presumably along the Memphis & Charleston) until he reached Bridgeport to "strike the enemy's communications" (meaning, destroy the vital rail trestle across the river there). This would force Sherman back across the Tennessee, "and thus we should at least recover our lost territory" (north Georgia, or what remained of it, after a half-year of war).

---

[18] General Orders No. 1 and untitled proclamation, October 17, *OR*, vol. 39, pt. 3, 824–26; Hood to Taylor, October 20, *OR*, vol. 39, pt. 3, 835; Connelly, *Autumn of Glory*, 483; McMurry, *Hood*, 162.

Writing six months later, Beauregard made a point of stating that Hood told him "that his general plan had been submitted to and approved by General Bragg"—which was, at least the latter half, blatantly false (Hood had informed Bragg of his intentions, but had neither sought nor received approval of them). Beauregard had reservations about Hood's plan, but he gave his assent for two important reasons. First, Hood told him that Sherman's troops were short on rations. (They were: General Slocum at Atlanta had to send out four foraging expeditions during October to bring in food for the men and animals of his XX Corps.) Second, given the effectiveness of his break in the Western & Atlantic, Hood promised more mayhem by aiming to destroy not only the Bridgeport trestle across the Tennessee, but also the Nashville railroad bridge across the Elk River at the Alabama state line. Beauregard added that Hood also aimed for the Duck River bridge: that meant the one nearly fifty miles into Tennessee, where the Tennessee & Alabama Railroad crossed the Duck near Columbia.[19]

Not mentioned by Beauregard, but an obvious reason for his approval of Hood's plan, was that it proposed to operate on Sherman's railroad *supply system*, not on Sherman's *army itself*. For John Bell Hood, battlefield hero of the war's earlier years, this was a pronounced turn indeed. Let us review the evolution of Hood's planning—assuming that *extemporizing* is not the right word.

September 4: Hood thought Sherman would not rest at Jonesboro but push south. If he did so, Hood would launch an attack on his flank and rear.

September 6: he proposed to the president an offensive movement, to "place our army upon the communications of the enemy, drawing our supplies from the West Point and Montgomery Railroad."

September 22: Hood announced his idea of crossing the Chattahoochee and taking defensive position near enough to the Western & Atlantic to threaten it and thus compel Sherman "to drive me off."

October 8. "Induced to somewhat change my original plan to draw Sherman to the Alabama line and then give battle," Hood wired Bragg that he would "destroy his communications from Kingston to Tunnel Hill," then "move to the Tennessee River."

---

[19] Roman, *Beauregard*, 2:286; T. H. Williams, *Beauregard*, 243; Beauregard report, April 15, 1865, *OR*, vol. 45, pt. 1, 647; Hood report, February 15, 1865, *OR*, vol. 39, pt. 1, 802; S. Davis, *What the Yankees Did to Us*, 350.

October 15–16. At Cross Roads, Hood claims in his memoirs that he "conceived the plan of marching into Tennessee," although we have seen much evidence that he and Davis had discussed such an option as early as September 25.

October 21. Hood and Beauregard discussed a route that would take the army to Bridgeport, then presumably after a westward march on the north bank of the river, into Tennessee. At this point, as Williams observes, "it is difficult to summarize Hood's design."[20]

If the historian has a hard time figuring out Hood's "plan," try being Cump Sherman. "He is an eccentric," Sherman wrote of Hood to John Corse, "and I cannot guess his movements as I could those of Johnston, who was a sensible man and only did sensible things." "It is hard to divine his future plans," he admitted to Grant. Gradually Sherman began to conclude that Hood intended to march into Tennessee. General Corse telegraphed from Rome on October 11 that he had just interrogated a captured Mississippi cavalryman. "He says the men were told that they were en route for Tennessee and Kentucky." "I have a Montgomery paper of the 12th," General Slocum informed headquarters on October 17. "The dispatches from Hood as well as the editorials state that Beauregard is with Hood and that the army is going to cross the Tennessee River."

For Sherman, this was great news, for it allowed him to further press Washington for permission to march across Georgia. "I propose we break up the railroad from Chattanooga, and strike out with wagons for Milledgeville, Millen, and Savannah," he wrote Grant on the 9th. The next day he added that Thomas could defend Tennessee against a raid by Hood. "He will have an ample force when the re-enforcements ordered reach Nashville." Finally, Grant began to relent. "On reflection, I think better of your proposition," he wired on the 12th. "I believe Sherman's proposition is the best that can be adopted," Grant wrote Stanton the next day.

The president and his chief of staff had yet to be won over. Secretary Stanton confided to Grant that "the President feels much solicitude in respect to General Sherman's proposed movement." By marching into

---

[20] Hood to Bragg, September 4 (1 P.M.), *OR*, vol. 38, pt. 5, 1018; Hood to Davis, September 6, *OR*, vol. 38, pt. 5, 1023; Hood to Bragg, September 22, *OR*, vol. 39, pt. 2, 862; Hood, *A & R*, 258, 264; Hood to Bragg, October 8, *OR*, vol. 39, pt. 3, 805; Roman, *Beauregard*, 2:289–91; T. H. Williams, *Beauregard*, 243.

the interior of the enemy's country, Lincoln feared that Sherman might be ravaged by guerrilla raiders, run out of supplies, and be stranded. Stanton telegraphed Grant that the president feared that "a misstep now by General Sherman might be fatal to his army"—all at the time when Lincoln faced a worrisome re-election challenge that fall, with the Democrats sure to pounce on every defeat from the battlefield. And if Sherman wanted to head for salt water, General Halleck thought it should be the Gulf of Mexico, not the Atlantic Ocean, for a march through Alabama toward Mobile offered more strategic benefit.

Nonetheless, Sherman used Hood's apparent movement toward Tennessee to push his plan for a march across Georgia. "I want the first positive fact that Hood contemplates an invasion of Tennessee," he wired General Schofield, then at Chattanooga. "Invite him to do so. Send him a free pass in." "Damn him," Sherman spluttered, "if he'll go to the Ohio River, I'll give him rations. Let him go north. My business is down south."

Grant would end up being a hard sell. As late as November 1 he was still suggesting to Sherman, "If you can see the chance for destroying Hood's army, attend to that first." But Sherman persevered, and the farther Hood marched his army westward into Alabama the stronger his case became. If Hood advanced into Tennessee, Sherman predicted to Thomas on October 17th, "we can all turn on him and he cannot escape." But by "we," Sherman did not mean himself, only Thomas. Sherman had led his several corps as far as Gaylesville on October 20, but beyond that he refused to go. "To pursue Hood is folly," he wrote, "for he can twist and turn like a fox, and wear out an army in pursuit." Yet "pursuit" is the word Lieutenant Colonel Warner employed in his telegram to Secretary Stanton on October 21, adding, "Hood retreating at the rate of twenty-five miles a day"—using the obverse term for what General Hood would have called an *advance.*

Now that Hood had left Georgia and Thomas was preparing to deal with him in Tennessee, Sherman felt further liberated from Grant's concern that he should deal with Hood's army once and for all. "I now consider myself authorized to execute my plan to destroy the railroad from Chattanooga to Atlanta, including the latter city," he informed Halleck on the 19th, and "strike out into the heart of Georgia." "I will leave General George H. Thomas to command all my division behind me," he promised Grant three days later. The next day he affirmed to General Slocum, "We are piling up men in Tennessee, enough to attend to them

[Hood's forces] and leave me free to go ahead." Then on the 24th he explained to Halleck that despite Hood's apparent aim to force him to give up Atlanta and the initiative, "I dare him to the encounter, but am not willing to chase him all over creation."

Sherman spent a bit more than a week at Gaylesville—"dozing," as Horn contemptuously puts it ("one thing appears certain, if inexplicable: he was not anxious to engage Hood in a decisive trial at arms"). Sherman did not even learn that Hood had left Gadsden until October 26, and by then the Rebels were at Decatur. Marching back into northwest Georgia, he augmented Pap Thomas's force. He had already sent Corse's and Newton's divisions northward. Then on October 26 he also sent back Maj. Gen. David Stanley's IV Corps, which caused General Stanley to observe dryly, "Thomas was expected to beat Hood with one corps and two divisions. Sherman had failed to do this, with six additional corps during an entire summer."[21]

---

[21] Sherman to Corse, October 7, *OR*, vol. 39, pt. 3, 135; Sherman to Grant, October 22, *OR*, vol. 39, pt. 3, 395; Corse to Sherman, October 11 (4 P.M.), *OR*, vol. 39, pt. 3, 210; Slocum to Sherman, October 17, *OR*, vol. 39, pt. 3, 326; Sherman to Grant, October 9 (7:30 P.M.), *OR*, vol. 39, pt. 3, 162; Sherman to Grant, October 10 (12 m.), *OR*, vol. 39, pt. 3, 174; Grant to Sherman, October 11 (11:30 P.M.) and 12 (1 P.M.), *OR*, vol. 39, pt. 3, 202, 222; Grant to Stanton, October 13 (3:30 P.M.), *OR*, vol. 39, pt. 3, 239; Stanton to Grant, October 12 (8 P.M.), *OR*, vol. 39, pt. 3, 222; Flood, *Grant and Sherman*, 265 ("misstep"); Perret, *Lincoln's War*, 393; Ambrose, *Halleck*, 182; Sherman to Schofield, October 16 (5 P.M.), *OR*, vol. 39, pt. 3, 311; B. Davis, *Sherman's March*, 22 ("Damn him"); Grant to Sherman, November 1 (6 P.M.), *OR*, vol. 39, pt. 3, 576; Sherman to Thomas, October 17, *OR*, vol. 39, pt. 3, 333; Warner to Stanton, October 21, *OR*, vol. 39, pt. 3, 387; Liddell Hart, *Sherman*, 323–24; Kennett, *Sherman*, 258–59; Hirshson, *White Tecumseh*, 247; Sherman to Halleck, October 19, *OR*, vol. 39, pt. 3, 357–58; Sherman to Grant, October 22, *OR*, vol. 39, pt. 3, 395; Sherman to Slocum, October 23, *OR*, vol. 39, pt. 3, 406; Sherman to Halleck, October 24, *OR*, vol. 39, pt. 3, 413; Horn, *Army of Tennessee*, 381; Welcher, *The Union Army*, 2:583 (Corse and Newton), 584 (IV Corps to Chattanooga, October 26); Sherman to Thomas, October 28 (8 P.M.), *OR*, vol. 39, pt. 3, 484 ("I have already sent the Fourth Corps"); Carpenter, *Slight Demonstration*, 28 (quoting Stanley).

Henry V. Boynton was one Northern writer who criticized Sherman for chicanery in winning Grant's approval of the proposed march across Georgia: "By understating Hood's forces, and largely overestimating those proposed to be left with Thomas, Sherman obtained the desired permission." For instance,

Hood's army stayed a day or so at Gadsden so stores, especially of clothing and shoes, could be brought up and distributed to the men. Major General Elzey, back at Jacksonville, was ordered to send forth all the commissary and quartermaster supplies he had accumulated. With everyone—Beauregard, Taylor, Elzey—issuing orders, it is remarkable that anything got done ("the whole Confederate chain of command had become, in fact," as Robert Black notes, "hopelessly tangled"). Even General Hood was aware, as Major Mason acknowledged: "He fears that in the many orders that have been given and countermanded there may be delay." Nevertheless, the stuff got through. Col. Ellison Capers's South Carolina regiment alone received at Gadsden 112 pairs of trousers, 82 pairs of socks, 74 pairs of shoes, 46 pairs of drawers, 44 jackets, 37 shirts, and 21 blankets. "This issue by no means supplied our necessities," Capers recorded, "but relieved the most needy." As Hood prepared to lead his troops north from Gadsden toward Guntersville on the Tennessee, Beauregard stayed behind to attend to issues of subsistence and supply, of which (as Roman politely puts it, "General Hood was disposed to be oblivious"). Of particular concern to Beauregard was the task of moving the army's supply base from Jacksonville to Tuscumbia. Richard Taylor was instructed to see to the proper running and protection of the railroad, and supply officers were ordered to cooperate in the logistical transfer.[22]

Beauregard also played a key role in bringing Bedford Forest into Hood's operations. The Creole wanted Wheeler and his command sent to Georgia to watch Sherman and dog any advance he might make from Atlanta. This left the Army of Tennessee with only Red Jackson's mounted division until Forrest could join it. Beauregard therefore instructed Taylor to have Forrest report to Hood "as soon as practicable

---

Boynton charges that in his *Memoirs* Sherman wrote that Thomas had 82,000 to 88,700 men, when the latter had 55,472. For these and other reasons, Boynton explained that he was writing a critique of Sherman's postwar recollections "to show wherein the *Memoirs* of General Sherman fall short of presenting the correct history of many events of which they treat" (Boynton, *Sherman's Historical Raid*, 3, 146–47). On Boynton and his purpose in contesting Sherman's memoir, see Rubin, *Through the Heart of Dixie*, 126–27.

[22] Mason to Elzey, October 20, *OR*, vol. 39, pt. 3, 834; Black, *Railroads of the Confederacy*, 264; Capers report, *OR*, vol. 45, pt. 1, 733; Roman, *Beauregard*, 2:286, 288; Beauregard to Taylor, October 22, *OR*, vol. 39, pt. 3, 843; Beauregard to Cooper, October 24, quoted in Roman, *Beauregard*, 2:289–91.

after executing his present instructions." At the time, Forrest was in the field, leading a raid in middle Tennessee. He did not even get the order until a week after Beauregard had wired Taylor.[23]

During the 20th and 21st, the men received twenty days' rations and ammunition. The artillery received 106 rounds per gun (which was seriously short of the usual 400). Despite the distribution of clothing and shoes, many men remained unshod. Nevertheless, spirits soared when the commanding general had an announcement read to the troops that they were heading north for Tennessee. The news thrilled the men in the ranks. On the night of October 20, jubilant soldiers with regimental bands serenaded Hood and other generals at their quarters and asked for speeches. "Gen. Hood did not say he would cross the Tennessee," one wrote home to his newspaper, "but we all understand it." "Last night there was a perfect jubilee," wrote Lt. Marcus Ely of the 54th Georgia to his wife on October 22. "Genl's Beauregard, Cheatham, Wood, Butts and all the generals were serenaded by the different bands & of course all made speeches." "When we get to Tennessee," he added, "we expect to whip Sherman and wrest Tennessee from the Yanks and establish our independence." "Although the army was greatly reduced in numbers," remembered Tom Walker of the 9th Tennessee, "the idea of again entering on Tenn soil made us feel at least we Tennesseans bonant [buoyant] and happy."

"From the different encampments," Hood remembered, "arose at intervals that genuine Confederate shout so familiar to every Southern soldier." O'Connor goes further: "the Army of Tennessee, for the first time since Hood took command, cheered en masse when they heard the news of the proposed movement North." In Horn's words, the announcement to the troops "recharged them with their old-time enthusiasm." Noel Carpenter writes, perhaps too colorfully: "the loud cheers and rebel yells which erupted from the men every few minutes could be heard for miles.... As the Tennessee Campaign began, the morale of the army was the highest it had been since Hood took command."

Then the troops marched off on October 22 for Guntersville, the point on the Tennessee River Hood had told Beauregard would be his first objective (Beauregard wired General Cooper that the army "left today for vicinity of Guntersville"). But Hood changed his mind and or-

---

[23] McMurry, *Hood*, 163; Connelly, *Autumn of Glory*, 485; Brent to Taylor, October 23, *OR*, vol. 39, pt. 3, 845; Henry, *"First With the Most" Forrest*, 379.

dered his column northwest. On the 23rd he quartered at Brooksville; the next day the troops covered twenty-two miles to Blue Creek Church. On the 25th Hood and his staff set HQ at Somerville, thirteen miles southeast of Decatur. Hood later claimed that this drift westward was to facilitate his junction with Forrest, "an understandable position," observes Dyer. Beauregard remembered that Hood had learned that the enemy was heavily guarding the crossing point at Guntersville. The Creole was approaching Guntersville when he heard of the army's detour. No wonder he was angry when he finally encountered Hood at his headquarters two miles from Decatur on the 27th.[24]

Decatur is a town on the south side of the Tennessee and a station on the Memphis & Charleston Railroad. In spring 1864, Federal troops occupied it, evicted its civilian population, and built a line of earthworks surrounding it, buttressed with two artillery forts. When Confederates approached Decatur on October 26, they found a Union garrison of 1,200. Stewart's and Cheatham's corps deployed outside the Federal works, but Hood seemed without resolve either to storm the place or even to demand its surrender. When General Beauregard arrived at Decatur on the 27th, Hood came in for sharp criticism for the Guntersville miscommunication. He would have tried to explain his decision although we can only guess that he mentioned the need to rendezvous with Forrest and to avoid crossing at a heavily guarded river point. Hay adds a third possible reason for Hood not crossing at Guntersville: Sherman was at Gaylesville, two or three days' march away; a protracted river-crossing could expose his army to attack from the rear. Beauregard also questioned Hood's tactics at Decatur. In the words of Noel Carpenter, chronicler of the Decatur fight, Beauregard "told Hood he should have attacked Decatur by an all-out assault at

---

[24] McMurry, *Hood*, 163–64; Carpenter, *A Slight Demonstration*, 26 (106 rounds per gun), 27, 35, 42; [Shoup] journal, entries of October 22–25, *OR*, vol. 39, pt. 1, 807; "S.," "From the Front," *Atlanta Daily Intelligencer*, Nov. 2, 1864; McCardle, *A Just and Holy Cause*, 137; Cooling, *To the Battles of Franklin and Nashville and Beyond*, 284 (Tom Walker); Hood, *A & R*, 270; O'Connor, *Cavalier General*, 225; Horn, *Army of Tennessee*, 380; Beauregard to Cooper, October 22, *OR*, vol. 39, pt. 3, 841; Dyer, *Gallant Hood*, 282; Roman, *Beauregard*, 2:291–92.

dawn on the twenty-seventh rather than investing it, as Hood appeared to be doing, since the enemy was now prepared to resist."

They were. During the night of October 27–28, more Federals arrived by riverboat and across the pontoon bridge, bringing the garrison strength up to around 3,000. Brig. Gen. Robert Granger and the post commander, Col. Charles Doolittle, resolved to defend the place, but the Confederates, despite their numerical superiority, launched no all-out assault. Sorties by both sides led to captured rifle pits and prisoners. Cannoneers exchanged fire, although the Southerners never shelled the bridge across which Northern reinforcements kept coming. Indeed, on October 28, a charge by the 14th U.S. Colored Troops overran a Rebel shore battery. The Federals were spiking the guns when a counterattack by the 31st Mississippi drove them off.

Cold and hunger depressed the Southerners' spirits. "Today a wagon drives through the camp and issues two ears of corn to each man," wrote an obviously dissatisfied Capt. Samuel Foster. Some Tennesseans got only one roasting ear. Meanwhile Confederates took casualties not only from the forts' artillery but also from cannon of Northern gunboats that appeared on the 28th. That night even more Northern reinforcements crossed, so that on the 29th Federal strength at Decatur rose to 5,000. Hood broke the stalemate the only way he could, by ordering withdrawal of his troops. In this case, Sherman correctly predicted that Hood would walk away from the fight. "Allatoona and Resaca beat him off," Cump wired Thomas on the night of the 28th, "and neither was as strong as Decatur." The Army of Tennessee would obviously not cross the Tennessee at that point. Hood ordered a march westward for another crossing downriver.

In his campaign report Hood fails to mention the several days his army spent at Decatur. In *Advance and Retreat* he states merely, "I ordered a slight demonstration to be made against the town till our forces passed safely beyond." As Carpenter observes, this is not a convincing explanation: "The implication that he spent four days [October 26–29] using two army corps to neutralize the small fort while a third corps passed thirteen miles to the south has gone almost unchallenged." More to the point, even Hood's biographers have not explained it. Carpenter's conclusion is logical: Hood was hoping to hear from Forrest, and holding the army at Decatur offered more time for the cavalryman's arrival.

In their camps, the men talked about Hood's uncertain tactics. "The troops would have pitched into this place very willingly from all that I could see or hear," Lt. W. H. Berryhill of the 43rd Mississippi wrote his wife. "We serounded the place and staid there three days, though we did not try to take it," wrote A. L. Orr of the 18th Texas Cavalry to his sister. "We could of taken it but would of lost a grate many men." "We knew that this place surrounded with ditches & a strong network of wire where they had destroyed the telegraph would be as impregnable as Gibraltor," Lt. Marcus Ely informed his wife; "I think Gen Hood concluded that the sacrifice would be too great." As it was, Confederate casualties for the "slight demonstration" at Decatur totaled probably 250–300; Stewart reported 135 in his three divisions alone. General Granger counted 113 men lost on his side.[25]

On October 29 Hood established headquarters at Courtland, twenty miles west of Decatur. From there he sent to Stewart and Lee a pair of conflicting orders that showed the commanding general's uncertainty as to plan. At first the two corps commanders were told to march their troops to Leighton, ten miles west of Courtland, then to head north for a place on the Tennessee called Bainbridge Crossing, at which the engineers would lay down a pontoon bridge. Lee, in the advance, was to cross the river and guard the bridge. Stewart was to follow with "a good day's march, but not a hard one." Then, sometime in the afternoon Hood changed his mind, sending his infantry farther westward toward Tuscumbia, on the river's south bank and the Memphis-Charleston Railroad. According to Roman, the engineers sent ahead reported the crossing of the river at Bainbridge would be

---

[25] Carpenter, *Slight Demonstration*, 3, 12–13, 42, 64, 75 ("3,000 men"), 97, 102, 106–110 (14 USCT vs. 31 MS), 112 (Foster), 115, 116 (Berryhill), 119, 121, 127, 131 (Orr), 149; Hay, *Hood's Tennessee Campaign*, 58; Roman, *Beauregard*, 2:292–93; Sherman to Thomas, October 28 (8:30 P.M.), *OR*, vol. 39, pt. 3, 484; Hood report, February 15, 1865, *OR*, vol. 39, pt. 1, 802; Hood, *A & R*, 271; O'Connor, *Cavalier General*, 225; Dyer, *Gallant Hood*, 282–83; McMurry, *Hood*, 164; McCardle, *A Just and Holy Cause*, 138.

difficult, so Hood chose to march the army farther west. Hood arrived at Tuscumbia himself on the 30th.[26]

---

[26] [Shoup] journal, entries of October 29–30, *OR*, vol. 39, pt. 1, 807–808; Mason to Lee, October 29 (1 P.M.) ("cross your command and take position to protect the bridge"), Mason to Stewart, October 29 (1 P.M.) ("toward the Bainbridge Crossing"), James Cooper to Stewart, October 29 ("instead of moving to Bainbridge Crossing"), all *OR*, vol. 39, pt. 3, 866; Roman, *Beauregard*, 2:293; Beauregard report, April 15, 1865, *OR*, vol. 45, pt. 1, 648 ("it was now judged proper by General Hood to attempt a passage, first, at Lamb's Ferry, that failing, then at Bainbridge, or, finally, at Tuscumbia"); D. W. Sanders, "Hood's Tennessee Campaign. Chapter IV," 251; Carpenter, *Slight Demonstration*, 119 ("to cross at Bainbridge").

Chapter 3

November 1864

Hood would spend the next three weeks at Tuscumbia, awaiting the delivery of supplies and the arrival of Forrest. In both ways he would be vexed by delay. As for the first, the east-west railroad from Corinth ran only to Cherokee Station, fifteen miles west of Tuscumbia. It was being repaired, but the work would not be finished until November 21. In the meantime, stores had to be brought by wagons over a poor road, made worse by rains. "Want of supplies and bad weather have delayed forward movement," General Beauregard wired Cooper in Richmond on November 3.

The military division commander was getting anxious. Part of the problem involved Bedford Forrest. When Taylor summoned him to join Hood's forces, Forrest was in the field. He only completed his mission, destruction of a huge Federal supply depot at Johnsonville, on the Tennessee River, around November 7. During that time Hood could only wait. "As I had not a sufficient cavalry force without his to protect my trains in Tennessee," Hood later wrote, "I was compelled to delay the crossing and move further down the river to meet him." It was not until November 14 that Forrest finally arrived to meet Hood, but it would take another week still for all of his command to join the Army of Tennessee.[1]

---

[1] Vandiver, "Hood as Logistician," 11; Beauregard to Cooper, November 3, *OR*, vol. 39, pt. 3, 880; C. H. Brown, "Forrest's Johnsonville Raid," 57; Brooksher and Snider, "Devil on the River," 19 (Forrest "netted 4 gunboats, 14 steamboats, 17 barges, 32 guns, 150 prisoners, and over 75,000 tons of supplies"); Hood report, February 15, 1865, *OR*, vol. 39, pt. 1, 803; Henry, *"First With the Most,"* 382; Hay, *Hood's Tennessee Campaign*, 62 ("it was the 21st before his entire command arrived, though he himself reached Hood's headquarters on the 14th").

At Johnsonville on the Tennessee, Forrest opened his artillery on Federal gunboats helplessly moored along the riverbank. The Confederate gunners had a field day, with even General Forrest joining in the fun. "Elevate the breech of

Hood's regrettable delay in getting his army marching into Tennessee gave General Thomas valuable time to prepare to meet the Rebels. By November 4 Stanley and his IV Corps were gathered at Pulaski, on the Nashville and Decatur Railroad thirty miles south of Columbia. Sherman decided to send the XXIII Corps to Thomas as well. General Schofield arrived at Nashville on November 5, ahead of his troops. On the 11th Thomas placed Schofield in charge of the forces at and near Pulaski, which by then included the two divisions of his XXIII. Schofield arrived at Pulaski on the night of November 13 and assumed command the next day. Thomas also received some twelve newly minted regiments, and had scattered garrisons at Murfreesboro, Chattanooga, and in north Alabama. "What Thomas needed most of all," observes Eric Jacobson, "was time to consolidate his troops and every day mattered." With Schofield's and Stanley's veteran corps, Thomas expressed confidence to Halleck that he could "drive Hood back," but he still asked for reinforcements from Missouri that Sherman was arranging. These troops (two divisions of the XVI Corps under Maj. Gen. Andrew Jackson Smith) would not arrive for a good while; Confederate Sterling Price had led a raid across Missouri in October, and Smith's 9,000 infantry helped push him out of the state. Thomas anxiously awaited their arrival.[2]

Meanwhile, Hood seemed unwilling to embrace the military division organizational structure. (Stephen M. Hood disagrees, asserting that "Hood attempted to keep in touch with his multiple and changing superiors as much as possible.") On October 31, he telegraphed President Davis and Secretary Seddon; such correspondence should properly have gone through Beauregard's office. He also refused to involve Beauregard

---

that gun lower!" he ordered at one point, prompting howls of laughter from his men (Hurst, *Nathan Bedford Forrest*, 227).

[2] Welcher, *The Union Army*, 2:343–44, 585–86; Sherman to Thomas, October 28, *OR*, vol. 39, pt. 3, 484 ("on these two corps [IV and XXIII] you can engraft all the new troops"); Jacobson, *For Cause and Country*, 48, 50–51; Thomas to Halleck, October 31 (7 P.M.), *OR*, vol. 39, pt. 3, 534; SFO No. 135, October 22, *OR*, vol. 39, pt. 3, 401 (Schofield resuming command Army of the Ohio; Cox reverting to division); Sherman to Maj. Gen. William S. Rosecrans, October 29, *OR*, vol. 39, pt. 3, 494 (A. J. Smith's and Joseph Mower's divisions); Castel, *General Sterling Price*, 252 ("Smith arrived at Nashville in time to help Thomas crush Hood's army"); Smith report, January 10, 1865, *OR*, vol. 45, pt. 1, 433 (debarking at Nashville, November 30); Sword, *Embrace an Angry Wind*, 160.

in his communications within the military division. On the 30th he asked General Taylor for help in accumulating rations: "please give me all the assistance you can in obtaining them." Hood would not have needed to ask Taylor for this help if he had gone through Beauregard; the latter would have simply issued orders to Taylor. Similarly, Hood asked Forrest on November 2, "When can I expect you here or when can I hear from you? I am waiting for you." This kind of request also could have been reinforced as an order from either Beauregard or Taylor.[3]

At times Hood seemed to go out of his way to irritate his superior officer. When Beauregard on the 30th asked for "a brief summary of the operations of your army from the date of its departure from Jonesborough," Hood waited four days before replying on November 3. This caused Roman to note, "for reasons which cannot be explained this request was not readily complied with." Even worse was the nature of Hood's reply, in which he stated that he was too busy and unwell to prepare such a report, but that he would send copies of telegrams he had sent to Richmond.

Beauregard also had his A. A. G., Col. George Brent, write Hood on October 30, requesting "a concise statement of your plans for future operations" which could be sent to the War Department. Hood's answer of November 3 merely stated, "The army will commence the movement into Tennessee on Saturday morning next"—meaning November 5 (a deadline that came and went).

Hood and Beauregard met on the 3rd and discussed the forthcoming move into Tennessee. Beauregard gave his approval, but relations between the two generals soured again the next day when Beauregard had Brent write a letter again requesting "a brief report of your plan of operations from this point, for the information of the government," and asking also for a written summary of events since the army left Jonesboro ("a summary statement from your corps commanders would answer the purpose"). This latter request Hood simply ignored in a letter to Brent that day. As for "a brief report of your plan," Hood snapped back, "it is not possible for me to furnish any plan of my operations for the future, as so much must depend upon the movements of the enemy." Besides,

---

[3] Hood, *Rise, Fall, and Resurrection*, 104; Hood to Davis and to Seddon, October 31, *OR*, vol. 39, pt. 3, 870; McMurry, *Hood*, 165; Hood to Taylor, October 30, *OR*, vol. 39, pt. 3, 868; Hood to Forrest, November 2, *OR*, vol. 39, pt. 3, 879.

Hood added (as a deliberate shot at Brent), "the matter has been fully discussed between General Beauregard and myself."

The military division commander was not amused. On November 6 Beauregard sent a copy of this correspondence to the War Department, as if to show the administration the difficulty of trying to deal with an uncooperative, even petulant, army commander.

By this time the two generals, though both at Tuscumbia, were mainly communicating through written messages carried by staff officers. When Beauregard suggested that Hood issue a proclamation to the people of Tennessee as he prepared to march into their state, Hood did not even reply. On November 10, Colonel Brent wrote Hood, conveying Beauregard's wish that "roving or independent scouts" roaming the area be reined in; they were committing depredations on the people. Again Hood did not answer, exhibiting a surliness that Beauregard remembered years later. "These instructions," writes Roman, "appeared to produce an unfortunate effect upon General Hood, who began to chafe under the supervision exercised over him by General Beauregard, and to fear his superior influence with the army."[4]

Beauregard initiated another provocation, perhaps unwittingly, when he informed General Stewart that he would like to review his troops on November 12. Hood angrily shot back to Colonel Brent, declaring that "all orders, whether for review and inspection, or whatever kind, relating to this army, must pass through me, its Commanding General." He mockingly wondered how "one so familiar with the rule" as Beauregard should be, "should have overlooked it in this instance." But the military division commander had the last laugh—although relations between Hood and Beauregard were by now anything but a laughing matter—when the Creole reviewed Stewart's Corps on the 12th as planned.

Beauregard was on the scene with Hood, and by technical terms of President Davis's instructions, he could have taken personal command of Hood's army. Yet he refrained from doing so. "Although he was angry

---

[4] Brent to Hood, October 30, *OR*, vol. 39, pt. 3, 867; Hood to Brent, November 3, *OR*, vol. 39, pt. 3, 880; Roman, *Beauregard*, 2:296–99; McMurry, *Hood*, 165–66; Brent to Hood, November 4, *OR*, vol. 39, pt. 1, 800; Hood to Brent, November 4, *OR*, vol. 39, pt. 3, 887–88; Beauregard to Cooper, November 6, *OR*, vol. 39, pt. 1, 798–99; Brent to Hood, November 10, *OR*, vol. 39, pt. 3, 908.

over Hood's constant revisions in planning," Connelly notes, "Beauregard avoided taking command." According to Roman, Beauregard feared that openly interfering with Hood's authority "would bring upon him the censure of the War Department."[5]

Despite all the acrimony, Beauregard tried to put the best face on things to the authorities in Richmond. On November 3 he wired President Davis, "General Hood having conducted present movement successfully," and explained that he would not assume direct command of the Army of Tennessee. Yet the "present movement" was progressing much too slowly to suit the military division commander. "Want of supplies and bad weather have delayed forward movement," Beauregard wired General Cooper on the 3d. "I will assist [as] far as practicable," he assured the president.

As a means of assistance Beauregard assigned a task to Maj. Gen. M. L. Smith, the Army of Tennessee's chief engineer, who on October 17 joined the Creole's personal staff (allowing Lieutenant Colonel Presstman to return as at least acting chief engineer for the Army of Tennessee). Smith was sent ahead to check crossing points of the Tennessee River near Savannah, "as a change of base from this place [Tuscumbia] to Purdy may become necessary" once—assuming—the army marched into Tennessee.[6]

For the time being, making efficient use of the army's current base of supply was vexing enough. Principal problem was the railways, on

---

[5] Roman, *Beauregard*, 2:299 ("censure"), 606 (Hood to Brent and Brent to Hood, both November 12, not in the *Official Records*); Connelly, *Autumn of Glory*, 487.

Richard McMurry astutely conjectures that Hood dictated two responses to Brent that were "even harsher" than the one he eventually sent through him to Beauregard (McMurry, *Hood*, 223 n.45). All I see in Hood's dispatch book are two entries of November 12 addressed "Colonel Geo W Brent—void" (Book IV, Field Dispatches, Hood Papers, National Archives).

[6] Beauregard to Davis, November 3, *OR*, vol. 39, pt. 3, 879; Davis to Beauregard, November 1, *OR*, vol. 39, pt. 3, 874 (reaffirming Beauregard's authority to "exercise immediate command while there"); Beauregard to Cooper, November 3, *OR*, vol. 39, pt. 3, 880; General Orders No. 1, October 17, *OR*, vol. 39, pt. 3, 824 (announcing Beauregard's staff, including Smith); Presstman to Hood, December 22, *OR*, vol. 45, pt. 2, 721 ("Lieutenant-Colonel and Actg. Chief Engineer, Army of Tenn."); Roman, *Beauregard*, 2:604 (Beauregard to Smith, November 9); Sanders, "Hood's Tennessee Campaign. Chapter V," 289.

which everything depended. "The Mobile & Ohio railroad, north of Okolona to Corinth, was in a miserable condition, and could scarcely be operated," according to Major Sanders of French's staff. "The Memphis & Charleston railroad, from Corinth to Tuscumbia, was much worse, and there were only seven miles of passable roadbed between the two points." Hood tried to help, even though, as Stephen M. Hood has recently written, "civilian-owned railroads were out of Hood's control." Seeking twenty days' rations, Hood called upon the superintendent of the Mobile & Ohio, one Major Fleming, asking if M. & O. cars could also run on the Memphis & Charleston line, to "avoid transshipment at Corinth." Unloading (and reloading) of trains at the junction of two different lines was the usual business practice, yet "time is of the utmost importance," and every delay had to be eliminated. The fifteen-mile wagon-haul from Cherokee to Tuscumbia further impeded the supply flow. Instead of piling up a surplus, the logistical line barely fed Hood's men and animals. "The army," Sanders observed, "was barely furnished with the meanest rations. The weather changed, and it rained incessantly. The dirt road to Cherokee was almost impassable, and the teams of the army, instead of being rested and recruited for the march into Tennessee, were heavily worked and well-nigh exhausted, in hauling stores and supplies from Cherokee."[7]

Major Sanders was right: the weather in northern Alabama was a decided deterrent to the forward movement of Hood's army. Col. Edwin J. Harvie recorded in his journal, "November 1 to 10.—Remained in camp; raining most all the time." Cold rain pelted the soldiers and even flooded their tents. As a sign of progress for the campaign, Randall Gibson's Louisiana brigade paddled across the river in pontoon boats on October 31 and drove off Yankee cavalrymen harbored in warehouses on the north bank. Engineers laid a pontoon bridge and on November 2 Lee's Corps marched across to the north bank and the town of Florence. But a few days later the rain-swollen river began to rise and the pontoon bridge was partly submerged. Most of the army stayed encamped on the south bank while Hood waited for Forrest, rations, forage, and dry weather.[8]

---

[7] Hood to Superintendent of Mobile & Ohio Railroad, October 31, *OR*, vol. 39, pt. 3, 871; Sanders, "Hood's Tennessee Campaign. Chapter V," 289–90; Hood, *Rise, Fall, and Resurrection*, 95.

[8] "Itinerary of the Army of Tennessee," *OR*, vol. 39, pt. 1, 809; Jacobson, *For Cause & For Country*, 46 ("camps became flooded"); Sword, *Embrace an An-*

Another cause of delay at Tuscumbia was the deplorable condition of the army's draft animals. On November 4, Maj. E. H. Ewing, Hood's inspector of field transportation, reported to Lt. Col. A. H. Cole, chief of field transportation in Richmond, that as result of the "long and very fatiguing march made of over a month from Palmetto…all of our animals are fagged down and many rendered utterly useless." "I think we may safely say," he wrote, "that 1,500 animals have been utterly used up on this march." Hundreds of jaded horses and mules would have to be sent to the rear; replacements would have to be found and purchased. To be sure, animals will always be "used up" in any hard campaign, but General Hood bore some blame for making a bad situation worse. Ewing told Cole that he had urged Hood to dismount some officers, but that the commanding general "has not yet made up his mind." In the past month, many animals had been captured from the enemy, yet "as far as I have heard none have been turned over," Ewing complained. "General Hood steadily refuses to let us have anything to do with them." The situation was so bad that one inspector, Col. E. J. Harvie, wrote Hood himself on November 3. "When property is captured from the enemy it finds its way into private hands, no one seeming to know or care what becomes of it," Harvie explained. "Applications are daily made by officers to purchase public animals," he added, "and there is no organized system by which it can be ascertained whether those applications should be granted." Harvie appealed to Hood to take some action in at least streamlining the army's inspector-general department; "unless some change is made your army will be injured and its movements seriously embarrassed." Yet there is no evidence that Hood acted. As Beauregard feared, these were the logistical concerns of which Hood tended to be "oblivious." Officials in Richmond chose not to interfere, even as they issued dire predictions. Major Cole forecast "the probable total loss" of the Army of Tennessee's field transportation. "Our stock of horses is being destroyed," rued Quartermaster General A. R. Lawton. When the correspondence and en-

---

gry Wind, 66; Lindsay, "Capture of Florence, Ala., Under Hood," 423; Hood to Seddon, October 31, *OR*, vol. 39, pt. 3, 870 ("the pontoon bridge is being laid down"); Hay, *Hood's Tennessee Campaign*, 62–63; Sanders, "Hood's Tennessee Campaign. Chapter IV," 250 ("did not rise until about the 4th of November"); Cubbison, "John Bell Hood and the Campaign in North Alabama and Middle Tennessee, October–November 1864," 66. (I thank Greg Biggs of Clarksville, TN, for a copy of this article.)

dorsement reached his desk, Secretary of War Seddon termed the whole mess "a lamentable exhibition of the losses encountered in a hasty movement"—with "hasty" casting aspersions on the officer in charge, J. B. Hood.[9]

Despite the shortages of food, the onset of cold and rain, and the boredom of camp when everyone itched to get on the march, the troops appeared to be faring well. The arrival and distribution of quartermaster stores may have been a contributor. "At Tuscumbia," writes John Lundberg, historian of Texans in the Army of Tennessee, "Granbury's men were issued new clothing and shoes, well supplying them for the northward advance." "The army is a unit, and buoyant with hope," wrote Colonel Harvie, the inspection officer, to Brigadier Robert H. Chilton on October 31. A week later, after all the hard marching over the past month, the returns showed a respectable 35,662 officers and men present for duty: 30,155 in the three infantry corps, 2,857 cavalry (Jackson's Division), and 2,650 artillerymen.

As always, there were officer appointments for Hood to attend to. Frank Cheatham had taken over from Hardee as corps commander, yet (unlike Stewart and Lee) he still held a major general's rank. Hood recommended his promotion to lieutenant general on the 31st. That day Hood also asked the president to promote Col. Joseph B. Palmer of the 18th Tennessee to brigadier, as he was commanding John C. Brown's brigade. Davis approved, the Senate confirmed, and Palmer was conferred his higher grade effective November 15 although he would not learn of his promotion till after the Tennessee campaign.[10]

---

[9] Ewing to Cole, November 4, *OR*, vol. 39, pt. 3, 888–90 (including endorsements by Cole, Lawton, and Seddon); Harvie to Hood, November 3, *OR*, vol. 39, pt. 3, 880–81; Roman, *Beauregard*, 2:286 ("oblivious").

The decline of the Army of Tennessee's field transportation had begun earlier. Based on documents in the National Archives, Edward Hagerman states that Hood was concerned during summer 1864 that "the deterioration of field transportation…made it impossible to forage the thinly settled countryside or to carry adequate supplies if available." On August 9 he issued an order reducing his army's transport standard to nineteen wagons per 1,000 men. Five days later he increased it slightly, and did so again on August 20, "but on November 3 he reduced it to approximately sixteen wagons per 1,000 men" (Hagerman, *The American Civil War*, 289).

[10] Lundberg, *The Finishing Stroke*, 67; Harvie to Chilton, October 31, *OR*, vol. 39, pt. 3, 871; abstract from field return, November 6, *OR*, vol. 39, pt. 3,

Meanwhile Hood had to keep putting off the start-date of his forthcoming raid.

> Florence is in our possession and the pontoon bridge is being laid down. I hope to be able to advance across the river so soon as supplies can be obtained.
>
> —Hood to Seddon, October 31

> Part of the army occupies Florence, but want of supplies and bad weather have delayed forward movement, which will be resumed soon as practicable.
>
> —Beauregard to Cooper, November 3

> The army will commence the movement into Tennessee on Saturday morning next [November 5].
>
> —Hood to Brent, November 3

> In consequence of non-arrival of commissary supplies, I cannot hope to be able to make a general movement of the army before Monday next [November 7].
>
> —Hood to Beauregard, November 4

> I hope to march for Middle Tennessee by the 8th or 9th instant.
>
> —Hood to Davis, November 6

A. A. A. G. James Cooper wrote to all three corps commanders on the 3rd that the army would march "on Saturday morning next," a date that came and went.

Meanwhile to the president Hood affirmed, "General Beauregard agrees with me as to my plan of operations." In this he was correct. Hood and Beauregard got together on November 3 and talked things over. "The plan referred to, which was discussed and approved on the 3d instant," Beauregard wrote General Cooper on November 6, had two main elements. First task, obviously, was "to remove as soon as possible from Florence into Middle Tennessee," toward Lawrenceburg, Pulaski, or Columbia. Second, Forrest was to join the army as it marched north while demonstrating against the enemy forces said to be moving from Nashville toward Pulaski.[11]

893; Hood to Davis, October 31, *OR*, vol. 39, pt. 3 870; Losson, *Tennessee's Forgotten Warriors*, 213, 215 ("promotion not acted upon by the Senate"); Neff, *Tennessee's Battered Brigadier*, 107, 130, 132.

[11] S. R. Watkins, "Reminiscences of Hood's Tennessee Campaign," 400 (reference to "Hood's raid into Tennessee"); Hood to Seddon, October 31, *OR*,

Even with Beauregard's apparent support, Hood's plan to launch a raid into Tennessee worried Jefferson Davis for an important reason: Sherman. Hood informed the president that according to information from Wheeler, "Sherman is moving one corps to Tennessee," but three were marching south toward Atlanta—a signal that he planned a thrust deeper into Georgia. Sherman was indeed planning his march, to begin after he and his three corps (XIV, XV, and XVII) marched back to Atlanta from northwest Georgia. "By the 10th," he wrote to Grant on November 6, "I...will have the troops all paid, the presidential election over and out of the way." Indeed, November 8 was presidential election day in the Northern states, with Abraham Lincoln facing the Democratic challenger, Union Maj. Gen. George B. McClellan. Lincoln handily won re-election, with 86 percent of Sherman's soldiers casting their votes for him. The vote was a blow to Confederates, who had hoped for a possibility of negotiated peace with a new Democratic president. "Lincoln is elected we will have to fight him four years longer," was the resigned comment of Pvt. William Young of the 15th Texas, in Granbury's Brigade.

Wheeler wrote Hood on November 2, "Prisoners and scouts now say Sherman will march to Savannah via Augusta." Hood seemed to show no anxiety at this news. "Should he move two or three corps south from Atlanta," he expressed to Davis, "I think it will be the best thing that could happen for our general good." The president did not agree and seemed more interested in Hood engaging Sherman in battle, apparently unaware that 200 miles separated the two armies (from Tuscumbia to Kingston, Georgia, where Sherman was at the time). "The policy of taking advantage of the separated divisions of Sherman's forces, by attacking him where he cannot reunite his army, is too obvious to have been overlooked by you," Davis wrote, with that tinge of sarcasm he had so often employed in his correspondence with Joe Johnston. "I therefore take it for granted that you have not been able to avail yourself of that advantage during his march northward from Atlanta, and hope the opportunity will be offered before he is extensively recruited," he added, apparently for-

vol. 39, pt. 3, 870; Beauregard to Cooper, November 3, *OR*, vol. 39, pt. 3, 880; Hood to Brent, November 3, *OR*, vol. 39, pt. 3, 880; Hood to Beauregard, November 4, *OR*, vol. 39, pt. 3, 888; Hood to Davis, November 6, *OR*, vol. 39, pt. 3, 890; Cooper to Cheatham, Lee and Stewart, November 3, *OR*, vol. 39, pt. 3, 881–82; Beauregard to Cooper, November 6, *OR*, vol. 39, pt. 1, 799.

getting that Hood almost a month before had given up on the idea of drawing Sherman into battle. Offering advice that eerily echoed Grant's suggestion to Sherman ("if you can see the chance for destroying Hood's army, attend to that first"), Davis counseled, "if, as reported to you, he has sent a large part of his force southward, you may first beat him in detail." *Then*, he added, "and subsequently without serious obstruction or danger to the country in your rear, advance to the Ohio River." Perhaps Davis was sending a vague hint that he was not so supportive of Hood's plan as he had once been. In any event, it was a peculiar message to the army commander who any day now was hoping to launch his campaign in the direction of that very Ohio River.[12]

"Movement into Middle Tennessee is still delayed by want of supplies, bad weather, and impassable roads," Beauregard wrote General Cooper on November 9. That day Hood added his own assessment to Secretary Seddon: "Heavy rains will delay the operations of this army for a few days." As further bad luck, the pontoon bridge over the river broke on the evening of the 9th and was carried downstream by the torrent. The engineers required three days to get new parts from Corinth and rebuild the span. At last, around the 10th the weather began to clear, and Hood could contemplate getting Cheatham's and Stewart's corps across the river. On November 10 Hood moved his headquarters across to Florence without informing Beauregard; a staff officer bearing a message for Hood brought the news back to military division headquarters.

Supply problems still held the army back. The efficient Maj. George Whitfield, overseeing the railroads supplying Hood's army, on November 6 pronounced the line to Corinth "in tolerably safe order," and from there the eastward railway to Cherokee "about the same." "If I had the hands could soon have the cars running to Tuscumbia," he claimed. As it was, though, "no stores have accumulated at any point so far." General Hood did not help when on the 8th he recommended to Beauregard that repair of the railroad from Cherokee to Tuscumbia be suspended, and that instead work be done to restore the track from Corinth to Purdy, Tennessee, "the most convenient point from which to draw supplies dur-

[12] Hood to Davis, November 5, *OR*, vol. 39, pt. 3, 891; Lundberg, *Finishing Stroke*, 68–69; S. Davis, *What the Yankees Did to Us*, 377 (soldier vote); Wheeler to Hood, November 2, *OR*, vol. 39, pt. 3, 878; Davis to Hood, November 7, *OR*, vol. 39, pt. 3, 896; Grant to Sherman, November 1 (6 P.M.), *OR*, vol. 39, pt. 3, 576.

ing the movement contemplated." This is important: the Army of Tennessee was struggling to be supplied on northern Alabama railroads, and already its commander was focusing on his logistical lifeline into Tennessee. This kind of handling of supply issues has led the biographer of Confederate brigadier Randall Gibson to judge Hood not unfairly as "impetuous and ineffective in logistical planning."[13]

Hood updated the president on November 12. The army "is now in excellent spirits and confident," he affirmed. Forrest was on his way but had not arrived. "This circumstance," he explained, plus "high water, and the fact that I had to draw supplies from and through a department not under my command, involving delay in their reaching me, have retarded my operations." But once Forrest's cavalry joined the army, "which will be in a few days," Hood predicted, "I shall be able to move forward." Cheatham's Corps finally crossed the river on November 13, and the army's supply trains rumbled over the next day. Beauregard too showed a little more optimism. "Weather is now good," he wired Richmond on the 13th; "roads improving." But moods swing with the weather, and for Beauregard both darkened when rain returned the next day. "Weather again bad, and river still high," he reported to Cooper on the 15th.

Beauregard, like other observers, knew that Hood's delay was giving Thomas valuable time to concentrate his strength. "Press dispatches report part of A. J. Smith's forces having arrived at Paducah," he added. This news, and doubtless the feeling that he had done all he could to get the campaign going, led Beauregard to inform Hood that he would soon be leaving Tuscumbia for Corinth to see to defenses there, as Smith's

---

[13] Beauregard to Cooper and Hood to Seddon, November 9, *OR*, vol. 39, pt. 3, 903–904; Shoup journal, November 10, *OR*, vol. 39, pt. 1, 808; Sword, *Embrace an Angry Wind*, 69 (bridge rebuilt evening of November 12); Mason to Brent, November 9 (6:30 P.M.), *OR*, vol. 39, pt. 3, 904; Mason to Lee, November 9 (6 P.M.), *OR*, vol. 39, pt. 3, 905; Jacobson, *For Cause & For Country*, 46; McMurry, *Hood*, 166; Whitfield to W. F. Bulloch, Jr., November 6, *OR*, vol. 39, pt. 3, 903; Hood to Beauregard, November 8, *OR*, vol. 39, pt. 3, 900; McBride, *Randall Lee Gibson of Louisiana*, 111.

Daniel C. Warren sees linkage between Hood's logistical shortcomings in north Alabama and his indifference to supply issues when he served in Virginia: he had drawn Lee's rebuke in November 1862, and at Suffolk six months later, "he could not see the value or necessity of Longstreet's foraging expedition of April 1863" (Warren, "John Bell Hood: A Bid for Fame," 113 [courtesy of Greg Biggs, Clarksville, TN]).

movement portended a possible enemy advance on the Mobile & Ohio road. In a letter of November 15 Beauregard explained to Hood his intent to continue to tend to logistical issues. "General Taylor and myself will always be anxious to aid you in your present campaign with all the means at our control," he emphasized. As example, despite Hood's suggestion that repairs of the Memphis & Charleston line to Tuscumbia be halted, Beauregard looked to their completion, "in from fifteen to twenty days," weather permitting.

Then there was Sherman, and Beauregard worried about him, too. "Reports of General Wheeler," he telegraphed Richard Taylor, "indicate that Sherman is about to move with three corps from Atlanta to Augusta or Macon." In preparation for the march Sherman authorized Slocum to send out four foraging expeditions east of Atlanta to gather food. General Geary reported having collected 2,400 wagonloads of corn; Slocum's "bummers," as one biographer has termed them, gathered in two million pounds of corn and fodder from the people of Georgia. These supplies would help provision Sherman's troops in their upcoming march to the sea. Before leaving Atlanta, Sherman would order his engineers to blow up and burn all buildings, shops, and facilities that could help the Rebels. Word spread among his officers and men that the torch would be put to the city. "It is sumed a round Camp we are gonto erbandon this city and disstroy it by fiar," wrote Cpl. Rodolphe Rey of the 102nd New York to his girlfriend on November 7.

After the Yankees set out on their march, Wheeler's cavalry could only harass them. An alarmed Howell Cobb at Macon called on Beauregard for reinforcements. All the latter could do was ask Taylor to assemble what troops he could spare in his department, get to Georgia, and take charge of all forces there. Instead to traveling to Corinth, a week after leaving Tuscumbia Beauregard was in Macon, which since the outset of Sherman's march on November 16 had become a greater point of danger.[14]

---

[14] Hood to Davis, November 12, *OR*, vol. 39, pt. 3, 913; Beauregard to Cooper, November 13, *OR*, vol. 39, pt. 3, 917; Jacobson, *For Cause & For Country*, 46; Beauregard to Cooper, November 15 (11 A.M. and 7 P.M.), *OR*, vol. 45, pt. 1, 1208–1209; Beauregard to Hood, November 15, *OR*, vol. 45, pt. 1, 1209–11; Beauregard to Taylor, November 15, *OR*, vol. 45, pt. 1, 1213 ("future danger of Corinth"); S. Davis, *What the Yankees Did to Us*, 351, 387–90; Melton, *Sherman's Forgotten General*, 183; Rodolphe Rey to Lizzie DeVoe, November 7, Ru-

Before leaving Tuscumbia on November 17, Beauregard had Colonel Brent send Hood one more exhortation to get moving: "he desires you will take the offensive at the earliest practicable moment and deal the enemy rapid and vigorous blows, striking him while thus dispersed, and by this means distract Sherman's advance into Georgia." Beauregard went so far as to authorize Hood to issue any orders in General Taylor's department necessary for his success in the coming offensive, "sending General Taylor, or whoever may be in command, copies of all orders." This was a big step—Beauregard's final effort to help Hood achieve success. One may assume that he was able to take it, essentially a whittling away of Taylor's authority, because he could count on Richard Taylor's team spirit. One may also assume that if Beauregard had whittled John Bell Hood's authority in similar fashion, the latter would not have manifested such spirit. As to "earliest practicable moment," in his reply to Brent Hood merely stated that he would move as soon as he could. Prospects for that happening did not seem bright when Hood added, "I have now seven days' rations on hand and need thirteen days' additional." To Beauregard's request that he send one of Red Jackson's brigades, and eventually Jackson's entire cavalry division, to help Wheeler fight Sherman, Hood answered, "unless the general should insist upon it, I would prefer to keep it for the present.... To send Jackson's division at this time would materially endanger the success of the operations of this army."

While the commanding general put off the start of his campaign, the men voiced their intent to get going. "With this army we can tough it out four years longer," wrote Texas Private Young to his sister on November 17. He added, "as for rations and clothing we have done very well since here." That was a matter of opinion. On the same day another Texan, Pvt. E. G. Littlejohn, wrote, "We have suffered more for the want of something to eat on this campaign than on any other previous one."[15]

---

dolphe [*sic*] Rey Collection, New-York Historical Society (transcription courtesy of Douglas Ullman, Jr., Centreville, VA); Cobb to Beauregard, [November 15], *OR*, vol. 45, pt. 1, 1216; Beauregard to Taylor, November 16, *OR*, vol. 45, pt. 1, 1213; Beauregard to Davis, November 24, *OR*, vol. 45, pt. 1, 1242 (his arrival in Macon).

Writing after the war, one Southerner wrote of "the fatal delay of a week on the Tennessee river," which allowed Thomas to build and concentrate his forces around Nashville (M. B. Morton, "Battle of Nashville," 19).

[15] Brent to Hood, Hood to Brent, and Hood to Beauregard, all November 17, *OR*, vol. 45, pt. 1, 1215–16; Lundberg, *Finishing Stroke*, 68–69.

Meanwhile Wheeler sent reports from outside of Atlanta on Sherman's activity. "Enemy have been burning something in Atlanta for the last two days," he wrote ominously on November 13. The next day he confirmed that the enemy's XV and XX corps plus Kilpatrick's cavalry were in the city; "camp rumors are that they will march to Augusta and Savannah." Hood came closer to launching his own march when Forrest finally arrived at Florence on the 14th. The Wizard of the Saddle brought with him some 3,500 men. Hood's Tennessee troops were especially delighted to have Forrest with them; on the evening of his arrival they serenaded him and asked for a speech. Capt. Theodrick (Tod) Carter of the 20th Tennessee doubled as a correspondent for the *Montgomery Mail* and recorded Forrest's remarks thus:

> Well, soldiers, I came here to jine you. I'm gwine to show you the way into Tennessee. My conscripts are going, and I know Hood's veterans can go. I come down here with three hundred and fifty men. I got thirty-five hundred conscripts. Since May I have fought in every county in West Tennessee. I fought in the streets of Memphis, and the women ran out in their night clothes to see us, and they will do it again in Nashville. I have fought a battle every twenty-five days. I have seen the Mississippi run with blood for two hundred yards, and I'm gwine to see it again. I've captured seventy-eight pieces of artillery and sixteen thousand Yankees, and buried twenty-five hundred of them!

Boasting, especially with exaggeration, was a durable staple of nineteenth-century American frontier society, and Forrest's speech fit right into the tradition. Nevertheless, his feats were actual (or believed by at least some Southerners to be). Moreover, Forrest was bringing to the Army of Tennessee a reputation as the Confederacy's most feared cavalryman (Sherman certainly thought so). Accordingly, on the 15th Hood put Forrest in command of all of his army's cavalry, three divisions totaling 5,000, maybe 6,000 horsemen.

Amidst all the preparation for the upcoming campaign, John Bell Hood, who had been baptized and confirmed as a member of the Episcopal church, heeded the public proclamation of another Episcopalian, Jefferson Davis. Ten times during the war, beginning in June 1861 and ending in March 1865, the president proclaimed a nationwide day of prayer and fasting. These calls were publicized by Southern newspapers, as when Davis on October 26, 1864, called for the people on November 16 to "unite in prayer to Our Heavenly Father that He bestow favor upon us." Hood dutifully followed with a field order of November 15 declaring that the morrow "will be observed as a day of prayer throughout the army, and all military duty not absolutely necessary will be dispensed with," so that the officers and men could convene for worship. They did. Texas Captain Foster attended a service in Florence, estimating some 700 soldiers present. "All seem quiet and attentive," he noted in his journal.[16]

Taking a whole day off from work did not make the situation of Hood's army any better. Colonel Brent wrote Hood that the road from Cherokee to Tuscumbia at one point "is almost impassable," and other

---

[16] Wheeler to Hood, November 13, *OR*, vol. 39, pt. 3, 918; Wheeler to Hood, November 14, *OR*, vol. 45, pt. 1, 1206; Beauregard to Forrest, *OR*, vol. 45, pt. 1, 1207; [Shoup] journal, November 14, *OR*, vol. 39, pt. 1, 808; Henry, *"First With the Most,"* 383–84; Andrews, *South Reports*, 474 (Carter's report of Forrest's speech); T. D. Clark, *The Rampaging Frontier*, 208 ("tall talk" of frontier westerners); Sherman to Grant, November 6, 1864, *OR*, vol. 39, pt. 3, 659 ("that devil Forrest"); Forrest report, January 24, 1865, *OR*, vol. 45, pt. 1, 752 ("making in all about 5,000 cavalry"); Groom, *Shrouds of Glory*, 130 ("six thousand troopers"); Silver, *Confederate Morale and Church Propaganda*, 64–65; Stout, *Upon the Altar of the Nation*, 475 n.5; Crist, et al., *Papers of Jefferson Davis*, 11:111 (proclamation excerpts, October 26); General Field Orders No. 31, November 15, *OR*, vol. 45, pt. 1, 1211; Lundberg, *Finishing Stroke*, 68.

Example of church services held on fast day were those conducted November 16 (a Wednesday) in Columbus, Georgia, by the Rev. Dr. Charles Quintard. It was Quintard to whom General Hood had confessed his "two objects in life" on the day of his Episcopal confirmation three months before. Subsequently Quintard had moved with his family to Columbus, where he took an appointment from the church. On November 16, Quintard officiated at St. Paul's Church in Columbus, preaching from the Scripture, "I came not to send peace, but a sword" (Matt 10:34, KJV). The Rt. Rev. William Mercer Green, bishop of Mississippi, was to have led services, but was ill. Bishop Green had nonetheless prepared his prayer for the country, and Quintard asked him to write it in his diary. In his text the bishop asked the Lord's blessing upon "all engaged in our defence," upon sick, wounded and imprisoned soldiers, as well as the Confederacy's civil authorities. Green even asked the Lord's forgiveness of the South's enemies: "turn their hearts, that they may depart from us, and no longer seek to destroy us." Finally was this: "And grant, Oh Gracious Lord, that we may soon become not only a separate and independent people, but a holy nation, a peculiar people, zealous of good works, and in all things devoted to thy service" (Elliott, *Doctor Quintard*, 88–89, 93, 176–77).

points would become so with further rain. On the 18th Beauregard learned that the railroad had empty cars but nothing to load onto them. "Where are your supplies?" he asked; Hood did not answer, as he could not have done so with any certainty. Yet the Creole still hoped that a quick advance into Tennessee might halt Sherman's march farther into Georgia. After Wheeler reported that Sherman had fully four corps, Beauregard wired Hood, asking if he could send any troops to help Wheeler and Cobb. Failing that, he asked, "could you…take the offensive immediately?" "I think it best to take offensive immediately," Hood replied, "and expect to move on Monday." Beauregard answered back, "Push on active offensive immediately."[17]

The time spent in Tuscumbia and Florence, despite its privations, allowed the Army of Tennessee to gain strength, if only slightly. Medical officers in hospitals serving the army screened their patients for those men capable of field duty and called them forth. Lt. Robert Collins of the 15th Texas Dismounted Cavalry had been wounded at Peachtree Creek and was "convalescing" in Corinth, "laying up in day time and chewing sugar-cane, and running around courting at night." He was sent to the front on November 11. Such efforts had results. For example, Stevenson's Division (Lee's Corps) on November 6 mustered 3,540 officers and men present. About two weeks later another count showed 3,766. Any increase in strength was important to an army about to launch an offensive. With the accession of Forrest's cavalry after mid-November, Hay counts some 41,000 men under Hood's command. Yet some soldiers would be left behind when the Confederates marched north. On November 19, Dr. Andrew J. Foard, medical director, issued a circular prescribing how medical officers would see that sick and wounded men would be cared for once the army left Florence.[18]

---

[17] Brent to Hood, November 17, *OR*, vol. 45, pt. 1, 1216; Beauregard to Hood, November 18 (7:30 [A.M.]), *OR*, vol. 45, pt. 1, 1219; Beauregard to Hood, November 18 (3 P.M.), *OR*, vol. 45, pt. 1, 1220; Hood to Beauregard, November 19, *OR*, vol. 45, pt. 1, 1225; Beauregard to Hood, November 20, *OR*, vol. 45, pt. 1, 1226.

[18] Lundberg, *Finishing Stroke*, 66; Jacobson, *For Cause & For Country*, 79 ("some new recruits had joined the army"); Abstract, November 6, *OR*, vol. 39, pt. 3, 893 (also vol. 45, pt. 1, 678); "abstract from tri-monthly return of Stevenson's division…for November 18, 1864," *OR*, vol. 45, pt. 1, 1224; Hay, *Hood's Tennessee Campaign*, 78; A. J. Foard circular, November 19, *OR*, vol. 45, pt. 1, 1225–26.

The army was scheduled to march forth on the morning of the 19th, but rain again got in the way. Hood acknowledged his and the army's logistical inadequacies on November 20 when, in a General Field Order, he announced to the officers and men that "on the march we are about commencing there may be a scarcity of the bread ration." He asked the troops to "meet this privation, should it come, in a cheerful, manly spirit." He predicted that any bread shortage would be brief and promised that the meat ration would be proportionately increased. He offered hope that "the fruitful fields of Tennessee" would yield sustenance to the troops (unless the enemy had too thoroughly ravaged the land). "Let the privation be to us," he closed, "an occasion for the exhibition of a most determined patriotism." The commanding general balanced this message of concerned warning with an uplifting proclamation on November 21: "Soldiers: You march to-day to redeem by your valor and your arms one of the fairest portions of our Confederacy." If the men faced the upcoming challenges bravely, victory would result with "lasting renown to yourselves."

Hood also addressed the troops, according to a Confederate veteran many years later. J. W. Simmons remembered "his famous speech," "in which he said he was going to advance, and that he believed the army would rather advance on parched corn than to retreat on full rations. Every one in hearing hallooed themselves hoarse in rejoicing over the idea of advancing. 'Advance on parched corn' became a common expression in the army, but little did we think it would become a reality…."[19]

Finally, A. P. Stewart's infantry marched over the pontoons on November 20, and the army headed north the next day. "We left camp this morning at sun up and started for Tenn.," recorded Capt. Samuel Foster of Granbury's Brigade. "Damp, cold, and misty," Colonel Capers wrote of the weather. "The roads are Said to be in a desperate condition, almost impassable," recorded Alabaman James M. Lanning in his diary. General Walthall noted that "it was almost impossible for the artillery to move at all," with the guns being pulled over bad roads by animals already fatigued. The Army of Tennessee was launching its second raid back into

---

[19] Lundberg, *Finishing Stroke*, 68 ("The crossing was scheduled for sunrise on November 19, but because of drenching rains and overcrowding, it was delayed until the next day"); General Field Orders No. 35, November 20, *OR*, vol. 45, pt. 1, 1227: "SOLDIERS," November 21, *OR*, vol. 45, pt. 1, 1236; Simmons, "Pathetic Burial at Midnight," 358 ("his famous speech").

its native state under circumstances far less favorable than it had encountered in its first, when in late summer 1862 it marched forth under Braxton Bragg. This time snow and ice had already arrived, the horses had no forage, and the men were issued rations consisting only of corn, according to Capers. "In spite of this," he declared, "the men were cheerful and dutiful" as they set out on their new campaign. "We expect to have a hard time and to have some fighting to do," Lt. Hamilton Branch (54th Georgia) wrote to his mother on November 21, "but we intend with Gods help to ruin Thomas' Army."[20]

From Florence the infantry of Hood's three corps took as many roads on their march north. Cheatham's Corps, the most westward, headed for Waynesboro. Stewart, farthest to the east, aimed for Lawrenceburg. Lee's Corps marched in between. Hood rode with Stewart's column. The troops covered ten to a dozen miles on the 21st. As they crossed into Tennessee, the men faced snow and freezing rain. One soldier of the 27th Tennessee remembered the march into Tennessee as being "the coldest day I ever felt, and the army seemed to suffer most intensely." One legend has it that Confederates saw a sign someone had posted: "Tennessee: a Grave or a Free Home."[21]

On the 22nd the Confederates made further progress; Cheatham's infantry covered eighteen miles despite the bitter cold. The Yankees felt it, too. "The storms," wrote Union General Cox, "were continuous and severe in Tennessee, alternating between rains and severe frosts, covering

---

[20] Jacobson, *For Cause & For Country*, 46, 52–53; McMurry, *Hood*, 166–67 (Capers and Lanning); Walthall report, January 14, 1865, *OR*, vol. 45, pt. 1, 719; Capers report, January 4, 1865, *OR*, vol. 45, pt. 1, 735; Joslyn, *Charlotte's Boys*, 288.

In calling Bragg's advance into Kentucky in fall 1862 a *raid*, as opposed to *invasion*, I am adopting the terminology of Hattaway and Jones, *How the North Won*, ch. 9, "The Maryland and Kentucky Raids Repulsed." Their logic follows the Confederates' intentions of launching only temporary incursions into enemy territory—without the objective of permanently holding it, once conquered, with armed forces. The opposite—invasion and occupation of Rebel territory—was the grand objective of Union forces during the war.

[21] Jacobson, *For Cause & For Country*, 55; Crawford, "The Mystery of Spring Hill," 102–103 (including map, march from Florence); Mason to Lee, November 20 (4:40 P.M.), *OR*, vol. 45, pt. 1, 1227 (Hood on Waynesboro road); J. M. Taylor, "Twenty-seventh Tennessee Infantry," 1:427; Gillum, *Twenty-five Hours to Tragedy*, 35; White *Let Us Die Like Men*, 40-41.

the roads with a frozen crust over deep mire, just strong enough to make the utmost obstruction, without getting the solidity necessary to bear up the wagons and teams." Impeding Stewart's progress were Federal cavalry under Brig. Gen. Edward Hatch, whose horsemen had been in contact with Forrest's troopers since the 20th. In one fight, the Southerners had caught the enemy in "a well devised trap," admitted the Union commander, who managed to get his men out of it amid "those hideous yells, such as only rebels can make." On the morning of the 21st, Hatch sent General Thomas word of the movement of the three Rebel infantry corps. The next night Hatch informed General Schofield, still at Pulaski, that the Rebels (Stewart's) had driven him out of Lawrenceburg. Stewart then turned northwest so that the three corps reunited at Henryville on the 24th. Clearly the Confederates were headed for Columbia and Franklin (and perhaps from there to Nashville). W. T. Crawford, an early writer on "the mystery of Spring Hill," as the ensuing events have come to be called, can only guess why Stewart marched northwest from Lawrenceburg to Henryville, when the shortest route to Columbia was northeast via Mount Pleasant. He figures it was because Hood, the commanding general, was accompanying Cheatham on the longer, most westward route. A better guess is that by the 24th the Confederates knew that Schofield's two infantry corps were marching from Pulaski to Columbia; they would have reasonably worried about the enemy assailing Stewart's flank.[22]

Each side was racing for Columbia, where the roads from Pulaski and Mount Pleasant converge on the Duck River. Hood recalled in *Advance and Retreat* that he hoped "by a rapid march to get in rear of Schofield's forces, then at Pulaski, before they were able to reach Duck River." General Thomas wired Halleck that he had directed Schofield "to move back gradually from Pulaski and concentrate in the vicinity of Columbia, so as to reach this place before Hood could." In case of a fight, Thomas's intention was for Schofield to stand on the defensive, if

---

[22] Jacobson, *For Cause & For Country*, 58; J. D. Cox, *The March to the Sea*, 64; Hatch to Thomas, November 20, *OR*, vol. 45, pt. 1, 961–62; Col. Datus E. Coon report, January 20, 1865, *OR*, vol. 45, pt. 1, 585 ("hideous yells"); Thomas to Halleck, November 21 (9:30 P.M.), *OR*, vol. 45, pt. 1, 970; Crawford, "Mystery of Spring Hill," 104–105; Hay, *Hood's Tennessee Campaign*, 80; J. P. Strange to W. H. Jackson, November 24, *OR*, vol. 45, pt. 1, 1243 ("the enemy are all leaving Pulaski").

only because his strength was inferior to the Rebels'. Schofield's XXIII Corps numbered some 10,500 troops, and Stanley's IV another 16,000 effectives. The cavalry, of which Brig. Gen. James H. Wilson had recently taken command, counted some 4,300 to 4,800 sabers. By this measurement, Thomas was justified in announcing, "I shall have to act on the defensive."

Hood, on the other hand, had gotten a different message from Beauregard. On November 17 the military division commander dictated to Brent his desire that Hood would "take the offensive at the earliest practicable moment and deal the enemy rapid and vigorous blows." Hood, known even to his critics (such as Joe Johnston) as "the Striker of Manly Blows," needed little such direction before he would consider attacking the enemy in Tennessee.[23]

Forrest spread his three divisions to cover the advance of Hood's infantry. At Henryville, between Waynesboro and Mount Pleasant, Confederate horsemen brushed with the enemy and took forty-five prisoners. Jackson preceded Stewart, with a detachment off to the east to screen against the Yankees. "Both these divisions had several engagements with the enemy," Forrest reported, "and were almost constantly skirmishing with him, but drove him in every encounter." As they fanned out, Forrest's troopers sought forage for their mounts. "They were also to glean the country of cattle, and set the mills of the neighborhood to grinding all the meal possible," write Jordan and Pryor, Forrest's early biographers. They did their job well. Hood was able to write Beauregard from Mount Pleasant on the 25th, "Plenty of mills now in our possession. I think I will have no difficulty about supplies." "We are getting plenty to eat now," wrote a Georgian in the 54th to his wife; "we draw bacon beef hard bread and sometimes flour. We have no cause to complain." The people of the countryside helped by offering their own victuals. "White

---

[23] Hood, *A & R*, 281; Hay, *Hood's Tennessee Campaign*, 78 ("Hood hoped to get in the rear of Schofield at Pulaski before he could withdraw to Columbia"); Thomas to Halleck, November 21 (9:30 P.M.), *OR*, vol. 45, pt. 1, 970; Crawford, "Mystery at Spring Hill," 105; Thomas to Schofield, November 21, *OR*, vol. 45, pt. 1, 970, 973 ("I have sent General Wilson out to take general charge of the cavalry, and directed him to report to you"); Longacre, *From Union Stars to Top Hat*, 166 ("4,300 mounted cavalrymen"); Wilson report, December 21, 1864, *OR*, vol. 45, pt. 1, 550 (Hatch, Croxton, and Capron totaling 4,800); Brent to Hood, November 17, *OR*, vol. 45, pt. 1, 12154; Symonds, *Johnston*, 334 (the "Striker of Manly Blows").

handkerchiefs are waved by the ladies in every house we pass that is inhabited," wrote one Johnny Reb. One good woman filled cups with molasses as the men passed. The spirit of homecoming perfused the ranks with good humor. A lot of soldiers still had deficient footwear, but as one expressed it, "we didn't care"; they were in Tennessee.

But Hood lost the race to trap the Yankees. "I hoped to be able to place the army between these forces of the enemy [those at Pulaski] and Nashville," he later wrote. Schofield, warned by Hatch, started marching from Pulaski on November 23 for Columbia, which Cox's division reached the next morning. Hood took the developments in stride. "The want of a good map of the country, and the deep mud through which the army marched," he reported, "prevented our coming up with the enemy before they reached Columbia."[24]

In Columbia, with his troops "pretty well intrenched" on the south bank of the Duck River, Schofield halfway thought Hood might attack him. He also knew that Thomas hoped for him to hold Hood south of Nashville as long as he could; A. J. Smith was expected from Missouri and every day helped. Nevertheless, Schofield was wary that the Rebels might attempt a river crossing above or below him. Wilson posted his cavalrymen at the various fords of the Duck east of Columbia. "If you cannot hold Columbia," Thomas cautioned, "you had better withdraw to the north bank of the river." Schofield began marching a couple of brigades to the north bank on the night of the 25th, but he hoped to hold his position on the south side in case Thomas sent him reinforcements.[25]

Hood's infantry reached Columbia on the morning of November 27, relieving Forrest's cavalry. Even though the Federals had a river at their back, Hood decided against launching an attack; his troops were tired from marching over muddy roads. More importantly, the situation

---

[24] Forrest report, January 24, 1865, *OR*, vol. 45, pt. 1, 752; Jordan and Pryor, *The Campaigns of Lieut.-Gen. N. B. Forrest*, 612; Hood to Beauregard, November 25 (12 m.), *OR*, vol. 45, pt. 1, 1245; McCardle, *A Just and Holy Cause*, 140 (Marcus Ely to Martha Ely, November 23); Bailey, *Chessboard*, 79; Hood report, February 15, 1865, *OR*, vol. 45, pt. 1, 652; Schofield report, December 31, *OR*, vol. 45, pt. 1, 341; Connelly, *Schofield*, 123 ("Cox's arrival on November 24"); Hood report, December 11, *OR*, vol. 45, pt. 1, 657.

[25] Schofield to Thomas (8 P.M.), November 24, *OR*, vol. 45, pt. 1, 1017; Connelly, *Schofield*, 124; Longacre, *From Union Stars*, 168 ("observing the Duck River fords east of Columbia"); Thomas to Schofield (3 P.M.), *OR*, vol. 45, pt. 1, 1017; Schofield report, December 31, *OR*, vol. 45, pt. 1, 341.

offered opportunity for maneuver. Hood developed the idea of leading most of his army across the river east of Columbia and gaining Schofield's rear while S. D. Lee, with two of his divisions and the army's artillery, were to keep up a demonstration, making Schofield believe he was about to be attacked. Forrest's cavalry was sent off to the east to find crossing points over the Duck. On the night of the 27th Hood called Lee, Stewart, Cheatham, and Forrest to his headquarters, the home of Mrs. Amos Warfield ("Beechlawn"), to explain his plan.[26]

At this point Hood's tactical objective is hard to ascertain. Everyone knew that the aim was to gain the enemy's rear. While two divisions of his corps kept the enemy occupied before Columbia, S. D. Lee later wrote, "the other two corps and Johnson's division [of his corps] would be crossed above and moved to the rear of the enemy in the direction of Spring Hill," a town about twelve miles north of Columbia on the road to Franklin. "We made a forced march to get in rear of the enemy," recalled General Stewart. Maj. Gen. Carter Stevenson, division commander in Lee's Corps, wrote that Hood's forces "pushed for the enemy's rear." "This was a strategic movement of Hood's to gain the Franklin pike in rear of the enemy," recorded General French, in the memoir based on his wartime journal.

But once he gained position behind Schofield, what did Hood intend to do? Two weeks afterward he wrote that his object was "to make a rapid march on roads parallel to the Columbia and Franklin pike, and by placing the troops across this pike at or near Spring Hill to cut off that portion of the enemy." Hood does not make clear whether he expected Schofield, once he realized he had been outflanked and began retreating northward from Columbia, to attack him, or whether he intended to attack Schofield. This allows James Lee McDonough to claim "the Confederate objective was general and somewhat vague." "Hood's corps leaders, Stewart and Cheatham, marched without any specific orders,"

---

[26] Forrest report, January 24, 1865, *OR*, vol. 45, pt. 1, 752 (relieved by infantry morning of November 27); Crawford, "Mystery of Spring Hill," 105 ("the Confederate infantry arrived on the morning of November 27"); Hay, *Hood's Tennessee Campaign*, 82 ("intrenched and awaiting enemy"; "muddy roads and were fatigued"); Jacobson, *For Cause & For Country*, 71–72; McDonough and Connelly, *Five Tragic Hours*, 34; Connelly, *Autumn of Glory*, 491.

Connelly charges, remarking on "the almost total lack of planning for the Spring Hill venture."[27]

After the war, General Lee declared there was an understanding that once the Confederates had gotten their position astride the Columbia pike in Schofield's rear, there would be fighting. "Before starting from Columbia," he wrote, "it was understood that the first troops reaching the pike at Spring Hill should cross it or commence fighting and would be reenforced by the other troops as they arrived in succession on the field. A lodgment was to be made on the pike and the march of the enemy stopped." Eric Jacobson absorbs the customary sources and adds a few new ones (Pvt. Edward McMorries: "pass the enemy and throw itself across the Franklin pike in front"; Surgeon Urban Owen: "Our Gen'ls all say that our route is to get between Thomas' army & Nashville"), deciding "it is not at all presumptuous to conclude that Hood's goal on November 29 was, as he stated, to gain control of the pike and force Schofield's hand in battle."

Clouding the situation further is Chaplain Quintard's record of conversations he had with Hood at the time. Quintard had traveled from Georgia to join the army and encountered the general on November 25 when his headquarters were at Mount Pleasant, a dozen miles southwest of Columbia. Hood greeted the good doctor warmly. "The General is in the best of health and spirits," Quintard recorded in his diary, "and full of hope as to the results of the present movements." Two days later, on the evening of the 27th—about the time Hood was outlining his plans to his corps commanders—Quintard had "much pleasant conversation" with Hood. In his diary that night Quintard wrote, "He detailed to me his plan of taking Nashville, calling for volunteers to storm the Key of the works about the city and etc." The next day Quintard wrote that Hood "told me just now confidentially that he would either beat the enemy to Nashville or make him go there at a double quick." "At all events our forces cross at daylight," he added, "and, as the General says, will 'go through the woods.'" Before they parted, Quintard said a short prayer,

---

[27] Lee report, January 30, 1865, *OR*, vol. 45, pt. 1, 687; Stewart report, April 3, 1865 *OR*, vol. 45, pt. 1, 712; Stevenson report, January 20, *OR*, vol. 45, pt. 1, 693; French, *Two Wars*, 291; Hood report, December 11, *OR*, vol. 45, pt. 1, 657; McDonough, *Nashville*, 54; Connelly, *Autumn of Glory*, 492.

asking God's blessings on Hood and his endeavors. "Thank you, Doctor," the general replied, "that is my hope and trust."[28]

In his postwar memoirs, Hood claims that after marching his forces to Spring Hill, he planned "to attack as the Federals retreated, and put to rout and capture, if possible their Army." Writers, including Thomas Robson Hay, have accepted Hood's statement ("he planned to attack Schofield"). "He would ambush his opponent," claims O'Connor. "It does seem Hood was trying to destroy Schofield, not just beat him to the Tennessee capital," judges Jacobson. Connelly, however, is skeptical of Hood's assertion. "There is evidence"—Quintard's diary entries—"that he was more interested in gaining the pike at Schofield's rear in order to outrace the Federals to Nashville." Others seem to hedge in their assessment. Horn writes that once the Confederates got in Schofield's rear, blocking the Columbia-Nashville pike, they "would have Schofield in a trap from which he would not easily extricate himself." McDonough and Connelly, in their treatment of Franklin, state simply that Hood "would cross the river and march on the Davis' Ford Road to Spring Hill," without saying what he would do there. McMurry offers the criticism that "Hood had started north from Florence with no more plan…than he had had at any time since he drew back from Allatoona in early October," yet judges that "his objective was probably Nashville, not the [enemy] force."

In *Advance and Retreat*, Hood couches his tactical situation at Columbia-Spring Hill by referring to Jackson's exploits in Virginia. At Chancellorsville, Stonewall had gained the enemy flank and attacked it. But at Second Manassas, once in the enemy rear he had taken the defensive and waited for Pope to attack him. Which of Jackson's stratagems did Hood seek to emulate on November 28–29?

The uncertainty as to Hood's plans reminds one of General Beauregard's cautionary warning about this campaign and Hood's manner of "planning" military operations, especially because "much had to be foreseen, and much prepared or created." Beauregard after the war believed that the whole movement into Tennessee "had been rather hastily undertaken," and that "a great deal had been left to future determination, and even to luck." This indictment—leaving his army's fate to unforeseen

[28] Lee, "From Palmetto to Nashville," 257; Jacobson, *For Cause & For Country*, 81–82; Elliott, *Doctor Quintard*, 95, 181, 183–84; Shoup journal, entry of November 25, *OR*, vol. 45, pt. 1, 670.

developments—seems to apply to Hood as he pushed farther north into Tennessee.[29]

Hood's flanking maneuver was set in motion on November 28. Forrest's cavalry led the way, crossing the Duck at several places, including Hardison's Mill, where the Lewisburg-Mount Carmel road crossed. There the Yankees offered some resistance before riding off to the north. One of Forrest's divisions was on the north bank of the Duck before midday. The Southern horsemen "forced back the Union cavalry on roads leading toward Spring Hill and Franklin," wrote Union officer Henry Stone. Meanwhile S. D. Lee performed his role after learning that Schofield had marched the rest of his troops across the river during the night of the 27th–28th, burning the railroad bridge behind them. Stevenson's Division of Lee's Corps entered Columbia before dawn; Clayton's also came up. The Southerners put on "such a continual and noisy demonstration," one of Schofield's men observed, that they "believed Hood's whole army" was in their front on the south bank. With the Federals dug in on the other side in a position Hood termed "quite strong in front," Lee achieved his purpose in holding their attention. The 28th was spent in "considerable skirmishing across the river and some artillery firing," Lee reported, "resulting in nothing of importance."[30]

Union General Wilson, falling back before Forrest, sent word at 1:30 P.M. to Schofield that Rebel cavalry were crossing the river east of Columbia, heading for Spring Hill. "You had better look out for that place," he cautioned. After that, however, Wilson left Schofield to handle things alone, as noted by J. P. Young, an early chronicler of the campaign. "General Wilson seems to have obtained the idea that General Forrest's objective point was Nashville via Franklin, and so firmly was he

---

[29] Hood, *A & R*, 282–83; Hay, *Hood's Tennessee Campaign*, 82; O'Connor, *Cavalier General*, 229; Jacobson, *For Cause & For Country*, 79, 81; Connelly, *Autumn of Glory*, 492; Horn, *Army of Tennessee*, 384; McDonough and Connelly, *Five Tragic Hours*, 35; McMurry, *Hood*, 169; Roman, *Beauregard*, 2:281.

[30] Wigginton, "Cavalry Operations," 40, 45 (W. S. Nye's map showing Chalmers's, Jackson's and Buford's crossing places); Longacre, *From Union Stars*, 168; Wyeth, *Life of General Nathan Bedford Forrest*, 539 ("forced back the Union cavalry on the roads leading towards Spring Hill and Franklin"); Mathes, *General Forrest*, 309; H. Stone, "Repelling Hood's Invasion of Tennessee," 4:444; Hay, *Hood's Tennessee Campaign*, 83; Jacobson, *For Cause & For Country*, 72–74; Hattaway, *Stephen D. Lee*, 134; Hood report, February 15, 1865, *OR*, vol. 45, pt. 1, 652; Lee report, January 30, *OR*, vol. 45, pt. 1, 687.

impressed with this idea that in retreating before the leader next day he rode entirely past the flank of Schofield's forces, retreating along the Franklin Pike in the vicinity of Spring Hill, leaving that hapless commander entirely uncovered."

"In his eagerness to withdraw," Edward Longacre writes, "Wilson made the most serious tactical error of his military career": he fell back up the Lewisburg pike, riding *away* from Schofield and the infantry near Columbia. This left Schofield guessing whether Rebel infantry were crossing the Duck behind Forrest. At 8:30 on the night of the 28th, Wilson sent word that Forrest's cavalry had crossed and were heading for the Lewisburg pike, but he mentioned no infantry. "If he [Schofield] assumed infantry and began a general retirement toward Franklin to avoid being flanked," writes a chronicler of Wilson's campaigns, "his brigades, strung out along the pike, would be vulnerable to an attack from both rear and flank if all this proved to be only a feint on Hood's part." "The question," he concludes, "was how long should he wait?" James L. McDonough adds more reasons for Schofield's hesitation as to what he should do. "He did not know if Hood, with most of his army, was heading for Spring Hill, about twelve miles in his rear, or coming westward up the north bank of the Duck River to attack him at Columbia." That was the first reason; McDonough adds another: "He also had to consider the possibility that Hood was trying to panic him into retreat in order to gain the road from Columbia to Franklin, the only macadamized route available, for the easier and more rapid transport of his artillery and baggage trains."

Complicating things was the instruction Schofield was receiving from his superior in Nashville, General Thomas. "If you are confident you can hold your present position," Pap wired at 8 P.M. on the 28th, "I wish you to do so until I can get General Smith here." Decades later Schofield defined his dilemma.

> I was then confronted with the grave question, How long might it be possible to hold Hood back, and thus gain time for Thomas to get up his reinforcements [A. J. Smith]? By holding on to the crossing of the Duck River at Columbia until dark that night, and thus preventing Hood from using the turnpike for the movement of his artillery and trains until the next day, we would practically gain twenty-four hours; for he could not move them readily over his mud road from Huey's Mill [Davis Ford]. To do this, I must not only head Hood off at

> Spring Hill, but defeat any attempt he might make to dislodge me from the north bank of the Duck River.

Schofield's dilemma, and his uncertain response to it, caused some stir among his officers. "It seems to me a little strange that General Schofield does not intimate what measures he proposes to adopt to protect ourselves and guard our trains, and still more strange that he does not initiate such measures at once, as the enemy, according to his own statement, has crossed the river in force," wrote Brigadier General Thomas Wood to his commander, Major General Stanley, sometime on November 28. "It is perfectly patent to my mind, if the enemy has crossed in force, that General Wilson will not be able to check him." Wood closed ominously, "It requires no oracle to predict the effect of the enemy's reaching the Franklin pike in our rear."

Schofield's inactivity worked in Hood's favor. Lt. Col. Stephen Presstman, returned to position of the army's chief engineer with the reassignment of Major General Smith, had his crews working on the 28th preparing the riverbanks and bringing up the pontoons from Columbia. Then, working all night of the 28th–29th, the engineers laid down a pontoon bridge at Davis Ford, a crossing point some four miles east of Columbia that was unguarded by the Yankees. A local resident, G. Wash Gordon, pointed Presstman to the site. As with Jackson's flanking maneuver at Chancellorsville, here, a year and a half later, Southern sympathizers in the neighborhood helped Hood find direction for his troops' march.

Hood's flanking maneuver seemed on the verge of success. "As November 28 drew to a close," writes Jacobson, "Hood's plan was working splendidly." "John Bell Hood appeared to have gained a decided advantage," concedes Wiley Sword. The general was more subdued in his wire that night to Beauregard and Seddon. "The enemy evacuated Columbia last night and are retreating toward Nashville," he stated (not quite correctly). "Our army is moving forward."[31]

---

[31] Connelly, *Schofield*, 125–26; Wilson report, February 1, 1865, *OR*, vol. 45, pt. 1, 558; Wilson to Maj. J. A. Campbell, November 28 (1:30 and 8:30 P.M.), *OR*, vol. 45, pt. 1, 1112–13; J. P. Young, "Hood's Failure at Spring Hill," *Confederate Veteran*, vol. 16, no. 1 (January 1908), 26; Longacre, *From Union Stars*, 169; Crawford, "Mystery of Spring Hill," 110 ("Because of Wilson's incorrect interpretation of Forrest's movements, Schofield was deprived of his cavalry at Spring Hill on November 29"); Keenan, *Wilson's Cavalry Corps*, 51 ("The

Hood knew that November 29 would be the critical day of his maneuver. According to Chaplain Quintard, "The General [was] up at 3 A.M.," with couriers rushing around and aides bringing coffee with such frenzy, he added, "I did not get much rest." Recalling his disappointment at Hardee's handling of his previous "Jacksonian manoeuvre," Hood states in *Advance and Retreat*, "I resolved to go in person at head of the advance brigade, and lead the Army to Spring Hill." The Confederate infantry was also on the move early, beginning to cross on Presstman's pontoons at dawn, Lowrey's Brigade of Cleburne's Division in the lead. Hood and his staff rode with them. "In his uniform pocket was a regional map," writes Sword. "His local guide was present. Nothing seemed to be left to chance."[32]

Wilson's message of 8:30 P.M. closed with "Prisoners say that Hood is going to flank us on one side or other." This arrived at 2 A.M. on November 29. The distance between Wilson at Hurt's Cross Road and Schofield near Columbia was only a dozen miles, but to avoid capture couriers had to take a route that delayed messages at least four hours. Thus Wilson's next update, written at 1 A.M. on the 29th, did not reach Schofield till about dawn. From prisoners he learned that "the rebel infantry were then expecting every minute to march" across three pontoon bridges, then follow Forrest's route. "I think it very clear that they are aiming for Franklin," Wilson urged, "and that you ought to get to Spring Hill by 10 A.M." Adding impetus to his warning, Wilson wrote on the outside of his note, "Major-General Schofield by courier from Spring Hill. Important, Trot!!"

Schofield did not dawdle. After getting Wilson's report, he ordered Stanley and two divisions to march to Spring Hill, to "hold that place

---

question was how long should he wait?"); McDonough, "John McAllister Schofield," 14; Thomas to Schofield, November 28 (8 P.M.), *OR*, vol. 45, pt. 1, 1108; J. M. Schofield, *Forty-Six Years in the Army*, 213; Wood to Stanley, November 28, *OR*, vol. 45, pt. 1, 1115; Jacobson, *For Cause & For Country*, 74–75; Sword, *Embrace an Angry Wind*, 98; Hood to Seddon, November 28, *OR*, vol. 45, pt. 1, 1254 (same to Beauregard).

A detailed account of the life of Stephen W. Presstman, Hood's chief engineer, is to be found in Kundahl, *Alexandria Goes to War*, 197–206, noting that the engineer preferred to be called by his middle name, Wilson.

[32] Elliott, *Doctor Quintard*, 185; McMurry, *Hood*, 171 ("Hood was up at 3 A.M. on the twenty-ninth"); Hood, *A & R*, 283; Hay, *Hood's Tennessee Campaign*, 83; Sword, *Embrace an Angry Wind*, 98.

and cover the trains." Brig. Gen. George Wagner's division started marching at 9 A.M. with the army's train of 800 wagons and reserve artillery, about forty guns. Brig. Gen. Nathan Kimball's followed, but soon Kimball was ordered to halt at Rutherford Creek, some four miles north of Columbia, "to check any immediate movement of the enemy," Wood reported, in case the Rebels veered westward after crossing the Duck. Wagner's division marched on to Spring Hill, which it reached around 12:30. Schofield decided to keep back at Columbia his other three divisions, Cox's, Wood's, and Ruger's. "I did not apprehend any serious danger at Spring Hill," he later rationalized, "for Hood's infantry could not reach that place over a wretched country road much before night." (This sounds like Joe Johnston's assertion that he left Snake Creek Gap unguarded because he had calculated the Federals' rate of march through it.)[33]

General Lee saw the enemy activity in Columbia and decided to send troops across the Duck. Around 4 P.M. intrepid engineers under fire hauled pontoon boats to the riverbank so infantry could cross. Three Alabama regiments paddled over, attacked two enemy regiments and forced them back, allowing more Confederates to get across the river. Engineers completed their bridge by nightfall. In the fighting, Col. Robert Beckham fell mortally wounded, struck in the head by a rock sent flying by a Union shell. Into the afternoon Lee's artillery kept firing; Hood could hear it miles away as his infantry marched for Spring Hill. "Lieutenant General Lee performed his duty at Columbia, with great skill and fidelity which were crowned with entire success," Hood wrote later; "he attained the object of the demonstration, which was to keep the Federals in ignorance of our movements till sufficient time had been allowed the Army to reach the desired point."[34]

---

[33] Wilson to Campbell, November 28 (8:30 P.M.), *OR*, vol. 45, pt. 1, 1113; Wilson to Campbell, November 29 (1 A.M.), *OR*, vol. 45, pt. 1, 1143; Crawford, "Mystery of Spring Hill," 109 ("arrived near daylight"); Hay, *Hood's Tennessee Campaign*, 83 ("hold that place"); Wood report, January 10, 1865, *OR*, vol. 45, pt. 1, 122; Wagner report, December 2, *OR*, vol. 45, pt. 1, 229; Horn, "The Spring Hill Legend," 22 ("I did not apprehend any serious danger"); Johnston, *Narrative*, 315–16.

[34] Jacobson, *For Cause & For Country*, 102–104; Brewer, "How Lee's Corps Crossed the Duck River," 326; "Confederates Buried at Ashwood Church," 460 (Beckham mortally wounded near Columbia, November 29); Hattaway, *Stephen D. Lee*, 134; Hood, *A & R*, 291.

The morning of November 29 was one of movement and action. Red Jackson's cavalry assailed Wilson's rear, keeping the enemy horsemen riding north. With his main body, Forrest headed west toward Spring Hill, the town on the Columbia-Nashville road to which Wilson had urgently called Schofield's attention. At the time, Spring Hill was held by all of one Federal cavalry regiment, Lt. Col. Charles C. Hoefling's 12th Tennessee (U.S.).

The day promised milder temperatures than the Confederate infantrymen had seen in Tennessee. "The troops moved in light marching order," Hood reported, "with only a battery to the corps"; he had kept most of his artillery back with Lee in Columbia to reduce the marching column's *impedimenta.* Accordingly, the men left knapsacks and camp gear on the bank of the Duck River. They might need it all later, but first object was to whip the Yankees. Hood had with him a strong column of some 20,000 infantry and 5,000 cavalry. "As the infantry struggled northward on the Davis Ford road," explains David E. Roth of *Blue & Gray Magazine*, "it appeared Hood had engineered a great stroke."[35]

Having begun its march over the Duck on the morning of the 29th, Cleburne's Division, the lead, crossed by 7:30 A.M., pushing up the Davis Ford road. Stewart's Corps and Ed Johnson's division were across by 9 o'clock. From the ford to Spring Hill was only about a dozen miles by crow-flight, but the actual road-distance was seventeen. Besides, the road was narrow, rutted and at places washed out, slowing down the infantrymen's march. David Roth claims it was so bad that local farmers had even ceased to use it. One Confederate called the road from Davis Ford the "rockiest and roughest" he had ever encountered. Maj. Gen. John C. Brown remembered having to lead his troops "through fields and woods and over rough ground, adding greatly to the fatigues of the day." Brigadier General Featherston termed the troops' route "an unfrequented path." An apparently impatient Hood called upon a local resident, John Gregory, helping him as a guide, to confirm the accuracy of his map.

---

[35] Sword, *Embrace an Angry Wind*, 111, 114 (gear on the riverbanks); McDonough, "The Battle at Franklin, Tennessee, November 30, 1864," 21 ("Forrest turned the rest of his cavalry westward for a strike at the Spring Hill garrison"); Jacobson, *For Cause & For Country*, 87; Hood report, February 15, 1865, *OR*, vol. 45, pt. 1, 652 ("light marching order"); Hay, *Hood's Tennessee Campaign*, 85 (25,800 troops); Young, "Hood's Failure at Spring Hill," 26 (25,021); Roth, "The Mysteries of Spring Hill, Tennessee," 22.

Another factor was the pestiferous Yankees. Unhappy over the lack of news from Wilson, Schofield ordered Brig. Gen. Thomas Wood to send out a brigade (under Col. Sidney Post) to reconnoiter. The Federals got close enough to Hood's column of march that the Southerners around 10 A.M. had to form to the left against possible attack. General Brown later wrote that "by the orders of the Commanding General, in person, the whole of Gist's and about one-half of Strahl's brigade were detached for picket duty…necessarily causing more or less delay." Having spotted the Rebels, Post's troops carried the news of what they had seen back to Columbia. By this combination of causes, the pace of the march was frustratingly slow. "Hood, who exhibited buoyant spirits initially," writes one student, "quickly fell into an ugly mood." Near noon, when the men halted for rest and rations, "Gens. Hood and Granbury engaged in a rather heated argument about the overall speed of the march," according to Jacobson. One Confederate noticed that Cleburne became involved in the expressions of "chafing and dissatisfaction" at some of the troops' slowness. After lunch the men were set on their way again.[36]

Both sides were pushing troops toward Spring Hill. First Federals to reinforce Hoefling's mounted troops were two cavalry regiments sent by Wilson. The horsemen deployed on high ground east of the town, having heard that Forrest was heading toward them from Mount Carmel. The two sides clashed shortly after 11 o'clock, two miles east of Spring Hill. Forrest attacked, was at first repulsed, charged again and started pushing the outnumbered Yankees back to town. Forrest ordered General James Chalmers to charge into Spring Hill to capture the enemy wagons they could hear creaking into town—Schofield's trains. Meanwhile Wagner's lead brigade, Col. Emerson Opdycke's, arrived just in time. "I gave orders to Colonel Opdycke, commanding my leading brigade, to hurry up," General Wagner recalled. Wagner could see Forrest's skirmishers "in full view and moving upon the place from the east and not over half a mile distant." Opdycke's brigade, supported on its right by

---

[36] Jacobson, *For Cause & For Country*, 77–79, 82–83, 86, 97 (Hood and Granbury); Roth, "Mysteries of Spring Hill," 22; Jacobson, "Hood's Tennessee Campaign," 21 ("unfrequented path"); Cheatham, "The Lost Opportunity at Spring Hill, Tenn.," 537 (Brown's statement on delay); Sword, *Embrace an Angry Wind*, 114; Lundberg, *Granbury's Texas Brigade*, 210; Hood, *A & R*, 284 (denying that any time was lost when "Federal cavalry appeared on the hills to our left"); Sayers, "The Last Campaign," 244 ("chafing and dissatisfaction").

Col. John Q. Lane's, deployed, and the Federals were able to repulse Forrest's attack. According to one Union officer, Capt. John K. Shellenberger of the 64th Ohio, if Wagner "had arrived a few minutes later, he would have found Forrest in full possession at Spring Hill." Connelly aptly reasons, "Forrest probably did not know how close he had come to capturing Spring Hill."[37]

Nevertheless, Forrest saw the situation slipping away from him. At one point, probably late morning, he reported that "with a few men I moved to the left on a high hill, where I discovered the enemy hurriedly moving his wagon train up the Franklin pike." After his troopers had charged and again been repulsed, Forrest stated that "about this time I received orders from General Hood to hold my position at all hazards, as the advance of his infantry column was only two miles distant and rapidly advancing." The instruction missed the point: Forrest was not under attack; the enemy held Spring Hill and the Columbia pike; and Forrest was not strong enough to dislodge them. Indeed, by 2 P.M. it was Wagner's division, 5,500 strong, buttressed by artillery, which held Spring Hill. Stanley Horn places much importance upon this feat: "It is significant that General Forrest never referred to the Spring Hill affair as one of the Confederacy's 'lost opportunities,'" he writes. "Forrest's basic military maxim, it will be recalled, was that the way to win was to 'get there first with the most men.'" At Spring Hill "that was exactly how Schofield checkmated Hood," Horn reasons; "he got there first with 5,500 veterans, to back up the garrison troops' delaying action." By this assessment, John M. Schofield, who has been accused of being dilatory in the crisis of November 29, deserves credit for having beaten John Bell Hood to the tactical prize well before sundown that day.

It was not until 2:30 when Hood, riding at the head of his column with Frank Cheatham and Pat Cleburne, approached the Davis Ford

---

[37] Jacobson, *For Cause & For Country*, 87–92; Forrest report, January 24, 1865, *OR*, vol. 45, pt. 1, 753 ("Two miles from town the enemy's pickets were encountered and heavy skirmishing ensued"); Horn, "Spring Hill Legend," 23; Shellenberger, "The Fighting at Spring Hill, Tennessee," 101; Connelly, *Autumn of Glory*, 494.

The "battle of Spring Hill," as it has been called, was not much of one, certainly in terms of casualties. Both sides lost about 800 killed, wounded, and captured on November 29 (Gillum, *Twenty-five Hours to Tragedy*, 436). Another student breaks them down as 850: 500 among Forrest's cavalry and Cleburne's infantry, and 350 in Wagner's division (Sayers, "Last Campaign," 260).

road intersection with the Rally Hill pike near Rutherford Creek, 2½ miles southeast of Spring Hill. The odds at this point favored the Confederates, as Hay totals the numbers.

> At this time, General Hood was in front of Spring Hill with an army corps [Cheatham's] 9,800 strong and Forrest's cavalry, about 5,000 strong. A. P. Stewart's corps of about 8,500 men and Edward Johnson's division, of some 2,500 men, were nearby, close to the crossing of Rutherford's Creek and within easy supporting distance. No greater distance than five miles separated any of these troops. Thus it will be seen that, at this time, Hood had a force of some 25,000 men with which to strike Stanley's single division of some 5,500 men.

Hood was not focused on the town of Spring Hill, but on the turnpike south of it, apparently in the belief that his troops could get astride and block it against the enemy's northward march. In his reminiscences, he states that the Yankees were marching on the road when Cheatham's infantry approached. In his initial report, written not even two weeks later, Hood laid his first accusations against General Cheatham.

> About 4 P.M. our infantry forces, Major-General Cheatham in the advance, commenced to come in contact with the enemy about two miles from Spring Hill, through which the Columbia and Franklin Pike passes. The enemy were at this time moving along this pike, with some of their troops formed on the flank of their column to protect it. Major-General Cheatham was ordered at once to attack the enemy vigorously and get possession of this pike, and, although these orders were frequently and earnestly repeated, he made but a feeble and partial attack, failing to reach the point indicated. Darkness soon came on, and to our mortification the enemy continued moving along this road, almost in ear-shot, in hurry and confusion, nearly the entire night.

Three months later, in his final campaign report, Hood added, "Had my instructions been carried out there is no doubt that we should have possessed ourselves of this road." In October 1865 Hood was still talking about it. A newspaper reporter quoted him thus: "Maj. Gen. Cheatham was ordered to attack the enemy at once, vigorously, and get possession of this pike; and, although these orders were frequently and earnestly repeated, he made but a feeble and partial attack.... Had my instructions been carried out there is no doubt that we could have possessed ourselves of the road."

Writing his memoirs more than a decade later, Hood attempted to add color and emphasis to his accusation by reconstructing his alleged conversation with Cheatham. He claimed that, in company of both Cheatham and Cleburne, "I had in sight the enemy's wagons and men passing at double-quick along the Franklin pike." Then he turned to Cheatham.

> "General, do you see the enemy there, retreating rapidly to escape us?" He answered in the affirmative. "Go," I continued, "with your Corps, take possession of and hold that pike at or near Spring Hill. Accept whatever comes, and turn all those wagons over to our side of the house." Then addressing Cleburne, I said, "General, you have heard the orders just given. You have one of my best divisions. Go with General Cheatham, assist him in every way you can, and do as he directs." Again, as a parting injunction to them, I added, "Go and do this at once. Stewart is near at hand, and I will have him double-quick his men to the front."

Cheatham never submitted a report on the Tennessee Campaign, and the first substantial account he wrote of the events of November 29 appeared a year after the publication of Hood's posthumous memoir. In "The Lost Opportunity at Spring Hill, Tenn.—General Cheatham's Reply to General Hood," which appeared in the *Southern Historical Society Papers* in late 1881, Cheatham put a different cast upon the conversation. He asserts that he arrived around 3 and met General Hood at Rutherford Creek.

> At this point General Hood gave me verbal orders, as follows: That I should get Cleburne across the creek, and send him forward toward Spring Hill, with instructions to communicate with General Forrest, who was near the village, ascertain from him the position of the enemy, and attack him immediately; that I should remain at the creek, assist General Bate in crossing his division, and then go forward and put Bate's command in to support Cleburne; and that he would push Brown forward to join me.

As for Hood's recollection of their conversation at Rutherford Creek, Cheatham was adamant: "There is not a bit of truth in this entire paragraph." First, at 3 o'clock on the 29th, there were no enemy troops and wagons moving on the Columbia pike. Students from Thomas R. Hay to Eric Jacobson side with Cheatham on this: "there was no retreating enemy or wagon trains on the pike at that house," writes Hay; Wagner's division with the trains was already at Spring Hill (at least according to General Wagner, who reported that he arrived at Spring Hill "early in the afternoon"). "At 3 P.M. there were no Union troops or wagons moving along the pike south of Spring Hill," Jacobson echoes. Cheatham also argued that Hood could not have seen the Columbia-Franklin pike from their vantage point. "The turnpike was never in view, nor could it be seen until I had moved up to within three-quarters of a mile of Spring Hill," he claimed. "Only a mirage would have made possible the vision which this remarkable statement professes to record." On this point, too, the evidence favors the corps commander. Dave Roth of *Blue & Gray* personally visited the Davis Ford/Rally Hill/Rutherford Creek intersection, consulted a topographical map of the area, and confirmed that high ground blocked the view northwest to the Columbia pike. "Hood could not have seen the road from there," he concludes. Hay adds a further challenge of Hood's version in *A & R*: "Stewart's Corps was not 'near at hand,'" but south of Rutherford Creek.[38]

Regardless of technicalities, Cheatham understood his orders to "attack immediately." Accordingly, Cleburne's Division marched toward and crossed the creek, Mark Lowrey's brigade on the left, Granbury's on the right, and Daniel Govan's in reserve (James Smith's had been left behind to guard supply trains). Tyree Bell's dismounted cavalry brigade was to the right of Granbury, but the troopers were low on ammunition. Coming up on Cleburne's left was the division of "Grits" Bate, who recalled that Hood, "who was personally present," told him "to move to the turnpike and sweep toward Columbia," orders that indicate Hood apparently believed Cheatham's troops had gotten in rear of the enemy, despite his reference in *A & R* to "the enemy there, retreating rapidly to escape us."[39]

---

[38] Forrest report, January 24, 1865, *OR*, vol. 45, pt. 1, 753 (including "hold my position at all hazards"); Horn, "Spring Hill Legend," 32; Jacobson, *For Cause & For Country*, 96, 97 ("at around 2:30 P.M., he reined up southeast of Spring Hill"), 99–100, 102 ("Cheatham…never filed an official report"); Roth, "Mysteries of Spring Hill," 23 (2½ miles south of the town), 26 ("actual visit to the place"); Hood report, December 11, *OR*, vol. 45, pt. 1, 657; Hood report, February 15, 1865, *OR*, vol. 45 pt. 1, 652; "Gen. Hood….On Board Steamer Clinton, At Sea, Oct. 9, 1865," *New York Times*, November 2, 1865; Hood, *A & R*, 284–85; Cheatham, "Lost Opportunity at Spring Hill," 524–25, 529; Hay, *Hood's Tennessee Campaign*, 87.

[39] Jacobson, *For Cause & For Country*, 100–101; Lowrey, "Autobiography," 374; Hay, *Hood's Tennessee Campaign*, 87; Wagner report, December 2, *OR*, vol.

Finally, by 3 P.M. General Schofield concluded that S. D. Lee's smoke and fire were diversionary—he could hear artillery coming from Spring Hill. He ordered Ruger's division, then north of Rutherford Creek, to start marching for Spring Hill. Before setting out with Ruger, Schofield issued orders for his other three divisions, Kimball's, Wood's and Cox's, to follow after sundown. Schofield and Ruger would arrive at Spring Hill about 7 P.M., well after nightfall.

Around 4 P.M. Pat Cleburne's division, 3,000 strong (even without Smith's Brigade), started advancing westward toward the turnpike. Cleburne and Forrest rode together behind Govan's Brigade, holding their swords aloft, accompanied by General Govan. "Forward echelon!" Cleburne shouted, standing in his stirrups. Advancing a third of a mile, the Southerners encountered enemy pickets and turned them back before running into trouble. About 4:15 the Yankees—Luther Bradley's brigade of Wagner's division, which had wheeled south of Spring Hill—fired upon Lowrey's right flank, forcing him to readjust his lines, "which was done without confusion," Lowrey stated, in a letter written years later to Cheatham. Govan came up and helped drive the Yankees. "The routed Federals retreated rapidly and in confusion across the pike," General Govan recalled decades later. The retreating Northerners "were called on with loud oaths" by Cleburne's men (according to Captain Shellenberger), "charging them with a Yankee canine descent, to halt and surrender." General Bradley himself was wounded in the arm and removed from the field as his troops fell back on the wagon train and reserve artillery. Later, he sheepishly explained that his brigade had fallen back "in some disorder, owing to the fact that over one-half of the men of these regiments were recruits and drafted men who had never been under fire, neither had they been drilled." The Confederates "were on the verge of taking control of the pike," Jacobson writes, when Northern artillery stopped them. General Stanley had lined up two batteries of Napoleons on a ridge southeast of the town; their fire forced the Rebels to retreat and take cover in a streambed.[40]

45, pt. 1, 229; Bate report, January 25, 1865, *OR*, vol. 45, pt. 1, 742; Hood, *A & R*, 284.

[40] Jacobson, *For Cause & For Country*, 102–113; Schofield report, December 31, *OR*, vol. 45, pt. 1, 342; Connelly, *Schofield*, 127; Schofield's SFO, 166, November 29, *OR*, vol. 45, pt. 1, 1139–40; Hay, *Hood's Tennessee Campaign*, 87; Symonds, *Stonewall of the West*, 250–51; Cheatham, "Lost Opportunity," 536

On this late autumn day, the sun went down at 4:35, but Cleburne was determined to press his assault. After reforming his brigades, he was ready to order all three of them in an all-out charge to seize the turnpike when he received an order from Cheatham, delivered by a staff officer, to halt any forward movement until he got further orders. Capt. L. H. Mangum, A. A. G. for Cleburne's Division, described the event in a letter printed more than twenty years later. "Cleburne's command was now in full view of Spring Hill and not exceeding 300 yards from it," Mangum declared. General Cleburne sent him off to tell Granbury to bring his brigade up to a fence running parallel to the pike some two hundred yards to its east; Cleburne was to deliver the same order to Govan, and to Gen. John C. Brown, whose division, having followed from the Davis Ford, was taking position on Cleburne's right.

Given the many conflicting accounts by Confederates as to the "mysteries" of Spring Hill, I favor the straightforward narrative offered by Symonds.

> As Mangum galloped off to deliver Cleburne's orders to Granbury, Cleburne rode to talk to John C. Brown, whose division had followed his own and who was now on his right. Cleburne was in the midst of a conversation with Brown near five o'clock when Cheatham rode up to tell him that he and Brown were to renew the attack on Spring Hill, which Cheatham still considered the primary objective, "as soon as they could connect their lines." Brown was to initiate the attack, Cheatham said, and Cleburne would pitch in when he heard the sound of Brown's firing to his right. Cheatham would personally ensure that Bate came up on Cleburne's left, noting also that Hood had promised them the support of Stewart's Corps if necessary. Then Cheatham left to direct Bate's division into line on Cleburne's left.

Cleburne waited for the sound of Brown's guns, but it was not to be heard. General Brown believed the enemy was massed on his right flank and sent a message to Cheatham so stating. "It was full dark at 6:17 that evening," Symonds writes, "with Cleburne's three brigades still in alignment when Maj. Joseph Bostick, Cheatham's assistant inspector general,

---

(quoting Lowrey letter to him of November 8, 1881; Lowrey's campaign report does not appear in the *OR* for, as he states, "I lost my papers"); Hay, *Cleburne and His Command*, 272 (quoting D. C. Govan letter of June 1906); Horn, "Spring Hill Legend," 27 (map, situation at 4:30 P.M., showing Cleburne's rout of Bradley's brigade); Purdue and Purdue, *Pat Cleburne*, 397 ("Yankee canine descent"); Bradley report, December 5, *OR*, vol. 45, pt. 1, 269.

rode up to Cleburne with order from General Cheatham to remain "where he was and not to move on the pike until after further orders."

Cleburne's battle-blood had been up, but a nighttime attack, Symonds asserts, was not to his liking: "released now from the obligation to make an attack in full dark over unfamiliar ground, he ordered his troops to lay down on their arms in line of battle where they were, a half mile south of Spring Hill and a quarter mile east of the turnpike." Meanwhile, in the darkness, Cleburne's men heard marching troops off to the left. Some thought it was Bate's Division coming up, but it was enemy infantry—Ruger's division and a brigade under Walter Whitaker—slipping out of the trap Hood had hoped to spring on them. In Captain Mangum's opinion, it all had hinged on the order brought by Cheatham's staff officer: "But for the order brought by Colonel Bostwick [*sic*], Cleburne would have been on the pike and had possession of Spring Hill in less than ten minutes. Then there would have been no battle of Franklin." "The arrest of his movement on the turnpike was a bitter disappointment to General Cleburne," Mangum wrote two decades after the war, "and he expressed himself very forcibly in regard to the failure that forced it."[41]

Did Cheatham send his A. I. G., Major Bostick, to Cleburne with orders to cancel the attack? Captain Mangum's postwar recollection is the only documentary source. In addition to his account in the *Missouri Republican* (quoted by Hay) and *Kennesaw Gazette* (cited by Jacobson and the Purdues), Anne Bailey has found it in a local Texas history. Early sources on the campaign mention Bostick and Cheatham's order, such as Buck and Young. Some secondary sources, including Horn, cite it, but others (notably Connelly) do not. Winston Groom refers to Bostick's instructions to Cleburne but guesses that "Cheatham's reasoning for the delay seems to be that he was under the impression Stanley had a much larger force in Spring Hill than he did, and so Cheatham wanted to get all three of his divisions in line for a massed assault." Wiley Sword

[41] Jacobson, *For Cause & For Country*, 112, 117 (Mangum, "bitter disappointment" in *Kennesaw Gazette*, June 15, 1887), 126 (Ruger with Whitaker's Brigade); Young, "Hood's Failure at Spring Hill," 37 (Bate's attack on Ruger); Hay, *Hood's Tennessee Campaign*, 88, 231–32 n.35 (text of Mangum's letter as it appeared in the *Missouri Republican* of September 12, 1885); Symonds, *Stonewall of the West*, 252–53; B. H. Stewart, Jr., *Invisible Hero*, 316; Purdue and Purdue, *Pat Cleburne*, 398–99 (also quoting Mangum in *Kennesaw Gazette*).

agrees: "Cheatham ordered a halt in the attacks.... By Cheatham's specific order, Cleburne was to halt and await the concentrated attack that might overwhelm Spring Hill's defenders."

Hood was aware of the role of Cheatham's staff as he pondered Spring Hill in the months afterward. In April 1865, aware that Peter Stewart had felt stung by some implications of his campaign report, Hood wrote Stewart, "I regret that I did not make myself more clear in my report by going more into detail about the staff officer of General Cheatham." Hood could only have meant Major Bostick.[42]

Yet Cheatham's biographer, Christopher Losson, makes no mention of the Bostick-Cleburne exchange, possibly because on other events he catches his subject in a few questionable assertions. General Cheatham's main contribution to the campaign literature was his essay in the *Southern Historical Society Papers* in late 1881, "The Lost Opportunity at Spring Hill," which he wrote as reply to Hood's version of events published in *Advance and Retreat* the year before and read at a meeting of the Louisville Historical Society in December 1881. In his paper he emphasizes his determination to carry out Hood's verbal order, made at Rutherford Creek, to attack the enemy. On this issue he made four points: 1) he oversaw the creek-crossing of Bate's Division; 2) he directed Brown to move his division to Cleburne's right "and attack to the right of Cleburne"; 3) when Brown voiced alarm that the enemy overlapped his right, "I sent word to him to throw back his right brigade and make the attack"; 4) he rode up to Brown and Cleburne and ordered that "as soon as they could connect their lines, they should attack the enemy."

Losson, however, finds fault with at least two of these statements. First, despite Cheatham's claim, the biographer states "he was absent from his assigned post at the creek when Bate's division came up." Second, "Cheatham claimed that he rode up and directed Bate to move to Cleburne's left *and support an attack* but this does not seem likely" (emphasis added). In truth, it appears that Cheatham's order was intended to

---

[42] Hay, *Hood's Tennessee Campaign*, 231–32; Jacobson, *For Cause & For Country*, 117; Purdue and Purdue, *Pat Cleburne*, 398–99; Bailey, *Chessboard*, 85 (Lt. Learned H. Mangum, Cleburne's aide and antebellum law partner in Helena), 195 n.34; Hay, *Cleburne and His Command*, 267; Young, "Hood's Failure at Spring Hill," 32; Horn, *Army of Tennessee*, 387; Connelly, *Autumn of Glory*, 496; Groom, *Shrouds of Glory*, 146; Sword, *Embrace an Angry Wind*, 131; D. W. Sanders, "Hood's Tennessee Campaign. Chapter VI," 360.

halt Bate's assault. In his campaign report, Bate related that General Hood, "who was personally present, directed me to move to the turnpike and sweep [southward] toward Columbia." Bate accordingly marched his troops "so as to strike the turnpike."

> Caswell's battalion of sharpshooters, deployed as skirmishers, was within 100 yards of and commanded the turnpike, checking the enemy's movement along it in my front, and my lines were being adjusted for a further forward movement, when I received an order, through Lieutenant Schell, from General Cheatham to halt and join my right to General Cleburne's left. My main line was within 200 yards of the turnpike when Major Caswell's battalion fired into the enemy on the pike.

In a letter to Cheatham of November 1881, Bate recorded that after he had gotten Cheatham's order to form on Cleburne's left he rode to Hood's headquarters, which the general and his staff had set around 4 P.M. at Oaklawn, the home of Absalom Thompson, a half mile north of Rutherford Creek. "I informed the General of having about dark come near to, in line of battle and commanded, with my skirmish line, the turnpike south of Spring Hill, and caused a cessation in the movements of wagon, horsemen, etc." But he told Hood he had not swept the pike toward Columbia "because just at that time I received an order from my corps commander, General Cheatham, to halt and align the right of my division with the left of Cleburne's, which I declined to do until I received a second order to the same effect, and then I did so."

"Cheatham, it is to be assumed," Horn charitably judges, "did not realize that Bate was in actual contact with the enemy when he issued this order." "But if it had not been issued," he adds, "Bate might have stopped the movement of Schofield's force along the pike at least until morning."[43]

The bulk of Cheatham's "Reply to General Hood" in *SHSP* involves his refutation of at least seven points of fact, lengthily quoting the disputed parts of *Advance and Retreat* and following each with a detailed rejoinder. The points range from minute ("Lowry's not Granberry's brigade of Cleburne's division, was in front" of the corps column marching

[43] Cheatham, "Lost Opportunity," 524–26, 541 (quoting Bate to Cheatham, November 29, 1881); Losson, *Tennessee's Forgotten Warriors*, 205; Jacobson, *For Cause & For Country*, 127–28 (Oaklawn); Bate report, January 25, 1865, *OR*, vol. 45, pt. 1, 742; Horn, *Army of Tennessee*, 388.

toward Spring Hill) to the more important, such as the specifics of Hood's orders to him on the afternoon of November 29 (e.g., whether the two generals could see the enemy column from Rutherford Creek, as noted above).

The critical disagreements hinge upon Hood's and Cheatham's communications that day. After he left Cheatham at Rutherford Creek, Hood rode off to see to the movement of Stewart's Corps and Johnson's Division, after which in *Advance and Retreat* he claims, "I dispatched a messenger to General Cheatham to lose no time in gaining possession of the pike at Spring Hill. It was reported back that he was about to do so." A dozen years after the war, Isham Harris, Tennessee's governor traveling with Hood and his staff during the campaign, wrote James D. Porter, Cheatham's chief of staff, that later Major Mason told him that Cheatham was blameless for not having launched a night attack, as he had sent no such order from Hood.

Harris found himself in the middle of the Hood-Cheatham squabble. The governor was at Hood's headquarters at the Thompson house when he and the general heard gunfire from Cheatham's initial attack. After that, the sound dropped off. In conversation three and a half years later with Campbell Brown, General Ewell's son-in-law, Harris remembered Hood saying, "something was the matter" and sending Harris off to "find out the situation." Harris sided with Cheatham, stating that later that night he heard Hood telling Cheatham to "await and conform to the movements of his troops in his right, telling him he had ordered Stewart to move beyond Brown's right until he got across the turnpike." Others shielded Harris. William D. Gale, a member of General Stewart's staff, wrote his daughter in mid-January about Spring Hill. "Strange to say, we remained all night in sound of voices of the men as they retreated in the greatest of haste, and not a blow was struck, though orders were sent by Gen. Hood several times to attack at once," he asserted. "One time Gov. Harris himself carried the order to Gen. ___," Gale wrote, choosing not to name the officer (assuming he knew the name).

Surely the hottest point of argument was Hood's allegation that at some time in the evening Cheatham rode up to him as Hood was realizing that the turnpike had not been seized. "General, why in the name of God have you not attacked the enemy and taken possession of that pike?" Hood claims to have asked. Cheatham allegedly (in *A & R*) answered "that the line looked a little too long for him, and that Stewart should first form on his right." Cheatham's rebuttal in *SHSP* was pointed: "here again General Hood's memory proved treacherous," he charges; "the dramatic scene with which he embellishes his narrative of the day's operations only occurred in the imagination of General Hood."

Another purported incident occurred later that night. When Hood, by then at his headquarters at the Thompson house, received news that the enemy troops were indeed marching north on the pike to Franklin and out of his hoped-for trap, he states that he "sent anew to General Cheatham to know if at least a line of skirmishers could not be advanced, in order to throw the Federals in confusion, to delay their march, and allow us a chance to attack in the morning. Nothing was done." Cheatham acknowledges having received this note from Major Mason around midnight, and forwarded it on to General Johnson, whose division was posted to his left. Johnson would have refuted this statement, according to one of Lee's staff officers, Maj. James Ratchford. Writing a half-century after the war in his memoir, *Some Reminiscences of Persons and Incidents in the Civil War* (1909), Ratchford claimed, "I heard General Ed Johnson say that during the night he could have routed the entire army with only his division, but was not permitted to attack them."

Then there was the generals' nighttime meeting at the Thompson house. According to Stewart's report, written four months later, Hood sent one of Cheatham's staff officers to direct Stewart to bring his infantry up on Cheatham's right. But after learning that taking such a position would put him a good distance from the pike and remembering Hood's orders to block the road north of the town, Stewart ordered his troops to bivouac and rode to headquarters for clarification.

> On arriving at his headquarters I inquired of him if he had sent this officer of General Cheatham's staff to place me in position. He replied that he had. I next inquired if he had changed his mind as to what he wished me to do. He replied that he had not, "But," said he, "the fact is, General Cheatham has been here and represented that there ought to be somebody on Brown's right." I explained to him that in the uncertainty I was in I had directed the troops, who had been marching rapidly since daylight, and it was now 11 P.M., to be placed in bivouac, and had come to report. He remarked in substance, that it was not material; to let the men rest; and directed me to move before daylight in the morning, taking the advance toward Franklin.

Losson judges from this meeting with Stewart that only at this point did Hood apparently realize "for the first time...that the pike was not sealed off."

> Here Hood reacted curiously; he did not order Cheatham or Stewart to blockade the road, either north or south of the hamlet, and instead asked Forrest if his weary cavalry could barricade the pike. Forrest replied that he needed ammunition, but the supply wagons were still in front of Columbia with Lee. Hood thus depended on tired horsemen with a shortage of ammunition to insure that Schofield was trapped.

Such is Losson's interpretation of what possibly occurred late at night at the Absalom Thompson house. More importantly, the biographer does not place Cheatham at Hood's headquarters for the high-level meeting. Indeed, he shrugs, "Cheatham's whereabouts during the night" are hard to confirm. Losson even mentions that, according to a rumor passed along by Union cavalry general Wilson, Cheatham may have "whiled away the time at the home of the attractive Jessie Helen Peters."

Losson further refuses to accept Cheatham's assertion in *SHSP* that late at night he had conversation with Hood at headquarters. Cheatham claims,

> I found a courier with a message from General Hood, requesting me to come to him at Captain Thompson's house, about one and a fourth miles on the road back to Rutherford's creek. I found General Stewart with General Hood. The Commanding General there informed me that he had concluded to wait till morning, and directed me to hold my command in readiness to attack at daylight. I was never more astonished than when General Hood informed me that he had concluded to postpone the attack till daylight. The road was still open—orders to remain quiet till morning—and nothing to prevent the enemy from marching to Franklin.

Losson's conclusion: "other evidence indicates it was Cheatham, not Hood, who was most responsible for calling off the attack."

Indeed it does. A key source Losson cites is the postwar recollection of Maj. Joseph B. Cumming of Hood's staff. He remembered seeing General Cheatham at Hood's headquarters "remonstrating against a night attack." Major Cumming's statement comes from his "Recollections" reposing at Chapel Hill. Stephen Hood quotes more of it:

> General Hood sent me forward with an order to General Cheatham to attack at once. I delivered the order, and as I had ridden hard to deliver it I returned to Gen. Hood's headquarters at a slow pace expecting every minute to hear the sound of the attack on the pike. It was now getting dark. It was the 29th of November, chilly and driz-

> zling. When I reached Gen. Hood's headquarters, to my astonishment I found Gen. Cheatham there, he having out-ridden me by a different route. He was remonstrating with Gen. Hood against a night attack.

Losson further quotes an important postwar letter written by S. D. Lee to James F. H. Claiborne, in which Lee related the rumor that "Cleburne and Cheatham consulted, and decided between themselves that no battle should be fought at night." Judge Claiborne was writing a history of Mississippi, and Lee was helping him with a series of reminiscing letters. In one, Lee complained that the incident at Spring Hill "was a criminal affair of somebody's" which led to the bloody battle at Franklin the next day. In his letter of May 17, 1878, to Claiborne, Lee went so far as to criticize Cheatham and "his excuse...that he was outflanked on the right and did not wish to bring on a night attack."[44]

After the war, Maj. Joseph Vaulx, one of Cheatham's staff officers, wrote a letter in which he recounted the events of the evening of November 29. Cheatham had arranged his three divisions from right to left: Brown, Cleburne, Bate. According to Vaulx, "Cheatham told Brown that he would order Cleburne to attack on hearing his guns, and that as soon as Cleburne became engaged he would order Bate also to advance." Before too long a worried Cheatham, according to Vaulx, asked, "Why don't we hear Brown's guns?"

Years later, in the letter he wrote to Cheatham (included in Cheatham's *SHSP* article), General Brown explained that he was anxious about the enemy at his right flank. "I had neither artillery nor cavalry, and was left in a position where I must meet with inevitable disaster if I advanced on Spring Hill," he wrote, claiming that he had conferred with his brigade commanders and reached "a determination to suspend the advance" while he sought out General Cheatham.

---

[44] Cheatham, "Lost Opportunity," 524–32; Hood, *A & R*, 285–87; Jacobson, *For Cause & For Country*, 128–29 (Harris's recollection); [Gale], "Hood's Campaign in Tennessee," 4; Hood, *Rise, Fall, and Resurrection*, 122–23 (quoting Cumming), 125 (Major Ratchford); Stewart report, April 3, 1865, *OR*, vol. 45, pt. 1, 712–13; Losson, *Tennessee's Forgotten Warriors*, 207 (quoting Cumming and S. D. Lee to J. F. H. Claiborne, June 12, 1878, in Claiborne Papers, also at Chapel Hill); Hattaway, *Stephen D. Lee*, 208, 255 n.8 (citing Lee to Claiborne, April 22, May 17, and June 12, 1878, Southern Historical Collection at the University of North Carolina, Chapel Hill); Sword, *Embrace an Angry Wind*, 154 (quoting Lee to Claiborne, May 17, 1878).

It is evident that if Pat Cleburne's battle-blood was up, John Brown's was not. Earlier that evening, as Brig. Gen. James R. Chalmers, one of Forrest's brigade leaders, recalled to J. P. Young, having come upon Brown as his division (according to Young) "was standing idly facing the town," Chalmers asked him why he was not launching an attack. "General Brown," Chalmers wrote, "answered me, as I thought, very curtly, saying: 'I have no orders.' His tone somewhat nettled me, and I replied in much the same tone to him: 'General, when I was circumstanced as you are at Shiloh, I attacked without orders.'" Brown answered, "I would prefer to wait for orders," whereupon Chalmers rode off.

James Dinkins served as A. D. C. to General Chalmers and was on the scene. "It was nearly night," he wrote well after the war, "while Cleburne held the position he had gained. Soon the firing ceased, and General Cleburne rode back, I presume for orders." Dinkins then related what he remembered Cleburne telling Chalmers:

> "They are badly paralyzed. I rode within fifty yards of their works without danger." At that moment blood tricked out of a wound in his horse's neck. As we rode to camp, we met Brown's division, of Cheatham's Corps, standing in column. The writer heard a conversation between Generals Brown and Chalmers on the situation. General Chalmers said: "I believe that, if you will join General Cleburne, and make a vigorous attack, you can capture or rout the force in his front." General Brown replied: "I have no orders." General Chalmers said: "I would make that attack without orders, general. I am confident it is a great opportunity; the enemy is very uneasy, and could be whipped and captured before the balance of Schofield's army can come to his assistance." General Brown answered: "I will await orders."

Another staff officer in Brown's division adds his support to the Brown "no orders" episode. According to Young, Capt. H. M. Neely (A. A. G. in John C. Carter's brigade of Brown's Division) related to him that "the failure of the division commander to attack was a subject of much comment along the line while the command was waiting in line of battle." "With the enemy in full view and at a short distance," just before darkness set in, Neely stated that he and another officer rode to General Carter "and asked if he knew why an attack was not made, and he replied: 'No; I am not in command.'"

> A little farther to the rear we met General Brown, and told him of what we had seen and our impressions of the situation in our front, and put the same question to him, and his reply was: "I don't know; I have no orders." We then told him in a pleasant way that if he would take the responsibility of beginning the attack without orders he could safely count on a "new feather in his cap," as it would be a quick and easy matter to capture or destroy Schofield's corps in its present condition. At that he laughed and said: "No; I must wait for orders."

While Brown awaited orders, his officers saw their opportunity slip away. Well after the war, Young wrote Ellison Capers, then Episcopal bishop of South Carolina, posing various questions about Spring Hill. Capers recalled the events quite clearly. It was not yet dusk when Capers and his men (the 24th South Carolina, Gist's Brigade, Brown's Division), were aligned in full view of enemy infantry marching "in great confusion" on the pike ahead of them. "We could not understand why we did not attack"; he recalled hearing the question pass through his troops. At one point in the growing darkness he and General Gist rode toward the enemy, whose "tramp of horses, the rattle of wagons, the command of officers, could all be distinctly heard." Before riding away, Colonel Capers later said, "I emptied my pistol in the direction of the sound of their voices." Capers made a point of stating that at the time no Federals overlapped the Confederate right, leading him to believe "that we could take the road on their left (our right) and block their way." The Southerners did not act, however, and "we lost a grand chance at Spring Hill." W. J. McMurray, postwar historian of the 20th Tennessee, was equally restrained: "Cheatham's Corps lay within two hundred yards" of the retreating Federal column, "and heard them passing almost the entire night undisturbed...this failure to attack was most serious." "Rumor said it was the fault of the commander of the leading corps," was the way S. D. Lee put it well after the war.[45]

---

[45] Young, "Hood's Failure at Spring Hill," 33–35 (including Chalmers to Young, Captain Neely's and Colonel Capers's recollections); Cheatham, "Lost Opportunity," 538 (Brown to Cheatham, October 24, 1881); Losson, *Tennessee's Forgotten Warriors*, 207 ("Some of Brown's staff officers intimated that Brown still felt uneasy about his flank and refused to advance"); Dinkins, *1861 to 1865 by an Old Johnnie*, 232; Capers, *The Soldier-Bishop Ellison Capers*, 112; McMurray, "Causes of Failure at Spring Hill," 395; Lee, "From Palmetto to Nashville," 257.

So Cheatham did not hear the sound of Brown's guns (assuming he did order the echelon attack of his three divisions from the right). It all seemed to stem from Brown's reluctance to advance. "I discovered a line of the enemy thrown out of Spring Hill, across and threatening my right flank," he wrote years later to Cheatham. Brown sought to justify his decision: "a hasty consultation with my brigade commanders resulted in a determination to suspend the advance and confer with the corps commander." Sumner A. Cunningham, sergeant-major in the 41st Tennessee, wrote after the war that in his diary he recorded that "the failure to fight at Spring Hill was because of Gen. John C. Brown's unwillingness to make the charge without the support of his right wing." Sometime during the late afternoon Brig. Gen. Tyree Bell, of Forrest's command, chanced upon Brown talking with one of his brigadiers, George W. Gordon, and the three engaged in a "lively chat" over the proposed attack. Bell remembered saying that "the prospects for an infantry fight were pretty good." Brown and Gordon got into "a dispute" over possible tactics when orders arrived to bivouac their troops for the night. "I need not remind you," Brown wrote Cheatham, "that in a very few minutes you were on the field and fully approved of what had been done." Brown's statement further indicts Cheatham. Peter Stewart wrote W. O. Dodd in early 1881, "My impression is that Cheatham and his officers thought themselves in great danger of being outflanked."[46]

With the infantry failing to launch any offensive on the night of November 29, the situation thus devolved upon "tired horsemen with a shortage of ammunition," as Losson writes. When Hood asked Forrest if his troopers could block the turnpike, the cavalry chief said he would do the best he could. Forrest sent forward Red Jackson and his small division, which had captured some ammunition earlier in the day. Jackson struck the pike four miles north of Spring Hill at Fitzgerald's around 11 P.M., "just as the front of the enemy's column had passed," Forrest later reported. Nevertheless, Jackson chased the Yankees, killed horses and

---

[46] Cheatham, "Lost Opportunity," 538 (quoting Brown to Cheatham, October 24, 1881), 535 (quoting Stewart to Dodd, February 8, 1881); J. Simpson, *S. A. Cunningham and the Confederate Heritage*, 31; Remington, "Cause of Hood's Failure at Spring Hill," 570 (Cunningham on General Brown); N. Hughes, Jr., *Brigadier General Tyree H. Bell, C. S. A.*, 199–200.

mules, and burned some wagons before having to drop back. Such was the Confederates' interdiction of the Columbia-Franklin turnpike.[47]

The collapse of Hood's plan allowed the one event it had been devised to deter: the escape of Schofield's army. Yet it had been an extremely close call. General Schofield's delay during the 29th to get his troops safely out of Columbia and the Duck River line had imperiled their march northward, putting them at risk of attack and defeat in detail by Hood. Schofield's delay also contravened General Thomas's specific instructions. As early as 3:30 A.M. on the 29th, Thomas wired from Nashville, "I desire you to fall back from Columbia and to take up your position at Franklin, leaving a sufficient force at Spring Hill to contest the enemy's progress until you are securely posted at Franklin." To be sure, this message was captured by Confederates. In his own defense, Schofield later claimed that his army had never been in great danger, that he was alertly tracking Hood's march north, and that he had not been swayed by Lee's demonstration, a contention Donald Connelly, the recent biographer, calls "bosh": "He had been lucky, very lucky."

As prescribed by Schofield's order that afternoon, Cox held the riverbank until darkness, then started marching; his pickets would stay till midnight. Wood's division, then Kimball's, followed. Aside from the difficulties of a night march, four miles north of Columbia the lack of a good bridge over Rutherford Creek caused irritating delays. Kimball's troops did not cross till midnight. Nevertheless, Schofield sent a staff officer ahead to Franklin to wire General Thomas on the situation. "General Schofield's troops are pushing for Franklin as rapidly as possible," Capt. William J. Twining telegraphed at 10 P.M. The officers and men knew what was at stake. "The men, knowing they were moving near the enemy's positions, kept well closed up," recalled General Cox. "This night march was a very rapid one," recalled Capt. Levi T. Scofield. At one point near Spring Hill, Scofield stated that a Col. George Northrup from Kentucky put his finger to his lips. "Hist," Northrup cautioned, pointing to the enemy campfires visible on the hills off to the right. Near Spring Hill Schofield's escort even brushed away a Rebel picket who ventured too close to the marching column. But the Federals got past

---

[47] Losson, *Tennessee's Forgotten Warriors*, 207; Jordan and Pryor, *Campaigns of Forrest*, 623 ("he would do the best he could"); Hurst, *Bedford Forrest*, 233; Wills, *Battle from the Start*, 283–84; Forrest report, January 24, 1865, *OR*, vol. 45, pt. 1, 753.

Hood's army without being attacked after Cleburne's and Bate's brief but canceled sorties late in the afternoon. Cox's division marched into Franklin just before dawn on November 30; Ruger entered about 7 A.M. Wood's division, reaching Spring Hill around midnight, "passed rapidly and silently through the village," General Wood reported. Kimball's division got out of the town two hours later. Wagner's, which had been first to reach Spring Hill, joined Kimball's as the rearguard of the retreating Federal army.[48]

The next morning, November 30, Hood awoke to learn that the enemy had escaped to Franklin. "General Hood is mad about the enemy getting away last night," Major Vaulx recalled General Brown as saying; "he is as wrathy as a rattlesnake this morning, striking at everything." "He is going to charge the blame of it on somebody," Brown warned.

The wrathy rattlesnake, "striking at everything," even struck at Pat Cleburne, Hood's most dependable division leader. Years later John C. Brown wrote Cheatham about a conversation the very day after Spring Hill, as the army marched toward Franklin in pursuit of the enemy. Cleburne was angry, having heard "through a very reliable channel" that General Hood was blaming him for the enemy's escape. Cleburne, "deeply hurt," Brown wrote, told him he would seek an investigation to clear his name once the campaign had ended.

Despite Brown's remembrance, it was Frank Cheatham who came in for most of Hood's blame. Evidently events moved so fast that Hood did not have a chance to rebuke Cheatham in person. In his *SHSP* article Cheatham claimed that "the first intimation made to me, from any source that my conduct...was the subject of criticism, was the receipt of a note from General Hood, written and received on the morning of the 3d of December." Cheatham stated that in the note, which he later lost, Hood explicitly absolved him of any blame for Spring Hill, but subsequent events do not support his claim. Even in the midst of his cam-

---

[48] Thomas to Schofield, November 29 (3:30 A.M.), *OR*, vol. 45, pt. 1, 1137; Jacobson, *For Cause & For Country*, 240 (captured message); Connelly, *Schofield*, 132–33; Cox journal (entry of November 29), *OR*, vol. 45, pt. 1, 358; Wood report, January 10, 1865, *OR*, vol. 45, pt. 1, 123; Kimball report, December 5, *OR*, vol. 45, pt. 1, 177; Twining to Thomas, November 29 (10 P.M.), *OR*, vol. 45, pt. 1, 1138; Crawford, "Mystery of Spring Hill," 124–25; Scofield, *The Retreat from Pulaski to Nashville, Tenn.*, 21; J. Cox, *The Battle of Franklin Tennessee November 30, 1864*, 34 ("a hostile picket"); Hay, *Hood's Tennessee Campaign*, 117; Jacobson, "Hood's Tennessee Campaign," 24–25.

paign, on December 7, Hood wrote Secretary Seddon withdrawing his recommendation of Cheatham to be lieutenant general, which he had made more than a month earlier. The next day he followed up by asking that "a good lieutenant general" should be sent to his army to replace Cheatham.

By that time S. D. Lee had written to his fiancée, Lilly Harrison, about Spring Hill, "one of the most disgraceful and lamentable occurrences of the war." "I will mention no names," Lee wrote in his letter, dated December 6, but it is clear to whom he was referring. "The Enemy...made their escape to Franklin during the night without being struck a blow....Gen H. gave his Corps Commander 4 or 5 distinct orders to attack and it was not done—his excuse being he was outflanked on the right and did not wish to bring on a night attack."

But then something must have happened on the 8th to change Hood's mind about replacing General Cheatham. Most likely it was a visit by Cheatham to army headquarters. Cheatham later thought that it occurred on December 4. The date of the meeting is not nearly so important as what was said between the two generals. On this matter they were in total disagreement. According to Cheatham in *SHSP*, he told Hood that "a great opportunity was lost at Spring Hill, but you know that I obeyed your orders there, as everywhere, literally and promptly." Hood, on the other hand, in *A & R* states that Cheatham came to him and admitted his mistake. He claims that he sent the following to Secretary Seddon on December 11: "Major General Cheatham has frankly confessed the great error of which he was guilty, and attached much blame to himself. While his error lost so much to the country, it has been a severe lesson to him, by which he will profit in the future." The evidence favors Hood's version, for after having asked Richmond to send a good lieutenant general on December 8, apparently later that day he reversed himself with this telegram to Seddon: "Major-General Cheatham made a failure on the 30th of November, which will be a lesson to him. I think it best he should remain in his position for the present. I withdraw my telegrams of yesterday and to-day on this subject."[49]

---

[49] Young, "Hood's Failure at Spring Hill," 36 ("wrathy as a rattlesnake"); Cheatham, "Lost Opportunity," 532 ("first intimation"), 533 ("I obeyed your orders"), 540–41 (Cleburne to Brown); Hood to Seddon, December 7 (received the 18th), *OR*, vol. 45, pt. 2, 659; Hood to Davis, October 31, *OR*, vol. 39, pt. 3, 870; Hood to Seddon, December 8 ("a good lieutenant general," received the

Nevertheless, Cheatham's mea culpa, assuming he made one on December 8, ultimately failed to mollify Hood. "Major-General Cheatham was ordered at once to attack the enemy vigorously and get possession of this pike," Hood wrote in a preliminary report on December 11. "Although these orders were frequently and earnestly repeated, he made but a feeble and partial attack, failing to reach the point indicated." "Darkness soon came on," he concluded, "and to our mortification the enemy continued moving along this road, almost in ear-shot, in hurry and confusion, nearly the entire night." "Had my instructions been carried out," he added in his final report of mid-February '65, "there is no doubt that we should have possessed ourselves of this road." These sentiments were amplified in *Advance and Retreat.* "Had I dreamed one moment that Cheatham would have failed to give battle, or at least to take position across the pike and force the enemy to assault him," Hood wrote, "I would have ridden, myself, to the front and led the troops in action."

Hood's soldiers commenting on these events after the war had split opinions on whether their commander should have done this. A. P. Stewart wrote J. P. Young in 1895 that when Hood complained to him that Cheatham would not carry out his orders, he wanted to say, "You, General, ought to have taken command yourself and made the attack." This prompted ex-Capt. A. C. Jones a few years later to comment, "It seems a strange sort of military discipline that a commander in chief should issue his orders to his corps commanders and then go to the front to see those orders executed." Jones asked, "Was that the custom of General Johnston or any other great commander?"

Cheatham's response, as we read in *SHSP*, was strong and ample, even if his contradictions of Hood are at times open to question. Ex-Confederates tried to sort it all out. "It was either the fault of General Hood or of General Cheatham," wrote W. O. Dodd, who as captain had ridden with Forrest. Hood "did not escape censure," Dodd explained, and "in my opinion both were to blame." "But the principal fault," he concluded, "is at the door of General Cheatham"; "the majority of the

---

17th), *OR*, vol. 45, pt. 2, 665; Jacobson, *For Cause & For Country*, 177 (Lee to Harrison); Hood, *A & R*, 290 (excerpt of Hood to Seddon, December 11), 356 (full text of which letter I do not see in *OR*, vol. 45, pt. 2); Hood to Seddon, December 8 ("I withdraw my telegrams," received December 17), *OR*, vol. 45, pt. 2, 665.

army believed General Cheatham mainly responsible for the misfortune at Spring Hill." Dodd's paper was published next to Cheatham's in the *Southern Historical Society Papers* in a point-counterpoint. Cheatham lined up numerous supporters, printing as part of his *SHSP* essay letters from Governor Harris, generals Stewart, Lowrey, Brown, and Bate. Hood did not do so in *Advance and Retreat*, and he died before Cheatham's article fired his shot across the bow.

It is not hard, nevertheless, to line up supporters of Hood's position. "Who was to blame of the subordinate commanders, I cannot say exactly," Chaplain Quintard entered in his diary on January 20, 1865, "but General Hood's orders were positive and specific, conveyed by two of his own staff officers and lastly by Governor Harris to General Cheatham." After the war, other officers came forth in memoirs that seemed to substantiate Dodd's claim that "the majority of the army believed General Cheatham mainly responsible." Peter Stewart's remembrance is a case in point. Decades after the war T. G. Dabney, who as sergeant had served in Stewart's artillery, ran into the general in Biloxi. Conversation naturally centered on the war, including Spring Hill. Stewart remembered being halted by Hood at Rutherford Creek, then ordered forward to form on Brown's right. Stewart encountered Hood at a roadside fire, accompanied by a sole orderly. According to Dabney, "General Hood began to inveigh against Cheatham for not making the attack on Spring Hill as he was ordered to do." Then Stewart told Dabney, "It was on my tongue to ask Hood, 'Why did you not see yourself that your order was obeyed and the attack made?' but I thought that would appear disrespectful."

Later, having brought his divisions toward Brown's right, Stewart met one of Cheatham's staff officers "who informed me that General Hood had sent him to place me in position." It struck Stewart as "strange" that "the commanding general should send an officer not of his own staff on this errand." Stewart soon met General Brown and became worried that it would take all night to get into the desired position. So Stewart ordered his troops to bivouac while he rode to Hood's headquarters. "It was now 11 P.M.," he later reported, claiming that Hood told him "in substance, that it was not material; to let the men rest; and directed me to move before daylight in the morning, taking the advance toward Franklin."

All these high-level goings-on were lost upon Hood's junior officers, who nevertheless realized that allowing the enemy army to sneak past them at night had been a monumental blunder. "I was within a few hundred yards of the pike & could—went to sleep about twelve o'clock—hear everything that was going on on the road," recorded Lt. Richard Cuyler King in his journal a week later. "It was a criminal mistake on the part of someone in authority for not getting possession of the pike," he concluded. General French entered in his journal on the 30th that he encountered General Hood. "Well, Gen. French," Hood allegedly said, "we have missed the great opportunity of the war!" "Yes," I replied, "I am told the Yankees passed along all night and lit their pipes at our camp fires." "Of course my answer was a *little* figurative," French added wryly, "but some soldiers heard it, and taking it literally it soon spread among the ranks." French's humor, however, was exceptional. One Confederate recalled years later that "the deep mortification and shame for the blunder could be seen in the bowed head of every one" the next day. Writing a generation after the war, one of Granbury's Texans waxed philosophical. While generals Hood, Cheatham, and others argued "in trying to fix the blame for this disgraceful failure," Lt. R. M. Collins of the 15th Texas concluded, "the most charitable explanation is that the gods of battle injected confusion into the heads of our leaders."[50]

"The majority of the army believed General Cheatham mainly responsible for the misfortune at Spring Hill," Dodd opined in 1888, and two recently published sources add weight to his judgment. Both were written by S. D. Lee to General Hood more than a decade after the Tennessee campaign, in August 1875 and April 1879. Stephen "Sam" Hood discovered both in 2012; they had long been discreetly preserved by descendants of the general and were then in possession of a great-grandson in Philadelphia. In this phenomenal cache, Sam found the letter (never before published) written by Lee to Hood in August 1875. "I

[50] Hood report, December 11 and February 15, *OR*, vol. 45, pt. 1, 657, 652; Hood, *A & R*, 287; Young, "Hood's Failure at Spring Hill," 36 ("You, General, ought to have taken command"), 41; A. C. Jones, "Criticism of General Hood," 508; W. O. Dodd, "Reminiscences of Hood's Tennessee Campaign," 521; Elliott, *Doctor Quintard*, 220; Dabney, "Strong Topics," 32 ("it was on my tongue," repeated in Young, "Hood's Failure," 36); Stewart report, April 3, 1865, *OR*, vol. 45, pt. 1, 712–13; R. K. Brown, *Our Connection with Savannah*, 135 (King journal, December 6); French, *Two Wars*, 292; S. G. French, "Errors in Reports of Battle of Franklin," 230 (repeating story of the Federals "lighting their pipes at our camp fires"); Shapard, "At Spring Hill and Franklin Again," 138; Schloemer, "General John Bell Hood," 23 (quoting Lieutenant Collins).

met A. P. Stewart at Columbus about 6 weeks ago and profounded why was no battle delivered at Spring Hill," Lee wrote. "He replied in substance that Cheatham & Cleburne determined it was not best to bring on an engagement at night." Lee admitted that Stewart had not personally heard Cheatham and Cleburne talking about the possible assault but added that Stewart "believed such was the case and had heard so."

Since Sam Hood's *John Bell Hood: The Rise, Fall, and Resurrection of a Confederate General* (2013), he has published *The Lost Papers of Confederate General John Bell Hood* (2015), and in it the full text of Lee's letter of August 1875, including this longer excerpt:

> I met A. P. Stewart at Columbus 6 weeks ago and profounded, why was no battle delivered at Spring Hill? He replied in substance that Cheatham & Cleburne determined it was not best to bring on an engagement at night. He said he heard or believed that Cleburne regretted it immediately afterwards, and said no such weight should be on his mind for similar cause again & that feeling lost his life at Franklin soon afterward. He further said he thought you were in a measure to blame as you were on the ground & could have put the troops up yourself or seen it done. As for himself he was put in position on the Columbia side of a creek to cut off any movement in that direction & was not moved from that position till near dark. He did not state the fact as to Cheatham & Cleburne on his own positive information, but said he believed such was the case & he had heard so.

Lee added that he hoped Hood would write his memoirs, at least partly because Sherman and Johnston had recently published theirs ("certainly Johnston & Sherman have hurt themselves badly"). "I believe now, a correct statement as to your movements would do good," Lee concluded.

Four years later Hood was indeed finishing his memoirs. S. D. Lee again wrote Hood about Spring Hill in April 1879, but this time he was more cautious, not mentioning Cheatham or Cleburne at all. "The events at Columbia, Spring Hill and Franklin are fresh in my mind & can never be forgotten," he affirmed. "I said a great, great blunder was made by somebody…." (Note the *somebody.*) "On arrival I recollect your chagrin & mortification at what had happened. You said they have eluded me owing to ____conduct"—here, editor Hood inserts "Lee drew a line with nothing on it." One may infer that Lee, having shared with Hood Stewart's remark about Cheatham and Cleburne several years before, was now drawing back, not wishing to be quoted about the controversy. Lee, according to Hattaway, was "making sure that Hood said nothing detrimental about him in *Advance and Retreat.*" Lee in the same letter implied as much. "It is natural that I should be sensitive," he confessed. Indeed, referring to a speech recently delivered by Hood to the Louisiana Division of the Army of Tennessee Association, Lee added, "I would like to have known what you did say about me in your speech." Hood took the hint. Whereas he printed some two dozen letters from ex-Confederates in *Advance and Retreat* (including one from Lee written January 26, 1874), he did not publish Lee's letters of 1875 and 1879 concerning Spring Hill. Lee himself seemed to hedge on what he should put on record about Spring Hill. He wrote Hood in April 1879 that he hoped that Hood's memoir "will make clear the Spring Hill matter, for it is time for that mystery to be cleared up." Lee even added, "If you do not, I will feel it my duty to do so after your book comes out." Yet just a few years before, in an article for *Confederate Veteran*, when Lee considered who was to blame for Spring Hill, he meekly declared, "Rumor said it was the fault of the commander of the leading Corps" (meaning Cheatham). But then he equivocated: "His [Cheatham's] friends said Hood was near enough to have enforced his orders and to have seen that they were executed."

Stephen Hood has found another postwar letter in the recently discovered archival cache, which Hood also did not print in *Advance and Retreat.* It supports the assertion of Major Ratchford that General Ed Johnson complained that "he was not permitted to attack them." A dozen years after the war one former Confederate, William W. Old, who had served as a major on Johnson's staff, wrote Hood about a conversation he had just had with another of Johnson's staff officers, former Maj. E. L. Martin. "In regard to the attack at Spring Hill," Old explained, "Mr. Martin says that Genl. Johnson did go to Gen. Cheatham and <u>beg</u> him to let him attack with his division, stating that he did not even require any support, but that Genl. Cheatham refused, stating *finally* he was opposed to night attacks. Mr. Martin says he was present" (emphasis original). Underlined words are in the original letter, Sam adds in his *Rise, Fall, and Resurrection* (2013). Yet in printing Old's complete letter in *Lost Papers* (2015), he does not include the emphasis on "beg" and "finally."[51]

[51] Dodd, "Reminiscences," 523; Hood, *Rise, Fall, and Resurrection*, 123 (quoting Lee to Hood, August 25, 1875, "John Bell Hood Personal Papers");

Nonetheless, Old's letter fits in with the several other sources we have seen, indicting Frank Cheatham for having cancelled a night attack on the Columbia turnpike and the enemy column marching northward upon it:

> S. D. Lee to Lilly Harrison, December 6, 1864: "gave his Corps Commander 4 or 5 distinct orders";
>
> Chaplain Quintard's diary entry of January 20, 1865: "General Hood's orders were positive and specific...to General Cheatham";
>
> Major Cumming's manuscript recollections at Chapel Hill: "I reached Gen. Hood's headquarters, to my astonishment I found Gen. Cheatham...remonstrating with Gen. Hood against a night attack";
>
> General Lee to Claiborne, in the judge's papers at Chapel Hill: "Cleburne and Cheatham consulted, and decided between themselves that no battle should be fought at night";

Hood, *Lost Papers*, 115–117, 119 (Lee to Hood, August 25, 1875), 122–23 (Lee to Hood, April 16, 1879), 124–25 (Lee to Hood, April 25, 1879); Lee, "From Palmetto to Nashville," 257; McMurry, *Hood*, 201 ("By 1879 [regarding his memoirs] he seems to have about finished the task"); Hattaway, *Stephen D. Lee*, 209; Hood, *A & R*, 138–39 (Lee to Hood, January 26, 1874, on the "general effect of entrenchments upon an Army").

I am guessing that the speech Lee refers to is that made by Hood to the Louisiana Division. Editor Hood does not identify it. McMurry states that Hood addressed the group in spring 1879, adding that "only a short excerpt of the speech has survived" in a newspaper clipping at Tulane (McMurry, *Hood*, 201, 227 n.29). McMurry quotes the clipping, in which Hood reportedly declared, "They charge me with having made Franklin a slaughter-pen, but, as I understand it, WAR MEANS FIGHT AND FIGHT MEANS KILL" (*Hood*, 201).

McMurry does not note that Hood's remark echoes the famous if apocryphal statement of Nathan Bedford Forrest, "War means fighting, and fighting means killing" (Tsouras, *The Greenhill Dictionary of Military Quotations*, 220). Recent biographers have identified the original source: McAlister, "Forrest on Fort Pillow" 8:56. McAlister, a Union officer posing as a journalist, interviewed Forrest in Meridian, May 13, 1865, four days after he had surrendered. Speaking about the Fort Pillow "massacre," Forrest was quoted as saying, "When I went into the war, I meant to fight. Fighting means killing." See Hurst, *Bedford Forrest*, 262; Davison and Foxx, *Enigma*, 409.

General Stewart to W. O. Dodd in 1881: "My impression is that Cheatham and his officers thought themselves in great danger of being outflanked or crushed";

Sergeant Dabney's assertion that after the war Stewart remembered encountering Hood late on the 29th, saying, "General Hood began to inveigh against Cheatham for not making the attack on Spring Hill as he was ordered";

Lee to Hood, August 1875, relating Stewart's remembrance that "Cheatham & Cleburne determined it was not best to bring on an engagement at night";

Major Old to Hood, September 1877: Martin, General Johnson's ADC, said Johnson begged Cheatham for permission to launch an assault, "but that Gen. Cheatham refused";

Lee to Hood, April 1879, "a great, great blunder was made by somebody";

Stewart to J. P. Young in 1895 that he recalled "General Hood complained to me that General Cheatham would not obey his orders to attack"; and

Major Ratchford, *Some Reminiscences of Persons and Incidents of the Civil War* (1909): "I heard General Ed Johnson say…he…was not permitted to attack."[52]

The problem is that S. D. Lee's "somebody" in 1879 could have been not only Frank Cheatham but also John C. Brown or even John B. Hood. Within two weeks of the army's defeat at Nashville, Governor Harris wrote Chaplain Quintard, "I was there and know much…and yet, I cannot fix the responsibility upon any one officer." Christopher Losson concludes that "Brown, Cheatham, and Hood all deserve some blame." General Brown by his own admission had reached "a determination to suspend the advance," which explains why no one heard his guns.

[52] Jacobson, *For Cause & For Country*, 177 (Lee to Lilly Harrison); Elliott, *Doctor Quintard*, 220; Hood, *Rise, Fall, and Resurrection*, 122–23 (Cumming), 123–24 (Old), 125 (Ratchford); Losson, *Tennessee's Forgotten Warriors*, 207 (Lee to Claiborne); Cheatham, "Lost Opportunity," 535 (Stewart to Dodd); Dabney, "Stewart on Strong Topics," 32; Hood, *Lost Papers*, 115 (S. D. Lee, August 25, 1875), 122 (Lee to Hood, April 16, 1879); Young, "Hood's Failure at Spring Hill," 36 (Stewart to Young).

Cheatham's role and responsibility in the event of November 29 have been elaborated. But what about General Hood? Confederates later criticized him for not having been on the field after darkness to direct the night assault. "Hood is not free from blame," charged Lee in a letter to Judge Claiborne; "he could have personally directed the movement when he saw delay." The commanding general had retired to the Thompson house sometime in the evening and remained there. He himself admits, "had I dreamed one moment that Cheatham would have failed to give battle…I would have ridden myself, to the front, and led the troops into action." To be sure, Hood had been up since 4 A.M. that day, and the ride to secure flank attacking position across Duck River had been a long and trying one, so end-of-day exhaustion would be understandable—save for the reminder that General Jackson, similarly directing a long march to get at Howard's flank at Chancellorsville, was still in the saddle at 9:30 P.M., well after dark on that spring Virginia evening, personally leading his staff through the Wilderness to find a way to continue his attack into the night. Some Confederates apparently held Hood to Stonewall's standard. Among them was Lieutenant General Peter Stewart. "The truth is, the failure at Spring Hill was General's Hood's own failure," he wrote in an article published a month after his death in August 1908. "He was at the front with the advanced troops, or could have been, and should have been," Stewart reasoned, "and if he gave 'explicit orders for an immediate attack and occupation of the pike' and they were disobeyed, the remedy was entirely in his own hands." A century later (2009), historian James R. Knight agrees. Hood, he writes, "left the conduct of the battle to his subordinates." When he withdrew to the Thompson house around 4 P.M., Hood had not just left the field, he had left Cheatham in charge of events from then on.

Probably Hood's chief mistake at Spring Hill was his apparent uncertainty as to tactical objective. Hay implies that it was, or should have been, the town of Spring Hill itself, where "Hood had a force of some 25,000 men with which to strike Stanley's single division of some 5,500 men." On the other hand, as we have seen, Hood was directing Cheatham and others to strike the Columbia-Franklin turnpike south of town in the belief that his forces could block the road and "trap" Schofield, perhaps forcing him to launch futile attacks against the defending Confederates. In other scenarios, according to various observers, Hood seems to have been aware that Schofield's army was marching past him towards Franklin and presented itself to a damaging flank attack, of the kind General Bate claimed that he wanted to launch. After the war, Major Vaulx claimed that when Cheatham reported to Hood's headquarters that night, the commanding general ordered, "Do not attack, but order your troops to hold the position they are in for the night." Relating Vaulx's story, Henry Field commented that it showed that Hood "had himself an access [excess] of irresolution."[53]

Stanley Horn points to a graver error in Hood's generalship on the 29th: ignoring the presence of a large force of enemy infantry holding Spring Hill to his rear.

> The records show that soon after Hood himself crossed Rutherford's Creek about 3 P.M. he met Forrest, who reported to him his action against the enemy during the preceding three or four hours. But in spite of this clear evidence that there was now a substantial part of Schofield's force already entrenched at Spring Hill, Hood believed only what he wanted to believe: that S. D. Lee's demonstration at Columbia was holding all, or most, of the Federal force at Duck River. When he was giving his final orders to Cleburne (in Cheatham's presence) he was not directing an assault on the enemy troops already entrenched around Spring Hill, but seemed to think he was intercepting or attacking whatever force was supposed to be advancing along the pike. He did not say to Cleburne: "Go, and attack those troops in Spring Hill that Forrest has been fighting"; but said, "Go, take possession of and hold that pike at or near Spring Hill." And that the pike and not the troops in the village, was the assigned objective for Cheatham's [*sic:* Cleburne's] division is shown further by his subsequent instructions to Bate to "move to the turnpike."

Horn's rightful attention to Wagner's division at Spring Hill leads him to offer the conjecture that the Confederates' failure to reach and block the turnpike south of the town might have been good for them after all. "If both Cleburne and Bate had been able to get astride the pike before dark," he posits, "they would have been caught between the Fed-

---

[53] Losson, *Tennessee's Forgotten Warriors*, 213 (quoting Harris to Quintard, December 29, 1894, in Quintard Papers at Duke University); Cheatham, "Lost Opportunity," 538 (Brown statement); Hattaway, *Stephen D. Lee*, 208 (Lee to Claiborne, "Hood is not free from blame"); Hood, *A & R*, 287; Robertson, *Stonewall Jackson*, 727 (9:30 P.M.); A. P. Stewart, ""A Critical Narrative,"" 463; Knight, *The Battle of Franklin*, 45; Hay, *Hood's Tennessee Campaign*, 85; Bate report, January 25, 1865, *OR*, vol. 45, pt. 1, 742; Field, *Bright Skies and Dark Shadows*, 216.

eral force that had already arrived at Spring Hill and the remaining portion of Schofield's army advancing northward on the pike." Horn's conclusion is that "in this situation they would hardly have been in position to 'destroy or capture' the enemy; on the contrary, they might have found it hard to escape capture or destruction themselves."

John B. Hood's willful neglect of Wagner's 5,500 infantry holding Spring Hill, thus threatening the flank or rear of Cheatham's proposed assault on the turnpike, points to his failure to send help to Bedford Forrest after the latter's repulse. According to Capt. H. A. Tyler, in Forrest's command, south of Spring Hill the cavalry leader could only look on as Schofield's infantry marched past him. "—Hood!" Tyler recalled Forrest exclaiming, "had he supported me here as he promised, that whole army would have been our prisoners." After the war, J. H. McNeilly, chaplain for Quarles's Brigade, recalled hearing Forrest exclaim to Major General Walthall, "O, General, if they had given me one of your brigades, just one, to have flung across this road, I could have taken the whole shebang!" In this vein General Chalmers told the Southern Historical Society, "I think I risk nothing in saying if Forrest had been in command of our army, General Schofield would never have marched by Spring Hill." Despite all of these factors, Hood's apparent negligence of Wagner's division in his rear is not nearly so scrutinized by scholars as, say, Pope's belief that Jackson was retreating at Second Manassas August 30, or Hooker's notion that there were no Rebels on his flank at Chancellorsville May 2. But Horn in his *CWTI* article makes a good case that it should be.[54]

Behind it all, however, remains the persistent Confederate legend that at Spring Hill a military miracle awaited its realization. J. P. Young expressed it in *Confederate Veteran*:

> It must be patent to the most casual observer from what has been narrated above that but for the failure of Hood's subordinates to act promptly at Spring Hill Stanley would have been crushed before nightfall, the wagon train and reserve artillery of the army captured,

---

[54] Horn, "Spring Hill Legend," 27, 32; McDonough and Connelly, *Five Tragic Hours*, 57 ("If he [Hood] had succeeded in taking the pike south of the village, his troops would have been placed between the two portions of the Union army, possibly in a more dangerous position than that of their enemy"); Tyler, untitled reminiscence, 15; J. H. M'Neilly, "Bloody Franklin," 5; Chalmers, "Forrest and his Campaigns," 482.

> Schofield entrapped as he approached in the darkness with Ruger's division, and the remaining divisions of Cox, Wood, and Kimball left at the mercy of Hood at daylight next morning.

With so many of us focused on the virtual petri dish of the "mysteries" surrounding Cheatham and the aborted attack on the turnpike, McMurry widens the microscope. "Even had the Army of Tennessee secured the pike," he offers, "Schofield could easily have marched around to the west of Hood's roadblock to reach Franklin." He goes even further: "Indeed, had it been necessary, the Yankees could have bypassed Franklin entirely and marched directly to Nashville. The Confederates, however, may not have known of the routes west of Spring Hill." Schofield himself wrote after the war that "if the enemy had gotten possession of a point on the pike, the column from Duck River would have taken the country road a short distance to the west of Spring Hill and Thompson's Station and marched on to Franklin." When Schofield read the speech on Forrest's campaigns given by General Chalmers in August 1879 to the Southern Historical Society (particularly Chalmers's claim that "if Forrest had been in command of our army, General Schofield would never have marched by Spring Hill"), Schofield wrote in the margin that he could easily have marched cross-country to reach Franklin. McDonough and Connelly point to the railroad bed and Carter's Creek pike west of Spring Hill, routes they believe Schofield was aware of. The general's biographer, though, has his doubts: "The prospect of his army, burdened with heavy wagons, crossing the still soft fields in the dark strains credulity," argues Donald Connelly. "He would have lost more than a few wagons, and the potential for troops getting lost in the dark was quite high." "Schofield's army could have escaped," he concludes, "but not easily."[55]

As much as the Confederates themselves, historians have struggled with Hood's responsibility for Spring Hill. Hay early on staked out the eventual consensus that among Brown, Cheatham, and Hood "the responsibility for the failure was divided." First, Major General Brown should have attacked, as "the prize of success was great." Major General

---

[55] Young, "Hood's Failure at Spring Hill," 41; McMurry, *Hood*, 173–74; Horn, "Spring Hill Legend," 31 (quoting Schofield); Chalmers, "Forrest and his Campaigns," 482; McDonough and Connelly, *Five Tragic Hours*, 56; Connelly, *Schofield*, 132, 376 n.50 (Schofield's marginalia on Chalmers's *SHSP* speech, in Schofield's papers at the Library of Congress).

Cheatham bore responsibility for the absence of coordination among his three divisions (recall Brown telling Chalmers that he had no orders). To quote Hay:

> Cheatham's troops were brought into line without any concert of movement and it appears that, when darkness came on, Cleburne and Bate and Brown were all separated and Stewart was still at Rutherford's Creek. Cheatham's three divisions did not join or co-operate, but moved, each of them, in an independent manner. They moved without any apparent definite plan or direction and bivouacked where the darkness of night found them.

But Hay also comes down hard upon the commanding general, Hood. 1) "He seemed uncertain as to the exact disposition of his troops, relative to each other and to the opposing Federals." 2) "His orders to his subordinates were contradictory and misleading." 3) "Hood had no definite knowledge of the position of Schofield's troops" and made the common mistake of believing they were where he wanted them to be, i.e., "in front of Columbia instead of on the march to Franklin." 4) "Instead of remaining at his headquarters he should have gone to Brown's position, on the right, and personally superintended his advance." 5) Finally, "if, at the time, Hood believed Cheatham had failed, from any cause, to comprehend and to execute orders given him, it was Hood's duty to relieve him and to order him to turn over his command to General Cleburne, the next in rank." That he did not do so, or at least vacillated in his messages to the War Department, Hay reasons, reflected Hood's uncertainty as to Cheatham's actual culpability. It also involved morale and politics within the army: "Cheatham was from Tennessee and many of the officers and soldiers in the army were also from this state."

In the end, Hay finds fault in Hood's "own mental limitations and physical incapacity," which led to an "absence of the powers of concentration, grasp, and vigor" needed by a commander on his critical day of battle. To a certain extent that old catchall, "poor staff work," can be blamed also at Spring Hill. Example comes from the letter written by ex-governor Harris to James D. Porter in 1877. In it Harris remembered that late on the night of the 29th, he, Hood and Major Mason (the A. A. G.) were asleep when—improbably—they were awakened by a private soldier who had himself seen the Yankees "in great confusion" hurrying toward Franklin on a road clogged with guns and wagons. Harris claimed that Hood told Mason to order Cheatham to launch an attack, then went back to sleep. The governor's story was a tall one, but it ended with an indictment of the adjutant: Mason later told Harris that he never delivered such an order to General Cheatham—he fell back to sleep before writing it. All this reminds us of Connelly's well-known observation that as army commander Hood envisioned the kinds of tactical masterpieces that Lee and Jackson were able to accomplish in Virginia, but that in the Army of Tennessee he could not pull them off through the incapacities of his staff officers.

Nevertheless, Hay concludes, "Hood failed largely because of the neglect personally to see to it that movements were rapid and that his orders were promptly obeyed and correctly executed." Thus on November 30 John Bell Hood, already labeled "rash to the point of indiscretion," added, according to Hay, mediocrity and inefficiency in overseeing a critical maneuver as elements of his unhappy emerging historical reputation.[56]

Others are uncharitable toward Hood but for different reasons. Horn regrettably succumbs to a local legend that the general was drunk: "the befuddlement of intoxication might account for the otherwise inexplicable confusion and paralysis at Spring Hill." Groom trots out the old saw that Hood "might have been taking drugs such as laudanum or other opiates." So does Anne Bailey ("probably took some laudanum"). McDonough and Connelly boldly assert that the Spring Hill affair showed that Hood "was too emotionally unhinged to command." Against this kind of allegation, Wiley Sword's conclusion appears much more reasonable: "Hood's impromptu decision making and the lack of cohesive, prior planning in defining tactical objectives were at the heart of the difficulty" at Spring Hill. Moreover, Sword states that Hood retired from his lines too early that evening and allegedly went to sleep by 9 P.M. Still, generals Brown and Cheatham come in for their share of criticism; the latter's performance, Sword writes, "was absolutely woeful."

---

[56] Hay, *Hood's Tennessee Campaign*, 94–102, 234 n.62 ("We have seen that Cleburne was fighting for possession of the pike, when halted, and that he was not supported as he should have been and as was originally contemplated"); Porter, *Tennessee*, 148–49 (Mason: "General Cheatham was not to blame….I did not send him the order"); "Spring Hill and Battle of Franklin," 343 (Mason, "I did not send him the order"); Gillum, *Twenty-five Hours*, 430 ("I fell asleep again before writing it"); Connelly, *Autumn of Glory*, 502.

Hood is not absolved even by his biographers. O'Connor criticizes—properly—the general for writing in his memoir that "the Army, after a forward march of one hundred and eighty miles, was still, seemingly, unwilling to accept battle unless under the protection of breastworks." In his tactical planning, Hood tried to be Robert E. Lee and Stonewall Jackson, O'Connor observes, "but he had not brought with him a first principle of the two generals he adored: to blame the commander rather than the commanded for any failures in expectations of a battle plan." Dyer adds that Hood was sick: "He had been ill with rheumatism at Florence, and nearly two weeks in the saddle in cold, wet weather had done nothing to help it." In such a state, his inattentiveness toward the execution of orders could presumably be explained. McMurry summarizes several criticisms, from Hay's time to ours: "Hood's orders were not clear. His lack of knowledge of his own army's position and movements was probably the result of a nearly complete breakdown of his staff. The absence of reliable information about Schofield's movements indicates that there was no provision of reconnaissance. Hood had once more developed a sound plan but had again failed to execute it."

Lundberg directs sharper criticisms:

> Who was to blame for the Confederate reversal at Spring Hill? The responsibility primarily lay with Hood and his failure to grasp the importance of details, the logistics of his army. Hood came up with what, on paper, appeared to be a brilliant plan but failed to execute the plan in large part because he did not pay attention to detail. He had no idea of the distance between Davis's Ford and Spring Hill because of his failure to reconnoiter the terrain and the misallocation of his cavalry. In addition, once Hood arrived on the field, he failed to supervise the attack until his window of opportunity passed. The only possible explanation for Hood's sometimes erratic behavior at Spring Hill remains fatigue. It is not difficult to understand how a physically crippled Hood failed to take a more active role rather than remaining at Oaklawn.

Given the back and forth among historians over the past century, Stephen Hood's observation is both sound and understated: "It will be interesting to see what changes, if any, are made to Civil War scholarship's future portrayals of John Bell Hood and Benjamin Franklin Cheatham at Spring Hill."[57]

However unresolved are the "mysteries of Spring Hill," the legend always ends with the incontrovertible fact that the Yankees got away. On the morning of the 30th, an angry Hood dressed at the Thompson house, mounted up with his staff, and rode north a couple of miles to the plantation home of Nathaniel Cheairs, "Rippavilla." There he called a conference of senior officers who were nearby: Cheatham, to be sure, perhaps a few others. Generals Stewart and Cleburne apparently were not present, nor was Forrest, who at daylight had set off with his cavalrymen to pursue the retreating Yankees. We have no written record of the meeting and cannot be sure of those present. S. D. Lee had not yet arrived from Columbia. The absentees were lucky, for the angry commanding general fumed at those in the room, especially Cheatham, for having allowed the enemy to escape. Mrs. Cheairs, who kindly served biscuits and bacon for the officers, later remarked that the language she heard was not appropriate for a lady's ears. Eric Jacobson reasons that it was likely Cheatham who later told Major General Brown that Hood had been "wrathy as a rattlesnake."[58]

After all the accusations and finger-pointing, there was little else for General Hood to do but order his army to pursue Schofield's retreating column. Jackson's cavalry was still clinging close to the enemy's rear when Forrest rode up with Bell's Brigade. The Confederates kept up a continuous skirmishing until the Federals withdrew into Franklin and took up a defensive position.

Hood's infantry set out on their march with a determined spirit. "They felt chagrined and mortified at the occurrence of the preceding day," recalled Captain Dodd years later, "and each man felt a pride in wiping out the stain caused by a superior's fault." The superior in question, to Dodd at least, was the army's commander; he remembered that "we lost confidence in General Hood" over "the wretched affair" at

---

[57] Horn, *Army of Tennessee*, 392; Groom, *Shrouds of Glory*, 150; Bailey, *Chessboard*, 87; McDonough and Connelly, *Five Tragic Hours*, 59; Sword, *Embrace an Angry Wind*, 136, 147, 153; O'Connor, *Cavalier General*, 235; Hood, *A & R*, 290; Dyer, *Gallant Hood*, 288; McMurry, *Hood*, 174; Lundberg, "Errant Moves on the Chessboard of War," 60; Hood, *Rise, Fall, and Resurrection*, 128.

[58] Jacobson, *For Cause & For Country*, 188–91; Gillum, *Twenty-five Hours*, 416–17; Jacobson, "Hood's Tennessee Campaign," 26.

Spring Hill. Nonetheless some in the ranks looked forward to getting even with the enemy in battle ahead. Others sought to shore up their comrades' spirits; as Gibson's Louisiana brigade marched by, a band played hearty tunes.[59]

Hood's three corps marched in order of Stewart's, Cheatham's, then Lee's, the last with the trains and artillery from Columbia. The Confederates found the Columbia-Franklin turnpike strewn with debris from the Yankees' night march: burned wagons, dead horses and mules, the usual equipment and baggage discarded by troops marching in haste. "Every one or two hundred yards we would pass an abandoned wagon with the team shot down in the traces," recalled D. H. Patterson (Quarles's Brigade, Walthall's Division). The retreating Yankees chopped the spokes of the wagon wheels to disable their vehicles. Tress T. Smith, a private in the 15th Mississippi, remembered well after the war that the Northerners had not only abandoned wagons but also "knocked the mules in the head and left them in the road hitched to the wagons."[60]

But the Federals had gotten away. When General Cox and his division arrived in Franklin just before dawn on the 30th, he got orders from Schofield to position his troops so as to cover the town and the pike into it, which would be used by the army's trains. General Schofield himself rode into town at 4 or 5 A.M. With daylight, he began inspecting the railroad and wagon bridges across the Harpeth River, which flowed just north of town. Even though General Thomas asked him by telegraph if he could hold Hood at Franklin for a few days, it was obvious that the Federals would have to continue falling back on Nashville. The problem lay in the bridges. Schofield quickly learned that the pontoon bridge he had requested Thomas to send had not arrived, and that the Columbia pike bridge had been washed away by the swollen current. Schofield set his engineers working to repair it and laying planks on the railroad

---

[59] Jordan and Pryor, *Campaigns of Forrest*, 625; Dodd, "Reminiscences," 522 ("felt a pride"); Dyer, *Gallant Hood*, 289.

[60] Hood report, February 15, 1865, *OR*, vol. 45, pt. 1, 653; McDonough, "Franklin," 22; Sword, *Embrace an Angry Wind*, 157; D. H. Patterson, "Battle of Franklin," 116; W. C. Thompson, "A Confederate's Recollection of the Battle of Franklin," 24; Tress T. Smith recollections, February 1899, typescript in possession of Mr. Wes Bowden, Fort Oglethorpe GA (copy provided to me by Dr. Dent Purcell of Savannah, GA).

bridge. The tasks were completed by noon so that the Federal wagon train and artillery were able to begin crossing the river. The infantry were to follow at 6 P.M. if the Rebels did not launch an attack, in which case Schofield would have to defend his position. On the north bank Wilson's cavalry had arrived and were on watch for any advance by Forrest to cross the Harpeth and get in the Federal rear. Later, around 1 P.M. after its arrival, Wood's division was sent to reinforce Wilson.

Meanwhile Cox laid out his line. There were already some works south of town. After the Federals occupied middle Tennessee in 1862, a Union garrison had been stationed at Franklin, guarding the Tennessee & Alabama Railroad bridge across the river. In March 1863, Bedford Forrest had captured a 230-man garrison in the stockade near the bridge. After that, Federals built a large fort on the north bank of the river, named for Union Gen. Gordon Granger, and dug entrenchments on the south bank near the railroad and the Lewisburg pike. Upon their arrival on November 30, Cox's men strengthened these old works and set about extending them. They tore down the Fountain Carter cotton gin house for its boards. They dug ditches in front of their parapets and placed headlogs with firing slits on top of them. The area south of Franklin was not wooded, but from an Osage orange hedge and locust grove they fashioned abatis against an expected Rebel infantry charge. Their line's only vulnerability was along the Columbia pike, where the road cut through the parapet and which they left open, but with a retrenchment two hundred feet behind. The fortifications wrapped around Franklin; a curve in the Harpeth allowed the Northerners to anchor both ends of their horseshoe-shaped line on the riverbanks. Cox's (under Brigadier James Reilly), Ruger's, and Kimball's divisions occupied the line. Wagner's, last to arrive about 11 A.M., took position two miles south of Franklin on some hills called Breezy and Winstead, looking out for the Rebels' approach.[61]

---

[61] Cox report, December 2, *OR*, vol. 45, pt. 1, 348; Connelly, *Schofield*, 132–33; McKinney, *Education in Violence*, 392; Bailey, *Chessboard*, 95; Seabrook, *Encyclopedia of the Battle of Franklin*, 165 (Fort Granger built March–May 1863); Jacobson, *For Cause & For Country*, 185; Horn, *Army of Tennessee*, 398 (Schofield's infantry to leave Franklin at 6 P.M.); Jacobson, "Hood's Tennessee Campaign," 25–26, 59 (Dave Roth's map); Connelly, *Autumn of Glory*, 503 (the Federal line); Sword, *Embrace an Angry Wind*, 163; Wills, *Battle from the Start*, 107; McDonough, "Franklin," 23–24; H. F. Walker, "Bloody Franklin," 16, 18–19.

They were not long in coming. When Stewart's infantry, in the lead, approached Winstead Hill, Wagner's troops retired north toward the main Union line. General Hood rode up Winstead about 2 P.M. and was joined by his officers. All were able to view the enemy position, the fortifications south of Franklin, and Fort Granger north of it. The terrain approaching the enemy line was fairly flat and open except for a hedge of bushes near the railroad cut and a locust grove in front of the Carter housc. It was an ideal field of fire for the Yankees. But Hood was determined to launch as assault upon Schofield's line. Maj. Gen. John C. Brown later wrote that Hood, after viewing the enemy position, explained his rationale: "The commanding general, after surveying the field, remarked in substance, 'The country around Franklin for many miles is open and exposed to the full view of the Federal army, and I cannot mask the movements of my troops so as to turn either flank of the enemy, and if I attempt it he will withdraw and precede me into Nashville.'" Some of the officers present may have offered dissenting opinions, or done so later in the afternoon (the sources are vague). General Brown, however, recalled that "the policy of General Hood's decision was not discussed, and I cannot recollect any question propounded by him to any one present indicating a desire for an expression of opinion by any one." "We will make the fight," he announced to General Mark Lowrey, and that was it. After the war one Confederate veteran claimed that at some point Cheatham exclaimed, "It is a mistake, and it is no comfort to me to say that we are not responsible."[62]

The soundest justification for Hood's decision to attack at Franklin is naturally to be found in his own words from *Advance and Retreat*:

> I hereupon decided, before the enemy would be able to reach his stronghold at Nashville, to make that same afternoon another and final effort to overtake and rout him, and drive him in the Big Harpeth river at Franklin, since I could no longer hope to get between him and Nashville, by reason of the short distance from Franklin to that

---

[62] Jacobson, "Hood's Tennessee Campaign," 26; Jacobson, *For Cause & For Country*, 237 ("It has been thought that Hood was on Winstead Hill at 1 P.M., but the evidence points to him being there somewhat later"); Porter, *Tennessee*, 156 (undated Brown report, which I do not see in *OR*, vol. 45, pt. 1); Rees, "Cleburne's Men at Franklin," 54 (General Lowrey); Losson, *Tennessee's Forgotten Warriors*, 219.

city, and the advantage which the Federals enjoyed in the possession of the direct road.

In his campaign report Hood states, "I knew that it was all important to attack Schofield before he could make himself strong, and if he should escape at Franklin he would gain his works about Nashville. The nature of the position was such as to render it inexpedient to attempt any further flank movement, and I therefore determined to attack him in front, and without delay." General Beauregard would likely have approved Hood's commitment to give battle. "It is essential you should take offensive," he had wired from Macon on November 24 in a message that could have caught up with Hood by this time, "and crush enemy's force in Middle Tennessee [as] soon as practicable, to relieve Lee."

It is not recorded that a factor in Hood's decision was Schofield's dangerous position, fighting with a river behind him. Others, however, thought about it. "I am puzzled to know why as skillful a general as Schofield should have elected to make a stand south of the town with a stream at his back," wrote Major Buck after the war; the decision created for the enemy "an element of danger in case of disaster."

Hood's biographers do not ponder the possibility that he should have sent Forrest off to flank Schofield out of Franklin. O'Connor quotes the general's *OR* statement without comment. Dyer simply asserts, "he was through with flanking movements and shilly-shallying." McMurry concludes that Hood reasoned thus: "Perhaps his army was incapable of imitating the strategy of Lee and Jackson by making successful flanking marches, but possibly it could execute their tactics and make a frontal attack that, if as successful as the one at Gaines' Mill, would overrun the Federals at Franklin and sweep on to the Ohio." Indeed, in *Advance and Retreat* Hood states that the frontal attack he ordered at Franklin had "become a necessity as imperative as that which impelled General Lee to order the assault at Gaines's Mills." In his postwar memoirs General Schofield agreed: "Franklin was the last opportunity he could expect to have.... I believe therefore, that there can be no room for doubt that Hood's assault was entirely justifiable." James R. Knight, a recent writer on Hood's Tennessee campaign, offers agreement with Schofield. He writes that on the afternoon of November 30, "it's hard to see what other options Hood had in the time remaining before sundown."

Some writers (e.g., Hay and Sword) disagree and argue that Hood should have attempted another Spring Hill flanking maneuver—say, send Forrest riding around Schofield's flank, crossing the Harpeth, and getting into his rear. Such speculation is persuasively quashed by Stephen Hood, who makes more than a half-dozen excellent points.

By the time Forrest set out on the afternoon of November 30 he would have had at most only a couple of hours of daylight left, insufficient time to get around Schofield's army.

Crossing the Harpeth itself would have been difficult and time-consuming, as the river was up and swollen by recent rains.

Setting out from Hughes Ford, the likely crossing point on the river, the Confederates would have had to ride farther east still to stay out of range of the enemy artillery posted at Fort Granger.

Such a course would have entailed a ten- to twelve-mile route. Forrest's troops were mounted, but any accompanying infantry, especially on a night march, would have slowed the column down.

By the time the Confederates reached the turnpike in the enemy rear it would be near dawn; Schofield's army marching speedily on the turnpike would have already escaped.

Hood doubted that Forrest's cavalry alone could have successfully held off Schofield's two infantry corps for very long.

And that is assuming that Union General Wilson and his 5,000 cavalry, already north of the river, would not have detected and blocked the Rebels' mounted movement.

Sam Hood's eighth point is his most salient: "Hood was attempting to destroy Schofield's army, not seize and occupy the town of Franklin."

General Thomas at Nashville, eighteen miles away, had been finally reinforced by A. J. Smith. With Schofield he would have about 60,000 troops—more than twice the size of Hood's army. Any hope for success in Hood's Tennessee Campaign therefore hinged on destroying Schofield's two corps before they could get to Nashville. That meant attacking the Federals' position before they slipped away, which they could be expected to do during the night. "The closeness of Thomas at Nashville mandated that Hood destroy Schofield at Franklin if he could," agrees another student, Michael Klinger. "The battle had to be fought before the Union armies could join forces." Hood thus determined not to await the arrival of S. D. Lee's Corps and the army's artillery. With only two batteries on hand, there would be no softening barrage, as Lee had done at Gettysburg. Besides, there would be precious little daylight left for the assault itself.[63]

After surveying the enemy position on Winstead, Hood met with some of his generals at the William Harrison house, a half-mile south of the hill. It was here that Forrest, Cheatham, and Cleburne probably voiced concerns about the attack; as Eric Jacobson states, little has been written conclusively about the meeting. (A local legend that an angry Bedford Forrest physically threatened Hood is patently untrue, Jacobson argues.) "They are very formidable," Pat Cleburne is said to have remarked about the Yankees' works. "I do not like the looks of this fight," Frank Cheatham said; "the enemy has an excellent position and is well fortified." Hood allegedly replied, "I prefer to fight them here where they have had only eighteen hours to fortify, rather than to strike them at Nashville where they have been strengthening themselves for three years." According to Wyeth, Forrest declared that any assault on such a position would bring needless casualties. "I do not think the Federals will stand strong pressure from the front," Hood is said to have answered; "the show of force they are making is a feint in order to hold me back from a more vigorous pursuit." Forrest persisted. "General Hood, if you will give me one strong division of infantry with my cavalry," he said, "I will agree to flank the Federals from their works within two hours' time." Hood was unswayed and ordered the assault. At some point Hood and Cleburne conversed, with the commanding general issuing tactical instruction: "Give orders to your men not to fire a gun until you run the Yankee skirmish line from behind the first line of works in your front,

---

[63] Hood, *A & R*, 291, 296; Hood report, February 15, 1865, *OR*, vol. 45, pt. 1, 653; Beauregard to Hood, November 24, *OR*, vol. 45, pt. 1, 1243; Buck, "Military View of Battle of Franklin," 383; O'Connor, *Cavalier General*, 237–38; Dyer, *Gallant Hood*, 291; McMurry, *Hood*, 175; J. R. Knight, *Hood's Tennessee Campaign*, 66, 68; Hay, *Hood's Tennessee Campaign*, 133–34 ("should have continued his flanking movements"), 138 (Thomas's 60,000 men); Sword, *Embrace an Angry Wind*, 262 ("easily could have outflanked Schofield"); Hood, *Rise, Fall, and Resurrection*, 146–50 (including Hal Jesperson's thoughtful map, "Forrest's Proposed Flank at Franklin November 30, 1864"), 240 ("Franklin was Hood's final opportunity to destroy the trapped Schofield before he reached the relative safety of the Nashville fortifications and joined Thomas's gathering forces"); Jacobson, "Hood's Tennessee Campaign," 42 (a flanking maneuver "would not have been as simple as some have suggested"); Walker, "Bloody Franklin," 20–21; Klinger, "Gallant Charge Repulsed," 33.

then press them and shoot them in their backs as they run to their main line; then charge the enemy's works." Cleburne allegedly replied, "General, I will take the works or fall in the effort."[64]

Getting the troops in position took a couple of hours, roughly 2 to 4 P.M., Stewart's Corps on the right, Cheatham's on the left. Cavalry guarded both flanks. If the attack broke the enemy line, Forrest with the heavier force (Red Jackson's and Abraham Buford's divisions) on the right was to cross the river and chase the fleeing Yankees. The six infantry divisions aligned from the right: Loring—Walthall—French—Cleburne—Brown—Bate, eighteen brigades in all, approximately 18,000 troops. As James Knight reminds us, counting Forrest's 4,000 cavalrymen and Allegheny Johnson's division, which would arrive later and be thrown into the battle, some 25,000 men would take part at Franklin, as many as Lee had committed to Longstreet's assault at Second Manassas (Lee's largest attack of the war).

"The soldiers were full of ardor, and confident of success," S. A. Cunningham remembered after the war. There was even joshing around during the deployment. When one learned Confederate raised his voice to quote Admiral Horatio Nelson's famous cry at Trafalgar, "England expects every man to do his duty," an Irish wag in the 1st/4th Missouri (with its large component of St. Louis Irishmen) replied, "It's damn little duty England would get out of this Irish crowd." Lending lightness to the air, regimental bands played "Dixie" and "The Bonnie Blue Flag." G. C. Phillips, surgeon of the 22nd Mississippi, also heard "The Girl I Left Behind Me," and later noted, "this was the first and only time I ever heard our bands playing upon a battlefield and at the beginning of a charge." Another writer recalled that in addition to these three songs, he heard the sound of a tune called "Ben Bolt."

Cleburne rode along his lines, reminding his men to save ammunition until the very last. But just before, when Cleburne addressed his brigade leaders with orders for the charge, he seemed "greatly depressed" to one observer. Govan caught on that there would be many men lost today. "Well, General, few of us will ever return to Arkansas to tell the story of

[64] Jacobson, "Hood's Tennessee Campaign," 42; Jacobson, *For Cause & For Country*, 241–43; Hay, *Cleburne and His Command*, 280–81 ("I do not like the looks of this fight," Cleburne's remark on Winstead Hill, which I place at the Harrison house); Horn, *Army of Tennessee*, 397–98 (Cheatham's comment on Winstead Hill); Henry, *"First With the Most,"* 397; Wyeth, *Life of Forrest*, 544.

this battle," he remarked to Cleburne. "Well, Govan," came the reply, "if we are to die, let us die like men."[65]

At 4 p.m. a flag signal from Winstead Hill gave the order to advance. General Stewart remembered that a staff officer sent by Hood brought the word. To the left of his corps were Cleburne's and Brown's divisions marching forward to battle. "Bands were playing, general and staff officers and gallant couriers were riding in front of and between the lines, 100 battle flags were waving...while 20,000 brave men were marching in perfect order against the foe. The sight inspired every man," one of Brown's regimental commanders wrote. Even the Yankees held their breath. "It was worth a year of one's lifetime to witness these first steps of the assault," recalled one Northern soldier; "emerging from the woods in the most perfect order, two lines, I think nothing could be more suggestive of strength and discipline and resistless power than was this long line of gray advancing over the plain."

Moving forward in the Confederate left-center, Cleburne's and Brown's troops came under enemy fire first from two brigades of George Wagner's division, which were pitched forward of the main Union line (reminiscent of Sickles at Gettysburg). Wagner's artillery fired some rounds, then mounted up to withdraw back to the works. Not so the infantry, which, under orders from Wagner, stayed in their exposed position. They put up a short fight, but when they were overwhelmed by the charging Rebels, they broke and ran. Cleburne and Brown paused briefly to reform their lines then sent their troops forward, chasing after the Yankees. "We will go into the works with them!" was the cry. Cleburne's horse was killed some eighty yards from the works. The general got up and exhorted his men to follow, waving his kepi and sword. Thirty yards farther Cleburne was shot through the chest and fell dead.

---

[65] Jacobson, "Hood's Tennessee Campaign," 42 ("the next couple of hours the Confederate army began to form up for battle"), 59 (Roth's map showing array of the eighteen infantry brigades); Hay, *Hood's Tennessee Campaign*, 246 (number of troops engaged); Sword, *Embrace an Angry Wind*, 181–84; Knight, *Battle of Franklin*, 68; Hennessy, *Return to Bull Run*, 364, 458; S. A. Cunningham, "Battle of Franklin," 101–102; P. T. Tucker, *Irish Confederates*, 37; G. C. Phillips, "Witness to the Battle of Franklin," 262; B. W. Lane, *Where No Sorrows Come*, 180; Patterson, "Franklin," 117; W. A. Washburn, "Cleburne's Division at Franklin," 27; Sutherland, "No Better Officer in the Confederacy," 246; Barefoot, *Let Us Die Like Brave Men*, 227.

Wagner's men, running ahead of the Southerners in "indescribable confusion" (General Stanley's words), prevented some of their comrades in the main line from opening fire. The Confederates thus broke into the enemy line on the heels of Wagner's retreaters. For a while they occupied some two hundred yards of the works in the very center of the Union line, astride the Columbia pike near the Carter house and cotton gin. Fierce hand-to-hand fighting erupted. General Hiram Granbury was shot just below the right eye and died instantly about the same time as Cleburne. In Gist's Brigade, the flag of the 24th South Carolina was planted on the parapet just after General Gist was shot in the chest. Brigadier O. F. Strahl was struck in the head trying to hoist his men from the ditch to the top of the parapet so they could fire. One of his soldiers later recalled the scene:

> I was near General Strahl, who stood in the ditch, and handed up guns to those posted to fire them. I had passed to him my short Enfield (noted in the regiment) about the sixth time. The man who had been firing cocked it and was taking deliberate aim, when he was shot and tumbled down dead in the ditch upon those killed before him. When the men so exposed were shot down, their places were supplied by volunteers until these were exhausted, and it was necessary for Gen. Strahl to call upon others. He turned to me, and though I was several feet back from the ditch, I rose up immediately, and walking over the wounded and dead, took position with one foot upon the pile of bodies of my dead fellows, and the others in the embankment, and fired guns which the General himself handed up to me until he, too, was shot down.... I asked the General for counsel, he simply answered, "Keep firing." But just as the man to my right was shot, and fell against me with terrible groans, Gen. Strahl was shot. He threw up his hands, falling on his face, and I thought him dead, but in asking the dying man, who still lay against my shoulder as he sank forever, how he was wounded, the General, who had not been killed, thinking my question was to him, raised up saying that he was shot in the neck, and called for Col. Stafford to turn over his command. He crawled over the dead, the ditch being three deep, about twenty feet to where Col. Stafford was. His staff officers started to carry him to the rear, but he received another shot, and directly the third, which killed him instantly.

Brigadier John C. Carter, also in Brown's Division, was shot from his horse and borne to the rear. "Officers and men fell like dead leaves when forests are shaken," remembered a soldier in the 6th Tennessee, in Carter's Brigade.

A counterattack by Reilly's brigade, coupled with the fire of the 6th Ohio Battery, stopped the Confederates from charging farther. Emerson Opdycke's brigade, held in reserve for this sort of emergency, also counterattacked. Once the Federals had regained their works, the earthen protection gave them the advantage. Union General David Stanley took a bullet wound in the back of the neck but left the field under his own power. In the growing twilight those Confederates still standing began to fall back. Many, such as Brig. Gen. George Gordon, fell prisoner when they could not safely leave the ditch outside the Yankee works. Others clung to the ditch till darkness. Sumner Cunningham of the 41st Tennessee (Strahl's Brigade) remembered that the ditch "was almost full of dead men."

To the left of Cleburne and Brown, Bate's attack struck the Federal line later, around 5 P.M., virtually by itself. Federal cannon fire was particularly cruel against Bate's troops. As they did elsewhere that afternoon, some Confederates reached the enemy works, planting flags on the parapet. But they could not hold their position. Union General Ruger, in charge of that sector, saw the Rebels "gradually waste and disappear." Among the Southerners left on the field was Capt. Theodrick (Tod) Carter, whose family owned the land at the center of the battlefield. Severely wounded, he was found that night and carried to the Carter house, where he died.

On the Confederate right, the assault of Stewart's three divisions (Loring, Walthall, and French) was bloodied by artillery fire both from the enemy line and from the fort across the river. "At every discharge of the guns," one Union artilleryman later noted, "there were two sounds—first the explosion and then the crack of the bones." Merciless musketry also cut down the Confederates struggling to get through the Osage orange hedge and thick *abatis.* Those who did charged into the ditch in front of the enemy parapet only to be shot down there. Color bearers of the 3d and 22d Mississippi planted their colors on the enemy works. They were shot down, and their flags fell into enemy hands.

Brigadier General Adams rode straight for the Federal works and received a volley of musketry as he mounted them. He fell mortally wounded; his horse, Old Charley, was killed and lay atop the enemy parapet. Lt. Col. Edward A. Baker of the 66th Indiana described the general's death: "Gen. Adams rode up to our works and, cheering his men, made an attempt to leap his horse over them. The horse fell dead upon the top of the embankment and the General was caught under him pierced with seven, maybe nine bullets." "As soon as the charge was repulsed," Baker went on to write, "our men sprang upon the works and lifted the horse, while others dragged the General from under him. He was perfectly conscious, and knew his fate. He asked for water, as all dying men do in battle." Two Federals brought water and a clump of cotton as pillow for Adams. "It is the fate of a soldier to die for his country," are said to be his last words.

Federal Col. John Casement described the "terrible slaughter" his soldiers inflicted upon Francis Cockrell's Missouri brigade. General Cockrell was severely wounded; half of his brigade became casualties. Major General Walthall described the enemy rifle and cannon fire as the most deadly he had ever seen; nevertheless his men closed on the enemy works. "Numbers of every brigade gained the ditch and there continued the struggle with but the earth-work separating them from the enemy until late in the night," he reported. Winfield Featherston's brigade lost a third of its strength, taking enfilading artillery fire from the enemy fort across the river. Losses among Confederate officers were frightful. Brig. Gen. Thomas Scott suffered a concussion from an artillery shell so severe that he would never again serve in the field. Ultimately those Southerners not captured sought to escape back to their lines. General Loring rode among those fleeing and tried in vain to turn them around. The assault of Stewart's Corps had been decisively repulsed.

The fighting continued well past sunset. When Ed Johnson's division of Lee's Corps came up around 7 o'clock, it too was ordered to move forward despite the looming darkness ("a better word would be 'grope' forward," comment Connelly and McDonough). Nevertheless, the Confederates launched an assault in which Brig. Gens. Arthur Manigault, Zachariah Deas, and Jacob Sharp were all wounded. Sharp's Mississippians reached the enemy works and in hand-to-hand fighting captured three enemy flags. But they soon faltered and were forced to fall back. By 9 P.M., after five hours of hellish fury—when Confederate James Dinkins said "the devil had full possession of the earth"—the battle of Franklin was over.[66]

---

[66] Stewart report, January 20, 1865, *OR*, vol. 45, pt. 1, 708; Sword, *Embrace an Angry Wind*, 187 ("bands were playing"); J. N. Beach, "From the Union Side at Franklin," 239; Hay, *Hood's Tennessee Campaign*, 121–25; Hay, *Cleburne*

General Hood had watched it first from Winstead Hill, then from near the Neeley house to its north. "The troops moved forward most gallantly to the attack," he wrote in his campaign report, although it is doubtful he could have seen very much after his troops closed with the enemy, given the thick smoke hanging over the battlefield. "The field became blanketed in smoke after the battle began and was heaviest where the fighting was the worst," Jacobson has written. "There was little wind and the only time the smoke cleared was when artillery fire or musket volleys cut through the air." In this situation, the commanding general awaited reports from his field officers.[67]

The reports were bleak. Around midnight at his headquarters, Hood heard from his three corps commanders. Stewart told him that his was "all cut to pieces; that there was no organization left except with the artillery." Cheatham had just come from a dolorous torchlit ride across the darkened battlefield; according to one soldier, "great big tears ran down his cheeks and he sobbed like a child." He therefore appeared before Hood "despondent and gloomy." Only S. D. Lee could offer anything positive. "Are you, too, going back on me?" Hood is said to have angrily asked. Lee announced that while Johnson's Division was "badly

---

and His Command, 281–82; Jacobson, "Hood's Tennessee Campaign," 44; Jacobson, *For Cause & For Country*, 255–56, 268 ("Wagner refused to call the brigades in"), 281–82, 286 ("terrible slaughter"), 287, 290, 294, 303, 309, 314, 318–19, 323 (Dinkins quote), 326, 333–34 (Stanley), 339–40 (Gordon), 342–44 (Adams), 349 (General Carter), 353–54 (Strahl), 356, 363 (Tod Carter), 365, 381–83; "Death of Gen. Strahl," 31; Cartwright, "Franklin: The Valley of Death," 277 (Cleburne's death); Brennan, "The Battle of Franklin," 43 (Granbury's wound); Walthall report, January 14, 1865, *OR*, vol. 45, pt. 1, 721; Robison, "The Carter House: Focus of the Battle of Franklin," 16; Knight, *Battle of Franklin*, 74; "Gen. John Adams at Franklin," 300–301; L. R. Tucker, *Brigadier General John Adams, CSA*, 197–98, 217 n.24; B. Lane, "The Familiar Road," 46; B. W. Lane, *Tennessee Hero*, 150, 152; R. Gates, "Sixth Tennessee Infantry," 1:222 ("when forests are shaken"); Sword, *Embrace an Angry Wind*, 217 (flags of the 3d and 22d Mississippi); McDonough and Connelly, *Five Tragic Hours*, 150–51; Lee report, January 30, 1865, *OR*, vol. 45, pt. 1, 688 ("three stand of colors"); Dinkins, *1861 to 1865*, 236.

[67] Jacobson, *For Cause & For Country*, 261, 296 (smoke); Hood report, February 15, *OR*, vol. 45, pt. 1, 653; Connelly, *Autumn of Glory*, 505 ("little could be seen").

cut up," his other two, Clayton's and Stevenson's, could be used for any coming work.

Everyone must have been stunned to hear that Hood intended to continue the battle on the morrow. He ordered the army's artillery, now up, to swing into position for a morning barrage. Each gun was to fire a hundred rounds beginning at 7 o'clock, and the infantry assault—"which it was announced was to be made by the entire army," General Stevenson noted in his report—was to follow two hours later. Hood's decision clearly demonstrated his determination to damage or destroy the Union IV and XXIII Corps before they retreated back to the safety of Nashville.

General Schofield saved him the trouble; at 11 P.M. on the 30th his troops started quietly to withdraw from their positions, headed northward for junction with Thomas's newly formed army. Artillery wheels were wrapped in blankets to muffle their noise on the stone pike; padding was also laid across the wagon bridge. All the Federal troops were across the Harpeth by 2 A.M., when the rear guard set fire to the bridges. Flames from the railroad bridge alerted Confederates that the Yankees were leaving Franklin.[68]

---

[68] Groom, *Shrouds of Glory*, 208; Losson, *Tennessee's Forgotten Warriors*, 226–27; McDonough and Connelly, *Five Tragic Hours*, 155; Stevenson, "Report of Major General Carter L. Stevenson of the Tennessee Campaign," 163 (also in *OR*, vol. 45, pt. 1, 694); McMurry, *Hood*, 176 ("Hood planned to renew the attack in the morning"); Schofield report, December 31, *OR*, vol. 45, pt. 1, 344; Connelly, *Autumn of Glory*, 506; Thomas to Schofield, November 30, *OR*, vol. 45, pt. 1, 1171 ("send back your trains to this place at once, and hold your troops in readiness to march to Brentwood, and thence to this place"); Sword, *Embrace an Angry Wind*, 254–57.

Chapter 4

December 1864

Confederates began entering the town around 4 A.M. on December 1, and were in full control by dawn. Their first task that day was to gather up the wounded still on the battlefield. The men of Cheatham's and Stewart's corps had this task (Lee's Corps, with Forrest's cavalry, had been sent off in pursuit of Schofield). The injured were carried to all nearby buildings, which surgeons commandeered as field hospitals. So many amputations were performed that severed limbs were hauled away in wagons. The large McGavock home, Carnton, soon had hundreds of wounded throughout its rooms, veranda, and yard. The McGavocks gave up their bed linen, even personal clothing, for use in bandaging the soldiers' wounds. The house of Fountain Carter, in the middle of the battlefield, was similarly overwhelmed. It was here that a wounded Capt. Tod Carter was brought and here where he died. In the town itself churches, schools, and other buildings were made into hospitals—forty-four in all (including three for wounded Federals). The courthouse, for example, became Brown's division hospital.

After the wounded were cared for, the dead were tended to. Confederates owned the battlefield, and upon them fell the responsibility for its clean-up. Ditches—none too deep, maybe two feet—were dug near where Southern soldiers lay; at some places long trenches accommodated the bodies. There were so many that at some places they had to be laid in two layers. Dead Confederates, placed on their backs, got a piece of cloth or blanket to cover their faces; hands were folded, and soil was then shoveled over them. When known, headboards were placed with what information the burial details could glean. Dead Federals received less ceremony; their corpses were often thrown in the ditches of their works, with dirt dragged from the parapets to cover them. In other places, fallen Rebs and fallen Yanks were interred together, so quick did the work have to be done. All of this labor was as emotionally draining as it was physically exhausting. One Southerner, Capt. James L. Cooper, recorded that he and his comrades were "so disheartened by gazing on that scene of slaughter that they had not the nerve for the work before them." The dreadful labor took all day December 1 and, in some cases, even into the next.

As Hood rode into town—a boy years later recalled "his long, tawny mustache and whiskers"—he could see for himself all the dead soldiers lying on the battlefield. "His sturdy visage assumed a melancholy appearance," one soldier later said, "and for a considerable time he sat on his horse and wept like a child."

It would take some time for casualty reports to come in, but the identification of officers killed and wounded took priority. For this reason on December 1, after setting up headquarters in town at the home of William Sykes, Hood had his A. A. G. Mason (recently promoted to colonel) issue a circular asking each corps commander to send in a list of all division, brigade, and regimental officers who were casualties of the battle. In his first report to General Beauregard and Secretary Seddon, sent by wire on December 3, he named as killed Major General Cleburne, brigadiers John Adams, S. R. Gist, Otho Strahl, and Hiram Granbury; Major General Brown, brigadier generals Carter, Manigault, Quarles, Cockrell, and Scott wounded; and Brigadier General Gordon captured. (He omitted brigadier generals Zachariah C. Deas and Jacob Sharp; John C. Carter died of his wound on December 10.) In his first message to Richmond after the battle, Hood offered no numerical estimate of the army's casualties, stating only, "we have to lament the loss of many gallant officers and brave men."

They were many indeed. In addition to the fourteen general officers, fully fifty-four regimental commanders were killed, wounded, captured, or missing: twenty-three colonels, eleven lieutenant colonels, eleven majors, nine captains. Of these fifty-four, half were either dead or captured/missing and thus lost to the service; the other twenty-seven might presumably return for duty. Even this enumeration, ghastly as it was, did not include important officers who were not in regimental command; Lt. Col. J. S. Jones of the 24th South Carolina, Ellison Capers's second-in-charge, died in hospital a few days after the battle. In the words of Richard McMurry, "the leadership of the army had been shot away." The disastrous loss of leadership would have major repercussions, as Andrew Bledsoe points out: "the Battle of Franklin made a shambles of the Army of Tennessee's command structure, eliminating many of its finest combat leaders at a time when they were most urgently needed and leaving pitifully weakened for the remainder of the Tennessee Campaign."

"Our loss in killed, wounded, and prisoners was 4,500," Hood acknowledged in his campaign report, but the actual number was much higher, closer to 6,000. Of the seven infantry divisions participating in the battle, we have casualty reports from five:

DIVISION	REPORTED KILLED	REPORTED WOUNDED	REPORTED MISSING	TOTAL
Johnson	127	424	36	587
Loring	151	619	106	876
French	128	397	127	652
Walthall	128	304	148	580
Bate	47	253	19	319
Totals	**581**	**1,997**	**436**	**3,014**

After the war, Brig. Gen. George W. Gordon, who had himself been captured at Franklin, spoke on the battle and cited percentage losses for Cleburne's division (52 percent) and Brown's (31 percent) although he did not disclose how he got these figures. Using them, Jacobson calculates Cleburne's loss at Franklin as approximately 1,500 and Brown's as 1,150. Adding in 135 casualties in Forrest's cavalry, the Confederate total becomes a staggering 5,999. "About six thousand of the rank and file lay dead or disabled on the field at the close of that memorable day," declared General Gordon.

I am generally reluctant to accept the enemy's casualty counts, but General Schofield's claim is corroborative: After Federal forces reoccupied Franklin on December 18, they counted "1,750 buried upon the field, 3,800 disabled and placed in hospitals in Franklin, and 702 prisoners, making 6,252 of the enemy placed hors de combat." These figures are accepted by Eric Jacobson as "the Army of Tennessee's total loss." As Union General Cox later noted, the count of some 1,750 Southerners killed in several hours at Franklin exceeds such casualties in far longer battles—greater than Rosecrans's at Stones River (1,677, according to

Livermore) and Chickamauga (1,657). Besides his battle losses on November 30, Hood's army was weakened further by straggling on its march into Tennessee. A week and a half after Franklin, the army's statistical returns showed almost 8,800 fewer officers and men present for duty than the previous count taken two weeks before the army left Florence, a reduction in strength of a full one-quarter.

Schofield's casualties at Franklin were reported as 189 killed, 1,033 wounded, and 1,104 missing for an aggregate of 2,326. The key lay in the number of killed and wounded: for Confederates, 5,550; for Federals, 1,222. The gross disparity is best explained by the strength of the Northerners' fortifications; many Southerners died in the ditches in front of the enemy parapets, unable to climb out.

On the day after the bloody engagement, Southern officers and men who survived were shocked by the sight of so many corpses on the battlefield. "The ground about the Yankee works was literally piled with dead," observed Captain Cooper. "The place looked like one vast slaughter pen," he wrote, adding that "after gazing on it, I felt sick at heart for days afterwards." Gradually the butcher's toll began to be known. Cleburne's Division, according to General Govan in a letter of December 4 to his wife, lost "over half of its officers and men." Such high casualties, including the death of their beloved leader, depressed spirits throughout the division. When Brig. Gen. James Smith returned from detached duty on December 6 to Cleburne's Division, he did not find its "tone and morale...desirable, owing to the fearful loss sustained in that battle." The Confederates' repulse at Franklin and their enormous casualties led Captain Foster in Granbury's Brigade to record in his diary the day after the battle, "Gen. Hood has betrayed us (The Army of Tenn.). This is not the kind of fighting he promised us at Tuscumbia and Florence Ala. when we started into Tenn." Because of such comments, Emory Thomas has generalized that as a result of Franklin, "Hood's troops had lost almost all confidence in their commander." That may be a bit much. Hood, for his part, chose to recognize those officers and men who had performed the extraordinarily heroic act of reaching the enemy's breastworks. A week after the battle, he issued a general field order asking commanders to submit "the names of officers and soldiers who passed over the enemy's interior line of works at Franklin," so he could send them to the War Department in Richmond and there be added to the army's Roll of

Honor. There is no indication, however, that such a list was compiled—the army was in the field, and the officers had other pressing demands.[1]

---

[1] Sword, *Embrace an Angry Wind*, 260–61; Jacobson, *For Cause & For Country*, 415–20; McDonough, *Nashville*, 137; Bailey, *Chessboard*, 109 (three hospitals for Federals); Elliott, *Doctor Quintard*, 192 (courthouse hospital); Lowry, *Dark and Cruel War*, 583 ("wept like a child"); "List of division, brigade, and regimental commanders killed, wounded, missing, and captured in the battle of November 30, at Franklin, Tenn.," *OR*, vol. 45, pt. 1, 684–86; Figuers, "A Boy's Impressions of the Battle of Franklin," 7; Mason circular, December 1, *OR*, vol. 45, pt. 2, 629; Hood to Seddon (and Beauregard), December 3, *OR*, vol. 45, pt. 2, 643–44; H. D. Riley, Jr., "A Gallant Adopted Son of Tennessee—General John C. Carter, C. S. A.," 205; Capers report, January 4, 1865, *OR*, vol. 45, pt. 1, 738; McMurry, *Hood*, 176; Bledsoe, "The Destruction of the Army of Tennessee's Officer Corps," 78; Hood report, February 15, 1865, *OR*, vol. 45, pt. 1, 654; Johnson, Loring, French, Walthall, and Forrest casualty reports, *OR*, vol. 45, pt. 1, 691, 715–16, 726, 761; Bate report, January 26, 1865, *OR*, vol. 45, pt. 1, 743; "Confederate Monument at Franklin," 9 ("Address of Gen. Gordon"); Crownover, "The Battle of Franklin," 291 (Cox on casualties); Army of Tennessee returns of November 6 and December 10, *OR*, vol. 45, pt. 1, 678–680; Cox, *Franklin*, 211 (re-occupation of Franklin, December 18); Jacobson, *For Cause & For Country*, 417 (5,550 Confederate killed and wounded to 1,222 Federal); Schofield report, December 31, *OR*, vol. 45, pt. 1, 343–44; J. L. Cooper, "Service with the Twentieth Tennessee," 181; Smith report, January 23, 1865, *OR*, vol. 45, pt. 1, 739; Foster, *One of Cleburne's Command*, 151; E. M. Thomas, *The Confederate Nation*, 281; Gen. Field Orders No. 39, December 6, 1864, *OR*, vol. 45, pt. 2, 654.

Six thousand KWM has become the accepted number of Confederate casualties at Franklin, cited as early as 1908 in a history of the battle (R. W. Banks, *The Battle of Franklin November 30, 1864*, 27, copy courtesy of Gregory I. Wade of Franklin, TN). A basic source is always Livermore: 1,750 killed, 3,800 wounded, 702 missing, excluding cavalry. Acceptance of his calculation of 21,897 troops engaged leads to the figure of nearly 29 percent of the army lost (Livermore, *Numbers and Losses*, 132). Connelly, McDonough, and Weigley all accept Livermore (*Autumn of Glory*, 506; *Five Tragic Hours*, 157, Weigley, *A Great Civil War*, 414). Winston Groom accepts 1,750 killed as well, and on that basis asserts, "Never in any single-day battle during the entire war had that many Confederate soldiers been slain" (*Shrouds of Glory*, 220). The validity of that claim depends upon one's count of Confederate killed at Sharpsburg, which Sears places as 1,546.

Nevertheless, when adding 1,750 Confederate dead to Schofield's 189, one may safely conclude that November 30, 1864, was one of the five deadliest days in American war history, along with September 17, 1862; September 11, 2001; December 7, 1941; and June 6, 1944 (Sears, *Landscape Turned Red*, 295–96 [3,654 dead at Antietam]; Lipton, "Struggle to Tally All 9/11 Dead" [2,983]; Axelrod, *America's Wars* [2,403 dead at Pearl Harbor]; Ryan, *Longest Day* [1,465]).

Eric Jacobson notes that the number of Americans killed at Franklin—"just short of 2,000"—is greater than the nearly 1,500 Americans killed on June 6, 1944 (*The McGavock Confederate Cemetery*, 15). After the battle, Hood's soldiers buried the dead where they fell. A year after the war, the McGavock family donated two acres for a graveyard and oversaw the disinterment and reburial of almost 1,500 soldiers' remains (Webb, "The Confederate Dead at Franklin," 157).

John B. Hood was always very mindful of "guns and colors captured from the enemy in time of battle" as "the most valuable trophies of war." He was just as concerned when the enemy captured the flags of his own troops. It was characteristic, then, of Hood that a week after the battle he issued a circular to his corps commanders, asking them to report on the number of flags lost at Franklin and by what regiments. He apparently got replies for only three divisions concerning eight stands of lost colors, but eventually he came to conclude that his army had lost thirteen at Franklin. After he learned that the Federals were claiming the capture of as many as thirty Confederate battle flags ("we captured 702 prisoners and 33 stand of colors," Schofield asserted), Hood wired Richmond that the enemy's boast was untrue: "we lost thirteen, capturing nearly the same number." To be sure, in the hand-to-hand fighting at the enemy works, Southerners seized some Yankee flags. S. D. Lee made a point in his campaign report of commending Sharp's Mississippi brigade for capturing three stands of colors. But Hood probably exaggerated the number of flags ("nearly" thirteen) captured by his troops on November 30. David Fraley (former director of Carter House at Franklin, engaged at the time of this writing in the vexillology of the battle) believes the Confederates captured fewer than ten Union flags. Even more certainly, Hood downplayed the number of banners lost by his troops at Franklin, which Howard Madaus concludes was at least seventeen to twenty-two. John Keegan is one historian who accepts Schofield's boast of thirty-three stands of colors captured (Hood's Gen. Orders No. 40, March 16, 1864, *OR*, vol. 32, pt. 3, 644; Circular, December 7, *OR*, vol. 45, pt. 2, 659; Schofield report, December 31, *OR*, vol. 45, pt. 1, 342; Hood to Seddon, December 15, *OR*, vol. 45, pt.2, 690; Lee report, January 30, 1865, *OR*, vol. 45, pt. 1, 688; Banks, *Battle of Franklin*, 36 ["to General 'Jake' Sharp and his command the honor for their capture is solely due"]; David Fraley to the author, April 29, 2015; Madaus, *The Battle Flags of the Confederate Army of Tennessee*, 137; Keegan, *Military History*, 276).

After the war Mississippian James Binford of General S. D. Lee's staff recalled that Lee regarded the battle of Franklin "the most unnecessary loss of human life" he had known. Schofield's retreat during the night after the fight is a major reason for Lee's statement: if Hood had not ordered an attack on the afternoon of November 30, the Yankees would have vanished by the next morning anyway. We have already put forth that the destruction of Schofield's two infantry corps was Hood's objective, not forcing their retreat back to Nashville. Yet John Bell Hood has been blamed for far more for butchery than, say, another Confederate army commander who also ordered a frontal infantry assault in the face of daunting odds: Robert E. Lee. General Lee and his reputation have weathered a century of criticism for his decision to launch a magnificent, foredoomed attack on the third day at Gettysburg. Some students of the war have even justified it. The question for us to ponder is whether Hood's decision to launch a magnificent, foredoomed attack at Franklin was also justified.

First of all, we must dispense with the useless baggage. McMurry's observation, "we will never know what went through Hood's mind that day as his men deployed for battle," is a little off the mark, for we can pretty well outline the various options he worked through (e.g., sending Forrest off to the right flanking, waiting for the artillery to come up). More troubling still are the hypotheses and guesses of other writers who attempt to read Hood's mind. McDonough and Connelly boldly enter the realm of psychology in asserting that Hood's decision at Franklin "was the act of a tormented man driven by visions and dreams" of medieval chivalry and romance. In *Autumn of Glory*, Connelly boldly claims that at Franklin Hood "was too emotionally unhinged to continue to command." His "irrational mind would not entertain the idea" of Forrest flanking the enemy. "Worse," Connelly continues, "Hood now sought to discipline his army by means of a frontal assault." Stephen M. Hood

---

The flag of the 2nd/6th Missouri, Cockrell's Brigade, captured by Sgt. Alfred Ransbottom, Co. K, 97th Ohio, is now at the Missouri State Museum in Jefferson City (Jacobson, *For Cause & For Country*, 386 n.24). It is pictured (tattered and in color) in Rollins, *Returned Battle Flags*, 42.

rightly points out that there is absolutely no documentary evidence for these "eyebrow-raising statements."[2]

After that, comparisons and contrasts are in order between Gettysburg and Franklin, as Hood's and Lee's decisions share similarities both big and small. Both faced criticism by subordinates. Just as Frank Cheatham had warned, "I don't like the looks of this fight. The Federals have an excellent position, and are well fortified," so too did Porter Alexander observe that the sight of the enemy line on Cemetery Ridge struck him as "very unfavorable ground for us, generally sloping toward the enemy." Longstreet's objection to the assault of the third day is too well-known for further comment.

Afterward, the two generals' decisions came in for wide criticism. Hood's at Franklin was "one of the most questionable command decisions of the entire Civil War," according to Bradley Clampitt. Wiley Sword is much more critical: "one of the most controversial attack orders in the history of warfare on the American continent." Lee's decision for July 3 was "his worst of the war," write Gary Gallagher and his colleagues. Both infantry charges were disastrous failures, with Lee and Hood sharing the same criticism. "Both Hood and Lee committed cardinal sins of command," writes recent chronicler James R. Knight: "they made unreasonable demands on their own men, and they underestimated the strength and courage of their opponents."

At the same time, just as the charge of Pickett, Pettigrew, and Trimble at Gettysburg has earned magniloquent tributes for Confederate soldiers' bravery, so too has the charge of Hood's divisions at Franklin. "Don't you recall, Porter," Frank Cheatham remarked after the war, "that as they wheeled into battle in full view of the enemy, their precision and military bearing was as beautiful a sight as was ever witnessed in war?" Some have even ranked Franklin more highly. "'Pickett's charge at Gettysburg' has come to be a synonym for unflinching courage in the raw. The slaughter-pen at Franklin even more deserves the gory honor," asserts Horn. The fierce combat at Franklin was "unexcelled by any other in the war," exclaims W. J. Wood. Vincent Esposito asserts that the officers of the Army of Tennessee had a special motivation in making their charge. "Nettled ever since Atlanta by Hood's sarcastic references to how

---

[2] Jacobson, *For Cause & For Country*, 406; McMurry, *Hood*, 175; McDonough and Connelly, *Five Tragic Hours*, 68; Connelly, *Autumn of Glory*, 502–503; Hood, *Rise, Fall, and Resurrection*, 139–40.

fortifications had dulled the army's offensive spirit," he writes, "the corps and division commanders were incensed. Though logic dictated against a frontal assault, they vowed to demonstrate that their troops could storm fortifications as bravely as any." The result was bloody hand-to-hand fighting both around the Carter House and at the angle of the stone wall on Cemetery Ridge; charging Confederates both at Franklin and at Gettysburg temporarily overran the enemy line before being bloodily pushed back.

I am scarcely the first to consider Franklin in context with Gettysburg. More than a half-century ago, Stanley Horn noted three important differences between the two infantry assaults. First, Lee was able to order a preparatory artillery barrage before his infantry advance. Hood could not: most of his cannon were miles behind with Lee's Corps, and time was running out on the afternoon of November 30. Hay's statement that "Hood, most unwisely, did not precede his attack with artillery preparation" misses the point that most of the army's artillery did not arrive until after dark. By the time they got into position the next morning, the enemy had abandoned Franklin. Besides, there was no certainty that a barrage, even if Hood had been able to order one, might have tipped the tactical scales his way for the subsequent infantry charge. At Gettysburg the Confederate bombardment on July 3 knocked some Union cannon out of action but did no damage to other batteries that could enfilade the advancing Southern infantry. Moreover, firing high, the Confederate artillerists did nothing to break up or even demoralize the Union infantry waiting behind the stone wall that would be the focus of Lee's attack. Northern soldiers called the high-flying shells headed for their rear "quartermaster hunters"; Capt. Justus Scheibert, the Prussian observer in Lee's lines, called the entire cannonade *pulververschwendung*, a "waste of powder." Failure of the Confederates' bombardment at Gettysburg has usually been ascribed to their dwindling stock of ammunition; for this General Lee himself has recently come in for criticism. Peter Carmichael calls Lee's failure to check on his artillery ammo supply an "unconscionable oversight." Other students have called attention not so much to the quantity of the Confederates' artillery ammunition on July 3 as to its inferior quality. Porter Alexander wrote after the war that when Southern artillerists sought to fire over the heads of their infantry "premature explosions of shell & shrapnel" would inflict casualties among their own troops, to the point that the infantry refused to allow such indirect fire. Richard Rollins also points out that Lee's army did not carry into Pennsylvania enough rifled guns for long-range shelling. "The shortage of long-range rifled artillery was a major reason why the Confederate softening-up barrage just prior to Pickett's charge was, in the words of one Union artillery officer, 'the biggest humbug of the season,'" writes Charles. B. Dew, the historian of Richmond's Tredegar Iron Works. For all of these reasons, the Confederates' pre-assault bombardment at Gettysburg had little real effect. A similar pre-assault bombardment by Hood at Franklin, had he the guns and the time for it, might similarly have had little effect.

It is also worth noting that during the two assaults, Confederate artillery was much less engaged than the Federals' guns. At Franklin, Hood's charge was supported by only two batteries. Four of the six guns of Capt. Henry Guibor's Missouri Battery were placed to support the attack of Cockrell's Missouri brigade. But these, and the Confederates' other cannon, "fired very few shots and those at great distances," according to one commentator on the battle. Similarly, at Gettysburg, after the preparatory Confederate artillery barrage, only eighteen guns were moved forward to support Pickett's attack, so low was the supply of ammunition.[3]

For both charges, the ground to be covered by Confederates was similar. Between Seminary and Cemetery ridges, Stewart notes, "the land was low, offering no physical barrier." "Mostly flat and almost completely

---

[3] Losson, *Tennessee's Forgotten Warriors*, 218; E. P. Alexander, *Fighting for the Confederacy*, 245, 248; Eckenrode and Conrad, *Lee's War Horse*, 203 ("Longstreet did not think it possible for the assault to succeed"); Clampitt, *Confederate Heartland*, 117; Sword, *Embrace an Angry Wind*, 262; Gallagher, Engle, Krick, and Glatthaar, *Mighty Scourge*, 179; Knight, *Battle of Franklin*, 56; Horn, *Army of Tennessee*, 403; Wood, *Civil War Generalship*, 205; Esposito, *The West Point Atlas of American Wars*, 1:152; Hay, *Hood's Tennessee Campaign*, 134; Groom, *Shrouds of Glory*, 217 (noting that Lee's artillery barrage at Gettysburg was not effective); G. R. Stewart, *Pickett's Charge*, 137, 160; Carmichael, "'Every Map of the Field Cries Out about It,'" 281–83; Glatthaar, "Battlefield Tactics," 71 ("defective ammunition"); Rollins, "The Failure of Confederate Artillery in Pickett's Charge," 30 ("bad ammunition and fuses"); Rollins, "The Failure of Confederate Artillery at Gettysburg," 54 ("relatively low numbers of long-range weapons"); Dew, *Ironmaker to the Confederacy*, 196; Jacobson, *For Cause & For Country*, 113, 212–213, 257, 285, 307, 330, 362 (nine Union batteries, 38 guns engaged); Marshall, "Artillery in the Battle of Franklin," 101–102 (placement of the Federals' 38 guns).

clear of obstructions," Jacobson writes of Franklin. Horn and others note that the distance between the opposing lines was one mile at Gettysburg and two at Franklin, but this second tactical difference also proved of little consequence. For instance, the longer distance covered by Hood's men did not expose them to more extensive artillery fire. Loring's troops came under fire from Fort Granger when they were about a thousand yards from the enemy line; Pickett's left came under fire at about the same range, 1,100 yards. Moreover, the flat terrain of the Confederates' approach route south of Franklin and the absence there of thick undergrowth (of the kind, say, which had impeded their attack July 22 east of Atlanta) meant that the troops were not blown and winded for their final charge. The literature of Franklin has no counterpart to Brig. Gen. Richard B. Garnett of Pickett's Division, who called out, "Steady, men! Save your wind and ammunition for the final charge!"[4]

Horn's third tactical difference is much more important. Meade's infantry on Cemetery Ridge were virtually unfortified, only making use of the low stone wall, whereas Schofield's troops had built earthen parapets with headlogs, impeding ditches, and abatis. This factor does much to explain the disparity in casualties at Franklin: 1,200 Federals killed or wounded as opposed to some 5,550 Confederates. This ratio of about 4½-to-1 was the statistical norm for Civil War infantry attacking fortified defenders. Yet it is worth remembering that in terms of absolute numbers, Southern casualties in Lee's and Hood's two charges were about equal: 6,252 Confederates killed, wounded, and missing/captured at Franklin (Livermore, Weigley, Jacobson) and 6,200–6,400 K W M at Gettysburg (Stewart, Hess).[5]

What has generally not been commented upon is that Lee's casualties as percentage of numbers engaged were higher than Hood's: 6,200–6,400 of 11,700 in Lee's eleven attacking brigades amounted to 54 percent. Of Hood's forces, 6,250 of 21,000 (counting Edward Johnson's

---

[4] Stewart, *Pickett's Charge*, 12; Jacobson, *For Cause & For Country*, 238 ("land was low"); Sword, *Embrace an Angry Wind*, 216; Eicher, *The Longest Night*, 774 ("over two miles of open ground" [as opposed to one mile at Gettysburg]); Frey, *Longstreet's Assault—Pickett's Charge*, 126, 136.

[5] Horn, *Army of Tennessee*, 403; Livermore, *Numbers and Losses*, 131–32; Weigley, *Great Civil War*, 414; Jacobson, *For Cause & For Country*, 417; Stewart, *Pickett's Charge*, 263; Hess, *Pickett's Charge—The Last Attack at Gettysburg*, 333–35.

division) was 30 percent. On the other hand, an important exception involves general officer casualties. The Army of Tennessee lost at Franklin fourteen generals killed, wounded, or captured. The Army of Northern Virginia lost in the charge of Pickett, Pettigrew, and Trimble five generals. A key reason was that General Lee ordered no officers to ride into the charge on horseback; mounted officers always draw enemy fire. Despite the orders, though, several of Lee's officers did, notably Brig. Gen. Richard Brooke Garnett, who was killed in the attack. But Hood issued no orders against horseback, with the tragic result, as we have seen, of the deaths of Maj. Gen. Cleburne, brigadier generals Adams and Carter, and other mounted officers.[6]

Wiley Sword elaborates a number of Gettysburg-Franklin points of comparison, such as ground distance to be covered, artillery bombardment, fortifications, partial tactical success, and casualties. One that he does not bring up is that the numbers of Hood's and Schofield's troops engaged were about equal while Lee's attackers outnumbered the enemy two to one. Livermore counts Hood's infantry and artillery as 21,539 to

---

[6] Jacobson, *For Cause & For Country*, 418–20; S. Davis, "The Death and Burials of General Richard Brooke Garnett," 107; Coddington, *Gettysburg Campaign*, 792 n.52 (counting at least fifteen mounted officers on July 3).

Decades ago a legend developed that bodies of the five dead Southern generals were gathered from the field and placed on the back porch of Carnton, the home of John and Carrie McGavock (Catton, *The American Heritage Picture History of the Civil War*, 555). Since then, researchers have concluded that after the battle the McGavocks' porch held the bodies of generals Cleburne, Adams, Granbury, and Strahl, as well as Col. Robert B. Young, 10th Texas, and of Lt. John H. Marsh of General Strahl's staff (C. H. Brown, "The Myth of the 5 Dead Rebel Generals," 15; Jacobson, *For Cause & For Country*, 411–12). General Adams's body was carried to Pulaski, Tennessee, his wife's hometown. After General Gist died in a field hospital on the night of the battle, he was buried in a nearby family cemetery, but weeks later his remains were borne to Columbia, South Carolina. Generals Cleburne, Granbury, and Strahl were first interred at Rose Hill Cemetery in Columbia, Tennessee. But Chaplain Quintard objected to their having been placed so near dead Federal soldiers and soon saw that the three bodies were taken to Ashwood, Tennessee, several miles away, and laid to rest in the cemetery of St. John's Episcopal Church. But "rest" for each was temporary. After the war, Strahl's remains were moved and reinterred in Dyersburg, Tennessee; Cleburne's to his hometown of Helena, Arkansas; and Granbury's similarly to Granbury, Texas (S. Davis, "Franklin: Where Did Those Dead Generals Go?" 15).

Schofield's 21,897. Sword counts 22,000 Federals versus 23,000 Confederates. Hay counts about 20,200 Confederate infantry and artillery (seven divisions and two batteries) against 22,830 Federal infantry in the IV and XXIII corps. Lee's assault column at Gettysburg, on the other hand, had a two-to-one advantage. The Confederates' eleven assault and support brigades at Gettysburg numbered as many as 11,830 troops; Earl Hess counts 5,500 to 6,000 men in the eight Federal infantry brigades and five batteries engaged in repulsing the Rebel assault: "less than half as many Union infantrymen and roughly an equal amount of Union artillery power were involved, compared with their opponents."[7]

Russell Weigley states that Hood's troops had to cover two miles, "about twice the distance Pickett's Charge had to travel at Gettysburg," but he errs in stating that they advanced "on a narrower front." The point invites a further tactical contrast because the Confederates' attack front at Franklin was much *wider* than at Gettysburg. The point is important, for it meant that on July 3 Union infantry and artillery were able to deliver oblique and enfilading fire on the seven Confederate brigades that advanced farthest against the Union center. According to the Vitelli map in Coddington, their front extended a third of a mile. Indeed, Vermont and Ohio regiments advanced out into the open field against both flanks of Pickett's and Pettigrew's attackers. As Coddington puts it, "The shelling was bad enough, but now musketry! The combined fire was too much." The wilting of both flanks of the assault column was a major cause of failure for Lee's attack. On the other hand, Hood's attack front on November 30 extended virtually across Schofield's entire fortified line, two miles long. For the most part this prevented the kind of enfilade fire that crippled the Confederate infantry at Gettysburg. When Federals at Franklin were able to deliver a raking fire (as against Strahl's Brigade), they were able to do so because of the tactical configurations of their works, not because the length of their line overlapped that of the Rebels. Similarly, the enfilading cannonade that Loring's right took from the

[7] Sword, *Embrace an Angry Wind*, 269, 270; Livermore, *Numbers and Losses*, 131–32; Hay, *Hood's Tennessee Campaign*, 246–47; Hess, *Pickett's Charge*, 333–37.

In light of these figures, one must discount the hypothesis that in Lee's Pennsylvania campaign, two infantry brigades left behind at Richmond might have brought Pickett success had they been with their division on July 3. Philip Thomas Tucker repeats the assertion in *Pickett's Charge*, 91, 369–71.

guns across the river at Fort Granger was not nearly so hurtful to Hood's assault as was the enfilading Union artillery fire from Cemetery Hill. Pettigrew's exposed left flank presented for the Federals an "artillerists' dream," as Sears has phrased it: the opportunity to fire shot and shell into the flanks of advancing Rebel infantry. "I could sight down the entire length of their line," one Northern artillery officer recalled. Under such enfilade fire against the left of the assault wave, "Pettigrew's attacking wing was crippled before its men could fire a shot."[8]

There is yet another point to make: Lee did not use all his available infantry on July 3 while Hood very much did at Franklin. In addition to the nine brigades of Pickett's, Pettigrew's, and Trimble's divisions (plus Perry's and Wilcox's brigades of Anderson's division), Lee had in mind drawing support from Anderson's other three brigades, and more troops from three other divisions, according to Noah Andre Trudeau. "Lee's planning had provided for a second assault group nearly as strong as the first," another 11,000 soldiers from Longstreet who were not called into the battle on July 3. Wert concurs: "Lee apparently considered the fourteen brigades assigned to Longstreet's control as an integrated force that would be used in the assault." Cavalry brigadier John Imboden later related that General Lee on the night of July 3 told him that if Pickett, Pettigrew, and Trimble "had been supported as they were to have been...the day would have been ours." Lee may or may not have explicitly ordered such support troops for his attack, but there can be no doubt that at Franklin Hood used all his available infantry.

Jonathan M. Steplyk summarizes major points of contrast between the Confederate frontal assaults at Gettysburg and Franklin, contending that Hood's attack "surpassed the Gettysburg climax in size and scope":

> Pickett's Charge consisted of about twelve thousand men attacking across three-quarters of a mile, compared to twenty thousand veterans of the Army of Tennessee covering two miles of ground. The assault by the divisions of George Pickett, J. Johnston Pettigrew, and Isaac Trimble was preceded by a massive artillery bombardment, while the Army of Tennessee attacked with minimal artillery support. Moreover, Confederate forces in Pickett's Charge—attacking Union troops protected by only minimal field works—secured only a limited lodg-

---

[8] Jacobson, *For Cause & For Country*, 352 ("able to sweep the length of Strahl's Brigade"); Weigley, *Great Civil War*, 414; Coddington, *Gettysburg Campaign*, f.p. 505 (Vitelli map), 506–507, 515–16; Sears, *Gettysburg*, 424–25.

ment in the enemy position before their repulse, whereas the Army of Tennessee, through audacity and serendipity, created a dangerous breach in the heavily fortified Union line at Franklin, one which was fiercely contested for several hours and into the night before it was ultimately closed. In short, the Confederate charge at the Battle of Franklin was "larger, longer, and deadlier" than the final attack at Gettysburg.[9]

Both generals afterward put the best faces on their failed assaults. From Lee's campaign report:

> The battle recommenced in the afternoon of the 3d, and raged with great violence until sunset. Our troops succeeded in entering the advanced works of the enemy, and getting possession of some of his batteries, but our artillery having nearly expended its ammunition, the attacking columns became exposed to the heavy fire of the numerous batteries near the summit of the ridge, and, after a most determined and gallant struggle, were compelled to relinquish their advantage, and fall back to their original positions with severe loss.

And Hood's:

> We carried the enemy's first line of hastily constructed works handsomely. We then advanced against his interior line, and succeeded in carrying it also in some places. Here the engagement was of the fiercest possible character. Our men possessed themselves of the exterior of the works, while the enemy held the interior. Many of our men were killed entirely inside the works. The brave men captured were taken inside his works in the edge of the town. The struggle lasted till near midnight, when the enemy abandoned his works and crossed the river, leaving his dead and wounded in our possession.[10]

But note that Hood was able to end his statement with the fact that the enemy had abandoned his works and skulked away in the dead of night, leaving Confederates in control of the battlefield. Outcome is thus the most important difference between the battles of Gettysburg and Franklin. After his attack on July 3 was repulsed, Lee had to retreat if only because his army, deep in enemy territory, could not long subsist—

---

[9] Stewart, *Pickett's Charge*, 172–73; Trudeau, *Gettysburg*, 480–81; Wert, *Gettysburg Day Three*, 243; Steplyk, "Killing at Franklin: Anatomy of Slaughter," 81.

[10] Lee report, July 31, 1863, *OR*, vol. 27, pt. 2, 308; Hood report, February 15, 1865, *OR*, vol. 45, pt. 1, 653.

"the difficulty of securing supplies rendered it impossible to continue longer where we were," as Lee himself phrased it. On the other hand, Schofield's nocturnal retreat back to Nashville hours after the battle allowed Confederates to claim victory at Franklin. "Many believed the Battle of Franklin to be a great victory," Larry Daniel asserts. "We gained a complete victory," Samuel K. Vann, one of Hood's soldiers, wrote home after the fight. Edward N. Brown expressed to Fannie the day after the battle his regret at the army's "frightful" losses, "yet we gave the enemy a signal defeat." Texan Robert F. Bunting agreed in his diary entry, writing of "our great victory at Franklin." On December 1, Chaplain Quintard entered into his diary, "Our troops with the most determined gallantry charged and carried three lines of entrenchments routing the enemy who fled this morning in confusion towards Nashville." Quintard's exaggeration was exceeded by another of Hood's soldiers. After the battle, Hardin Figuers, a boy in Franklin, came upon a severely wounded Confederate unable to speak. When asked if he needed help, the soldier scribbled, "No; John B. Hood will be in New York in three weeks."[11]

In the long view, all of these tactical comparisons and contrasts tend to balance out. Lee had a pre-assault artillery barrage, but it was ineffective. Hood employed more troops, indeed all of his available infantry; Lee wanted to throw in another 11,000 men (according to Andy Trudeau and Jeff Wert), but couldn't—had he done so, his assault numbers would have equaled Hood's. Lee's Confederates advanced over open ground, in full view of the enemy; so did the Army of Tennessee's. Hood's attackers had a greater distance to cover, but they did not suffer proportionally. In both Confederate attacks, the enemy enjoyed advantages of position. The Federals' line overlapped Lee's shorter infantry-front, allowing them to pour a bloody enfilade fire on both flanks of his attack force. Federals at Franklin were more fortified, with entrenchments, abatis and ditch, obstacles Hood's troops could not overcome.

Considering all of the facts, one may therefore conclude that John Bell Hood at Franklin had about the same chances of victory in ordering his frontal infantry assault as Robert E. Lee did in ordering his, a year

[11] Hess, *Pickett's Charge*, 342 (Lee quote); Daniel, *Soldiering*, 146; Clampitt, *Confederate Heartland*, 118–19 (Vann and Brown); Grear, "'Our Peoples Are Depressed in Spirit,'" 206; Elliott, *Doctor Quintard*, 187; Jacobson, *For Cause & For Country*, 406–407 (Hood in New York).

and a half earlier. According to one Southern newspaper correspondent writing just days after the battle, Hood rode along his troops' lines just before the battle. "These lines must be broken boys," he shouted. "They are weak and cannot stand you." This story, printed in an Augusta newspaper a few weeks after the battle, is not to be seen elsewhere in the literature. It probably is not true; we do not see Hood on the battlefield after his observation from Winstead Hill. But if Confederate soldiers believed that they heard Hood exhorting them to break the enemy line with their charge, there was hope in the Confederate ranks that they could do just that.

Neither Hood nor Lee should thus be blamed more than the other, and both grieved over their lost soldiers. To Lee's famous lament, "Too bad! Too bad! Oh! Too bad!" was Hood's "O my God! This awful, awful day!" Perhaps the most telling observation comes from General Schofield. On a copy of the text of the address given by Confederate General Chalmers, printed in the *Southern Historical Society Papers* in 1879, Schofield wrote, "Was it not, in fact, such attacks as that of Franklin, Atlanta and Gettysburg, rather than any failures of defense, that finally exhausted and defeated the Confederate Armies?" Schofield's comment is echoed by that of McWhiney and Jamieson, in their argument several decades ago that Confederate generals depleted the South's manpower by favoring frontal infantry attacks in their battles: "Casualty lists reveal that the Confederates destroyed themselves by making bold and repeated attacks."[12]

A final contrast is in order: Hood and Lee at their respective battles learned lessons about their troops. "I thought my men were invincible," Lee later said; clearly, at least on July 3, 1863, they were not. Hood also

---

[12] Bohannon, "'A Bold Fighter,'" 274 ("These lines must be broken boys"); Trudeau, *Gettysburg*, 531; Bailey, *Chessboard*, 107 (Hood quoted); Weigley, *Great Civil War*, 415; McWhiney and Jamieson, *Attack and Die*, 9.

In terms of Lee at Gettysburg and Hood at Franklin, an historiographical distinction needs to be made: in the literature Hood lacks the charitable, even admiring absolution afforded by such scholars as Gary W. Gallagher, who has written of Lee, "He sometimes erred in his judgments, as when he asked too much of his exhausted army on September 16–18 at Sharpsburg or ordered the assaults at Gettysburg on July 3, but those errors and his eventual failure should not obscure his fundamentally sound approach to applying the Confederacy's resources under the most daunting of military and political circumstances" (Gallagher, *Lee & His Army in Confederate History*, 178–79).

learned a lesson about his soldiers. In his recollections Hood wrote that in late November he believed that his army "was still, seemingly, unwilling to accept battle unless under the protection of breastworks." He certainly learned otherwise at Franklin. Several months later, in the *Augusta Constitutionalist*, there appeared a long review of Hood's Georgia and Tennessee campaigns. In it, Confederate general Gustavus W. Smith comments that at Franklin "the courage and conduct of the army in this severely contested struggle proved, beyond a doubt, that its old fighting spirit was revived."[13]

After the battle of November 30, General Schofield claimed victory because he had repulsed the Rebel attack "at all points, and with very heavy loss," as he reported to General Thomas. But Schofield had skulked away after the battle, leaving Hood's army in possession of the battlefield. The situation is reminiscent of Antietam: Lee won a tactical victory in repulsing McClellan's assaults, but had retreated, giving the Federals a strategic victory. "Hood, of course, could scarcely claim a victory," Stanley Horn has written, "in view of his repulse and his crippling casualties." Yet that is exactly what Hood did. On the day after the battle Colonel Mason issued a general order congratulating the army "upon the success achieved yesterday." "The enemy have been sent in disorder and confusion to Nashville," it read, "and while we lament the fall of many gallant officers and brave men, we have shown to our countrymen that we can carry any position occupied by our enemy."

This last statement was quite a stretch, reflecting the commanding general's inclination to make the best of his troops' bloody repulse. One means of doing so was to downplay their casualties. On December 5 Hood wired Secretary Seddon, "Our loss of officers in the battle of Franklin, on the 30th, was excessively large in proportion to the loss of men," which could be taken to mean that the latter had not suffered severely in killed and wounded. To that point Hood added, "The medical director reports a very large proportion of slightly wounded men." Months later, in his campaign report, Hood maintained that at Franklin "the number of dead left by the enemy on the field indicated that his loss was equal or near our own," which was even more of a stretch. This up-

[13] Trudeau, *Gettysburg*, 542; Hood, *A & R*, 290; "G. W. Y.," "Atlanta and Nashville," *Augusta Constitutionalist*, February 5, 1865. The "G. W.Y. letter," as we call it, was written by Gustavus Woodson Smith; see McMurry, *Hood*, 183–85.

beat tenor was shared by some of Hood's staff officers. "Though our loss was heavy, everybody is in the finest humor," wrote Capt. Halsey Wigfall to his sister on December 5, "and ready for the fight again whenever Gen. 'John B.' gives the word." He added that another officer had carefully asked around, "enquiring the feelings of the men." His finding was that "the *morale* of the army was very much improved by the fight, and the men would go into the next with double vim and impetuosity."

Despite his fearful casualties, Hood's battle at Franklin was also reported as something of a victory in the Confederate press. The *Columbus Times* carried this message, issued from Hood's headquarters on December 8:

> About 4 P.M. on November 30, we attacked the enemy at Franklin and drove them from their outer lines and temporary works into their inner lines, which they evacuated during the night, leaving their dead and wounded in our possession, and retreated to Nashville, closely pressed by our cavalry.
>
> We captured several stands of colors and about 1,000 prisoners.
>
> Our troops fought with great gallantry.
>
> We have to lament the loss of many gallant officers and brave men.
>
> Major-General Cleburne, Brigadier-Generals Adams, Gist, Strahl, Granbury were killed.
>
> Major-General John Brown, Brigadier-Generals Carter, Manigault, Quarles, Cockrell and Scott were wounded.
>
> Brigadier-General Gordon was captured.

"We attacked him in position about 4 o'clock P.M. and successfully carried their two outer lines," was Governor Harris's summary of the battle, which was released to and carried by the Southern press. "We have captured about 1,300 prisoners and have picked up on the battlefield about 6,000 stands of arms," his message continued, as printed in the *Milledgeville (GA) Southern Recorder*; "We have also captured a large number of colors." From these statements, it would be hard indeed to infer that the battle had been a disastrous repulse. "The Army is in fine health and excellent spirts, and confident of success," Harris concluded; "the people are delighted and enthusiastic at our advance."[14]

Lee lost a third of his army at Gettysburg, and Hood lost a quarter of his at Franklin. After a day of holding his lines defiantly on the Pennsylvania battlefield, Lee ordered his army to begin its retreat back to Virginia. He had no other option. After Franklin, Hood at least had a few of them.

1. He could march back to where he had started, the south bank of the Tennessee River. But such a retreat, especially after gaining possession of the battlefield on December 1, was at odds with Hood's character and, he claimed, with his soldiers' spirits. He believed, as he put it in *Advance and Retreat*, that his army preferred to "face a decisive issue rather than retreat—in other words, rather than renounce the honor of their cause, without having made a last and manful effort to lift up the sinking fortunes of the Confederacy." Hay states that Hood believed his army "would disintegrate if he turned southwards." One can imagine what could have happened if Hood had retreated back to the Tennessee River after Franklin. His demoralized army would have slunk into winter quarters, and desertions would have further withered its strength. "A retreat would have invited wholesale desertions by Tennesseans," Fowler asserts. By a larger view, retreat from Tennessee would have in effect ended the war for the Army of Tennessee. "A retreat would mean admitting defeat and accepting inevitable surrender," Fowler adds. James Lee McDonough agrees: "to retreat, of course, would be an admission of defeat." Such line of thought leads one to speculate that if Hood's army had retreated, with no chance of reversing its fortunes, it would have indeed ultimately been forced to capitulate. For instance, if George Thomas with his big army had pursued Hood's retreat in December 1864, and brought on a big, decisive battle, it is not impossible that the Army of

[14] Schofield to Thomas, November 30 (7:10 P.M.), *OR*, vol. 45, pt. 1, 1171; Horn, *Tennessee's War 1861–1865*, 319 ("accurately it [Franklin] should be classified as a drawn battle"); General Field Orders No. 38, December 1, *OR*, vol. 45, pt. 2, 628; Hood to Seddon, December 5, *OR*, vol. 45, pt. 2, 650; Hood report, February 15, 1865, *OR*, vol. 45, pt. 1, 654; Wright, *Southern Girl*, 212; "Newspaper account of General John Bell Hood," December 8 and "Newspaper account by Isham Green Harris," December 5, *Supplement to the Official Records*, vol. 7, 676–78.

Tennessee could have surrendered in spring 1865 before the Army of Northern Virginia.

2. He could march around the enemy forces at Nashville, advancing into Kentucky, but his army was too small and theirs was too strong. "If Hood had attempted to flank Nashville and go into Kentucky, Thomas, with a numerically greatly superior force, would have been right at his heels," Hay believes, and an open-field battle defeat for Hood would almost certainly have followed. Moreover, Hood could not expect to increase his numbers by recruiting in Tennessee, much less Kentucky. Major General French claimed that Hood had written on his tent walls, "An army that can obtain no recruits must eventually surrender."

3. That left the third option: "to take position, entrench around Nashville, and await Thomas's attack which, if handsomely repulsed, might afford us an opportunity to follow up our advantage on the spot, and enter the city on the heels of the enemy," as he wrote in his memoir. Historians have been forced to agree. James Lee McDonough concludes that "acceptable alternatives did not exist" for Hood after Franklin, other than to advance to Nashville. This is what Hood decided to do. Orders for the march went out in a circular from Colonel Mason on December 2.

I have argued that Hood's decision to order his infantry assault at Franklin was justifiable by several standards, including comparison with Lee's order at Gettysburg. But Hood's decision to march his army to Nashville, dig in, and wait to be attacked is not so easily justified. Hay has laid out the reasons. 1) His army had received no reinforcements and could not expect any, while his enemy, already outnumbering him, grew stronger every day. 2) Poor roads and railroads hampered his army's supply, while the enemy's supply system to Nashville was operating at peak efficiency. 3) His men were scantily clad, poorly shod and without sufficient shelter to withstand the December's freezing weather in middle Tennessee, while his enemy's troops enjoyed comfortable encampments in the suburbs of Nashville. 4) As for morale, Hood's officers and men still reeled from the bloodletting at Franklin, while his enemy's confidence rose with each day that Pap Thomas spent preparing them for the big battle to come. 5) When it came, and *when* (note my choice of word) Thomas's attack broke his lines, Hood's retreat might be blocked by the enemy gunboats known to be patrolling the Tennessee River. "Under any circumstances Hood's plight was hopeless," Hay rightly concludes.

The best one may say—indeed, it is about the only thing to say—is that the farther John Bell Hood carried his army into Tennessee, the fewer strategic and tactical options he possessed. After Franklin, he had three choices of action. Even the least objectionable of them, as Hay contends, augured disaster. At the same time, one may reason that if Hood were going to lose his army by any of his three choices, the one he decided upon offered at least a slim hope of defensive tactical success—"a brave and vigorous effort…to save the country," as he put it in his memoirs. Stephen M. Hood agrees: after ruling out the other two options (bypassing Nashville, retreating south), "Hood had good reason for continuing the campaign" with his march on Nashville. Lt. Col. Harry Houchens, U.S. Army historian, adds broader context, as if in further justification: Hood's advance in Tennessee "was the South's 'Ardennes Offensive.'"[15]

Another way to view the exhaustion of Hood's options after Franklin is to count the number of battles in which Hood during the course of the war had ordered his troops to take a defensive position and await the assault of the enemy. Answer, zero:

Near Newport News, July 12, 1861. Attack.
Eltham's Landing, May 7, 1862. Attack.
Gaines' Mill, June 27, 1862. Attack.
Second Manassas, August 30, 1862. Attack.
Sharpsburg, September 17, 1862. Attack.
Gettysburg, July 2, 1863. Attack.
Chickamauga, September 20, 1863. Attack.

---

[15] Sears, *Gettysburg*, 469, 498 ("General Lee recrossed the Potomac on July 13–14 with just under two-thirds of the men who had marched north across the river with him in June"); Hood, *A & R*, 299–300; Hay, *Hood's Tennessee Campaign*, 146, 148 (Hood's three options); Fowler, "Finishing Stroke," 151–52; McDonough, *Nashville*, 136–37; O'Connor, *Cavalier General*, 242 (the three options, first of which "could hardly be considered by a man of his temperament"); Dyer, *Gallant Hood*, 296 (adding a fourth option, "attack Thomas at Nashville," so obviously unrealistic as not to be seen elsewhere); McMurry, *Hood*, 176 (reducing Hood's options to two, retreating or moving toward Nashville); French, *Two Wars*, 304; "Confidential Circular," December 2, *OR*, vol. 45, pt. 2, 640–41; Hood, *Rise, Fall, and Resurrection*, 178, 182; Houchens, "The Making of General John B. Hood: A Study of Command," 24 (copy courtesy of Greg Biggs, Clarksville, TN).

Resaca, May 14–15, 1864. Attack.
Kolb's Farm, June 22, 1864. Attack.
Atlanta, July 20, 22, 28, 1864. Attack.
Jonesboro, August 31, 1864. Attack.
Franklin, November 30, 1864. Attack.

You get the point. Hood was constitutionally averse to sitting on the defensive. For him to have decided to do so at Nashville—for the first time in the war—shows the poverty of the Confederates' tactical situation in December 1864.

In his campaign report Hood does not elaborate reasons for his decision, writing only that "we moved forward toward Nashville, Forrest with this cavalry pursuing the enemy vigorously." The latter statement implies that pursuit of the enemy was the only logical thing to do after the battle at Franklin. Besides, the idea of Hood's army marching north continued to have the administration's support. Hood did not know it, but on November 30 President Davis wrote Beauregard on how they should deal with Sherman's march across Georgia. They could do nothing to deter him, Davis acknowledged, "until Hood reaches the country proper of the enemy."

The country proper of the enemy was that north of the Ohio. General Grant worried Hood was headed there. Between retreat back into Alabama and marching toward Chicago, though, was a huge tactical span. Hood's decision to move on Nashville was essentially a halfway measure between the two, an "offensive-defensive step," as Horn puts it. That does not mean it is not open to criticism. "He knew of the overpowering numbers being gathered in Nashville," Horn writes. "He knew the shattered and battered condition of his own army; he knew that the careful Thomas would never come out till he felt reasonably sure he could crush him." Hood nevertheless determined to dig in south of the capital city and await attack. "Seldom has an inferior military force so obligingly waited nearby," Horn shrugs, "while the adversary completed preparations to demolish it." Stephen M. Hood tries to cushion General Hood_from such criticism by pointing out that while he "was fully aware of the heavy odds facing him," before Thomas's attack he laid contingency plans for his army's retreat.[16]

[16] Hood report, February 15, 1865, *OR*, vol. 45, pt. 1, 654; Roman, *Beauregard*, 2:303; McFeely, *Grant*, 192 ("across the Ohio River"); Horn, *Army*

There was no possibility that Hood would attack Thomas's army, for two important reasons. First of all, there were a lot more Yankees. General Schofield rode into Nashville before midnight on the 30th–1st; his troops arrived in the city the following day. On December 1 Thomas had the additional good fortune of greeting Gen. A. J. Smith, finally arrived from Missouri. One Union officer claimed that Pap "literally took him in his arms and hugged him." And no wonder: with Smith and Schofield, Thomas had an additional thirty thousand troops, bringing his strength at Nashville to more than 50,000. Hood knew that Thomas outnumbered him, although he would not have known by how much. The Confederate army after Franklin numbered 23,000 against 55,000 Federals. Hood was thus outnumbered by more than two to one.

Second, Thomas's army was strongly fortified in defensive works that the Federals had been building for more than two years. Forts crowned hills south and west of the city, whose north and east sides were protected by the Cumberland River. Fort Negley, on St. Cloud Hill near the city's south suburbs, was the key position, sprawling over fifty-one acres. Outside its stone and dirt walls, trees and buildings were chopped down to create a field of fire. (The position was renamed Fort Harker after Union Gen. Charles Harker's death at Kennesaw Mountain, but the name of Negley stuck.) A mile to its west was another fort, renamed for Gen. Dan McCook, also a casualty of Kennesaw Mountain. Miles of continuous entrenchments connected these forts and others in two lines that stretched from riverbank to riverbank. The lines were studded with artillery embrasures and protected with abatis. Once Hood entered Tennessee and seemed headed for Nashville, Northern engineers speeded up their work to make "fortress Nashville" even stronger. At the end of November, more than 10,000 black soldiers were at work on lines that included twenty-three forts, redoubts, and fortified bridges.[17]

South of the city, Confederates approached: Forrest's cavalry rode in on the morning of December 2; Lee's Corps, marching on the turnpike from Franklin, arrived mid-afternoon. Stewart's and Cheatham's troops,

---

of Tennessee, 405–406 ("even to Chicago"); Hood, *Rise, Fall, and Resurrection*, 180.

[17] McDonough, *Nashville*, 140–141; Horn, "Nashville: The Most Decisive Battle of the War," 6 (numbers); Beasley and Gotto, "Fortress Nashville," 25–26; Lovett, "Nashville's Fort Negley," 126, 129–33; Zimmerman, *Guide to Civil War Nashville*, 36–37.

their morbid work on the battlefield completed, came on the next day. Ladies on the roadside cheered, and the Tennesseans among the troops thrilled at the sight of their state capital, lost two and a half years before. "But between us and them," wrote one, "there bristles on every hill a fort, and long lines of rifle pits connecting them with the dark blue line of armed men."

On December 2, Hood issued orders deploying his army before the enemy in lines to be fortified. Lee's Corps, largely unbloodied three days before and therefore the army's strongest, would hold the center; Stewart would hold the left; Cheatham, the right.

> The entire line of the army will curve forward from General Lee's center so that General Cheatham's right may come as near the Cumberland as possible above Nashville, and General Stewart's left as near the Cumberland as possible below Nashville. Each position will be strengthened as soon as taken, and extended as fast as strengthened. Artillery will be placed in all favorable positions. All engineer officers will be constantly engaged in examining the position of the enemy and looking to all his weak points.

Unfortunately, Hood's orders called for a line that was impossible to construct and adequately man. At Franklin, his roughly 2¼-mile line, with cavalry on the flanks, had stretched between the banks of the Harpeth River above and below the town. Hood called for the same kind of line at Nashville, but the problem here lay in distance. If the Confederates at Nashville pegged their line a mile or so from the enemy's, it would stretch from the Cumberland riverbank above the city to below it for some ten miles—and there were 6,000 fewer troops after November 30 to man it. Hood's engineers did the best they could: their line running from the Nashville & Chattanooga Railroad to just over the Hillsboro Pike extended about four miles. Even then, as McDonough observes, the Confederate line "was too long to be adequately covered by some twenty thousand men." Cheatham's right flank was two miles from the river to the east; Stewart's left was almost four miles from the Cumberland bank to the west. This meant that both ends of the Confederate position were vulnerable to enemy envelopment. "It is not clear why Hood selected such a poor location," a perplexed Connelly writes; "it was a half-way job which not only failed to cover all approaches, but stretched Hood's strength to the breaking point." It was "an atrocious defensive position," affirms Woodworth. An already bad situation was made worse when Hood determined that the sector of his line across the Granny White

Pike was too close to the enemy works and had to be redug nine hundred yards to the rear. It so happened that a cold snap struck on December 9, bringing plunging temperatures, snow, and sleet. The men, most without overcoats and tents, suffered much in the terrible weather; having to work in it made things worse. Despite the bad weather, Hood on December 10 called for more construction past his line's left flank: "strong self-supporting detached works" were to be dug from the end of the line ("redoubt 1") southward. Four redoubts were sited on high ground for a distance of nearly two miles along Hillsboro Pike. Redoubts No. 4 and 5 were still unfinished when the enemy launched their attack on December 15. "Redoubts three, four, and five were located too far apart to support one another and there was no plan for an entrenched line to connect them," observes McDonough (Hood did not have enough troops to man such entrenchments, anyway). Such was the Confederates' short and incomplete line. (With it, for historians to say that Hood was somehow laying "siege" to Thomas in Nashville, as some have, is more than curious.)

The army settled into its semi-siege in the first week of December. At least there was enough nourishment. An Alabaman, writing home to his Selma newspaper on December 6, reported, "we are receiving full rations, the first since leaving Florence." The good people of Tennessee were offering up a "superabundance of rations," according to General Bate, who also put into use several nearby grain mills for bread and crackers. But "many of the men were barefooted," he noted. All had started out from Florence with some manner of shoes, but in the marches into Tennessee much footwear had fallen apart or been discarded as useless; someone estimated there were 3,500 barefoot men in the army. Clothing was a necessity also; Hood hoped that the tenuous railroad line from Corinth could alleviate the shortages. As they waited around for something to happen, Hood's soldiers had plenty of time to ponder what might happen (the Yankees attack? we attack? or retreat?). "On the hills and in the vales adjacent to the city our troops took a position," recorded one Tennessee veteran, "Micawber-like, waiting for 'something to turn up.'"[18]

---

[18] Sword, *Embrace an Angry Wind*, 278–79 ("Hood's 'siege'"), 303–305; Mason "Confidential Circular," December 2, and Mason circular, December 10, *OR*, vol. 45, pt. 2, 640–41, 672; McDonough, *Nashville*, 136 ("city under 'siege'"), 141–43, 161 (Jim Moon, Jr. map), 181–82 (redoubts); Henry, *"First*

Something that would definitely *not* turn up was a reinforcement of Hood's army by troops from across the Mississippi. In *Advance and Retreat* he writes that he expected such a reinforcement from as far away as Texas. In truth, General Beauregard was trying to make this happen. On December 2, he wired Kirby Smith, reporting the arrival at Nashville of enemy forces previously stationed in Missouri. This development suggested that Smith should also send troops east, perhaps two divisions. Beauregard asked for "prompt and decisive action," but did not get it. The Trans-Mississippi commander did not even reply to the Creole's request until after Hood had already suffered his defeat and retreated from Nashville back into Mississippi (and even then it was a "no"). At least Beauregard tried to help Hood. Possibly in recognition of his efforts, Hood wired his superior, then at Macon, on December 7 that he would "be pleased" if Beauregard could come and pay a visit.

Beauregard did not visit the Army of Tennessee; in the first days of December he was in Montgomery. Nonetheless he tried to stay in touch with Hood. "Keep me advised of your movements," he wired on December 2. Had he visited, he might have heard that among some officers and men, morale was surprisingly high after the bloody battle at Franklin. One soldier wrote home on December 6, "We are in sight of Nashville and the Yankees are alarmed as to their safety, and well they may be." Also on the 6th, Junius Macon wrote to a "Miss Clifford" that "the troops are not only cheerful but enthusiastic." That same day William M. Polk wrote to his sister, informing her "Our army is in fine spirits." On the 11th Captain Wigfall wrote his sister that details were at work in each brigade making shoes for the barefoot men, "some twenty pairs a day in each brigade," in addition to footgear being brought from the rear. The railroad was operating from Pulaski to Franklin, bringing in stores; the commissary and quartermaster officers "are hard at work getting other supplies," Wigfall noted. "So you see we are getting on finely."

---

with the Most," 402 ("undertook to besiege a fortified city"); Connelly, *Autumn of Glory*, 506 ("lay siege to the city"), 508; Woodworth, *Davis and His Generals*, 301; Horn, "Nashville," 6–7, 9 (W. S. Nye map); Hay, *Hood's Tennessee Campaign*, 140–41; "From Hood's Army," *Macon Telegraph and Confederate*, December 26 ("full rations"); Bate report, January 25, 1865, *OR*, vol. 45, pt. 1, 747; H. C. Irby, "Ninth Tennessee Infantry," 1:279; McPherson, *Battle Cry of Freedom*, 813 ("like Micawber, Hood seemed to be waiting for something to turn up").

At the same time, the army's recruitment efforts in the state whose name it bore were failing. Hood himself never expressed any real hope that thousands would flock to the colors. (Only politicians did that; Tennessee senator Gustavus Henry boasted in the Confederate Congress on November 29 that "fifty thousand…men…will be added as fresh recruits to our army from Kentucky and Tennessee.") On December 13, one of Hood's staff reported that since entering Tennessee 164 men had volunteered to join the army. On the other hand, in Johnson's Division, 254 dismounted cavalrymen had deserted. Aware of the numerical odds against him, Hood tried in other ways to strengthen his army. As he had asked Sherman after the fall of Atlanta for an exchange of prisoners, so too on December 5 did he ask Thomas. The latter replied the same day that he could not. "I have had quite a large number of prisoners from your army," Thomas acknowledged, but "they have all been sent North, and consequently are now beyond my control." There could thus be no POW swap. On the 7th Hood asked Beauregard if "all men belonging to this army and any re-enforcements that can be spared be sent forward as soon as possible." On the 9th he asked his superior if two Kentucky brigades earlier lent out could be returned to the army. Hood also inquired of Beauregard on December 10 whether garrison troops in Alabama and Mississippi might be freed for service in his army if state troops could take their places. He tried more immediate ways as well. On the 7th he sent out a cavalry column to nearby counties "to break up and destroy the home guards," local defense units that had become notorious havens for draft-dodgers, and "to conscribe men liable to military duty." At the same time, Hood ordered measures intended to keep what soldiers he had with him, such as frequent roll calls to discourage desertion. On December 10, a circular to officers directed that the troops be "kept well in hand at all times."[19]

---

[19] Hood, *A & R*, 299–300; Beauregard to Kirby Smith, December 2, *OR*, vol. 45, pt. 2 639–40; Hay, *Hood's Tennessee Campaign*, 149; Hood to Beauregard, December 7, *OR*, vol. 45, pt. 2, 659; Beauregard to Hood, December 2, *OR*, vol. 45, pt. 2, 640; McMurry, *Hood*, 178 (request to Beauregard about garrison troops, which I do not see in the *OR*, vol. 45, pt. 2); Wright, *Southern Girl*, 214–15 (Wigfall, December 11); Hood, *Rise Fall, and Resurrection*, 73–74 (Senator Henry's speech, typescript of which Hood kept in his personal papers well after the war); William F. Miller to E. J. Harvie, December 13, *OR*, vol. 45, pt. 2, 685 (164 recruits); "Journal of the Army of Tennessee," *OR*, vol. 45, pt. 1, 672 (entry of December 7); Thomas to Hood, December 5, *OR*, vol. 45, pt. 2,

Adding to the Confederates' discomfiture was General Hood's decision to detach Bedford Forrest with two cavalry divisions, sending him off to raid blockhouses on the Nashville-Chattanooga railroad. During December 2–4 Confederate cavalry forced the surrender of three Federal stockades, bagging more than 250 prisoners, stores, and even capturing a train. Hood compounded his decision on the 2d by ordering Cheatham to send Bate's infantry division plus some cavalry and artillery (1,600 men in all) to Murfreesboro, thirty miles southeast of Nashville, "and direct them to destroy the railroad from Murfreesborough to Nashville, burning all the bridges and taking the block-houses and then burning them." On the 4th, Forrest was ordered to take his two divisions to Murfreesboro and join Bate there.

A number of scholars condemn this "side bet gambit" (Wiley Sword's phrase) as a blunder. Stanley Horn terms Hood's decision a "masterpiece of suicidal folly," especially as his purpose in ordering the Murfreesboro gambit appears vague. Colonel Mason stressed to Bate that his primary mission was to destroy the railroad, not to attack the 5,000 Yankees protected in huge Fort Rosecrans outside of Murfreesboro. In his campaign report, however, Hood mentions his hope "to capture the forces at Murfreesborough." On December 11, in an interim report, he writes of the possibility that if the Murfreesboro garrison marched out of its works, or if Thomas from Nashville sent a relief column to aid, it, "I hope to be able to defeat them." Donald Connelly surmises that Hood "may have hoped to capture isolated garrisons, which would thus enable him to claim some success, even if he were later forced to withdraw" from Nashville. Sword charitably surmises that between December 2 and December 11 Hood simply "modified his plans." Even more charitably, Stephen Hood deems the dispatch of Forrest "more than justified," even "prudent." Despite such justification, Horn's argument resonates: the Murfreesboro gambit was "one of Hood's greatest blunders"...especially as it achieved nothing.[20]

56–57; Hood to Beauregard, December 9, *OR*, vol. 45, pt. 1, 669; Horn, "Nashville," 6; Mason Circulars, December 8 ("frequent roll-calls") and 10 ("well in hand at all times"), *OR*, vol. 45, pt. 2, 665, 672.

[20] McDonough, *Nashville*, 143; Henry, *"First with the Most,"* 402–403; Forrest report, January 24, 1865, *OR*, vol. 45, pt. 1, 754–55; Bate report, January 25, *OR*, vol. 45, pt. 1, 744 (Mason to Cheatham, December 2), Sword, *Embrace an Angry Wind*, 281–282; Horn, *The Decisive Battle of Nashville*, 36; Mason to Bate,

On December 4–5, Bate's troops tore up a few miles of track, burned three blockhouses after the enemy had evacuated them, and torched the nearby bridges. On the 6th, Bate united with Forrest, who took command of the combined force. Two more infantry brigades, those of Claudius Sears and Joseph Palmer, arrived from Hood's main force, bringing Forrest's command to some 6,500 men. Bate learned from Forrest that tearing up track was not the main object, after all; the Confederates pressed skirmishers toward the enemy works near Murfreesboro and dug in. Inside Fort Rosecrans, Union Maj. Gen. Lovell Rousseau decided to send out a reconnaissance in force to determine what the Rebels were doing. On December 7, this column of 3,000, led by Maj. Gen. Robert Milroy, attacked the Confederate line so forcefully that the Southern infantry gave way and "made a shameful retreat, losing two pieces of artillery" (Forrest's words). Both Bate and Forrest tried—but failed—to rally their troops. Only the timely recall of Milroy back to the fort by General Rousseau saved the Confederates from a worse defeat. Still, according to John Morton, chief of Forrest's artillery, the action at Wilkinson Pike near Murfreesboro, December 7, 1864, was Forrest's "hardest blow received during the war." "The infantry behaved badly," Hood was forced to acknowledge in his campaign report. Bate reported 19 K, 73 W and 122 M plus those two guns; Forrest did not account for his loss in the engagement. Milroy reported 30 killed and 175 wounded.

On December 9 Bate's Division and Sears's Brigade returned to the main force at Nashville. After that, Forrest's cavalry with Palmer's infantry brigade kept watch on the Yankees at Murfreesboro while Southern troopers tore up more track. Beyond that, "nothing of importance was accomplished," Hay judges. But as he also admits, "Forrest's absence from the army was unnecessary and in the end calamitous."[21]

December 2, *OR*, vol. 45, pt. 1, 744; Hood reports, February 15, 1865, and December 11, 1864, *OR*, vol. 45, pt. 1, 654, 658; Connelly, *Schofield*, 378 n.77; Hood, *Rise, Fall, and Resurrection*, 185; Horn, "Nashville," 6.

[21] Chalaron, "Hood's Campaign at Murfreesboro," 439–40; Bate report, January 25, 1865, *OR*, vol. 45, pt. 1, 744–46; Forrest report, January 24, *OR*, vol. 45, pt. 1, 755–56; McDonough, *Nashville*, 143–47; J. W. Morton, *The Artillery of Nathan Bedford Forrest's Cavalry*, 283; Hood report, February 15, *OR*, vol. 45, pt. 1, 654; Norris and Long, "The Road to Redemption," 38 (on General Milroy, seeking to efface the stain on his record from Ewell's rout of his command at Winchester, June 1863) and 56 (map, "Action at Wilkinson Pike");

The coming calamity was being carefully planned by General Thomas. The "Rock of Chickamauga," as Hay explains, "was calm, deliberate, and methodical in making ready to dispose of the Confederate army." "Old Slow Trot," as he was also known, would take his time to make sure everything was in place before he launched his *coup de main.* His stolid character had been observed as early as at West Point, where his friend William S. Rosecrans began to call him "General Washington." He further earned that dignified nickname after the first day's battle of Stones River in December 1862. Then, in a midnight council of war after a bloody day's doings, Rosecrans's generals urged that the army retreat. Thomas did not. As one officer recalled, he was "as always, calm, stern, determined, silent, and perfectly self-possessed, his hat set squarely on his head. It was inspiring to look at the man." By another account, though, Thomas dozed through most of Rosecrans's council, awakening only to say, "This army does not retreat."[22]

Similarly, two years later Thomas would stand solid at Nashville—not only against his enemy, but also his superiors. After his arrival from Franklin, General Schofield had suggested that the Union forces move against Hood's army, but Thomas declined. He was waiting for A. J. Smith's infantry, but he also wanted more to build up his cavalry arm. General Wilson was busy finding horses, and Thomas decided to give him time to do so. This did not please the general-in-chief back in Virginia. "Should he attack you all is well," Grant wired Thomas on December 2, "but if he does not you should attack him before he fortifies." Grant worried that Rebel cavalry would set to work on Thomas's railroads "if Hood is not speedily disposed of. Put forth, therefore, every possible exertion to attain this end." Thomas was unmoved. "I now have infantry enough to assume the offensive," he wired back that day, but he needed time to organize his forces, even referring to "my crippled condition." "In a few more days I shall be able to give him a fight," he promised.

Grant gave him those few more days but then telegraphed on December 5, "Hood should be attacked where he is. Time strengthens him, in all probability, as much as it does you." Old Slow Trot lived up to his

---

Wigginton, "Cavalry Operations," 42 ("Forrest's supporting infantry broke and ran, and could not be rallied"); Hay, *Hood's Tennessee Campaign*, 141; Neff, *Tennessee's Battered Brigadier*, 122 (Palmer left Murfreesboro December 16).

[22] Bobrick, *The Battle of Nashville*, 78; McDonough, *Stones River*, 160.

nickname. "As soon as I can get up a respectable force of cavalry I will march against Hood," he wired back the next day. Thomas was also promising General Halleck, "if I can perfect my arrangements I shall move against the advanced position of the enemy on the 7th instant." Grant then turned both peremptory and pleading. "Attack Hood at once," he ordered on the 6th; a few hours later he asked, "Why not attack at once? … Now is one of the finest opportunities ever presented of destroying one of the three armies of the enemy. If destroyed, he can never replace it." Meanwhile Grant on the 7th wired Secretary Stanton, "You probably saw my order to Thomas to attack. If he does not do it promptly, I would recommend superseding him by Schofield, leaving Thomas subordinate." "If Thomas has not struck yet," Grant informed Halleck on December 8, "he ought to be ordered to hand his command over to Schofield." Then came the bad weather of the 9th, worsening the situation. Thomas complained that "a terrible storm of freezing rain has come to-day," forcing a further postponement of his offensive operations. By that time General Halleck had confided to Thomas that Grant was most disgruntled: "If you wait till General Wilson mounts all his cavalry, you will wait till doomsday." Pap responded by offering his resignation. "If you should deem it necessary to relieve me," he wired Grant on the 9th, "I shall submit without a murmur." Indeed, Grant had already issued an order relieving Thomas from command but suspended it "until we should hear further." Grant's patience had come to an end, but he kept urging action. "If you delay attack longer the mortifying spectacle will be witnessed of a rebel army moving for the Ohio River…. Delay no longer for weather or re-enforcements," to which Thomas again promised, "as soon as we have a thaw, I will attack Hood." Meanwhile, on the night of December 10, Thomas called his senior officers together. With the exception of General Schofield, who remained silent, all sustained him in not ordering the assault amid such treacherous weather. Brigadier General Wilson remarked that if they tried to advance over an icy battlefield, the Rebels' first volley would send Thomas's troops sliding in every direction; the Rebels could defend their works "with nothing more dangerous than baskets of brickbats." "The whole country is now covered with a sheet of ice so hard and slippery it is utterly impossible for troops to ascend the slopes, or even to move over level ground," Thomas explained to Halleck on the 12th. The next day, the hoped-for thaw began to appear, as Thomas wired Washington on December 13. That allowed him to assure Halleck the next day, "the enemy will be attacked to-morrow morning." Thomas was fortunate indeed. Grant had sent Major General John Logan west to relieve him of command; Logan reached Louisville when he learned Thomas had finally launched his battle.[23]

Grant's impatience has certainly not endeared him to Thomas's champions. Wilbur Thomas calls Grant's peremptory attitude "one of the dirtiest, if not *the* dirtiest cutthroat jobs ever performed on a man of even remotely equivalent ability." Unfortunately, Schofield seems to have played a role in the job, possibly sending surreptitious telegrams to Grant complaining of Thomas's slowness. "There is considerable circumstantial evidence to support this charge," writes James L. McDonough. "As if Thomas didn't have enough to deal with!" exclaims Benson Bobrick; "he had a Judas on his staff!" To be sure, decades later in his memoirs, Schofield mildly criticized Thomas for having "made all those elaborate preparations to attack an enemy of less than half his own strength." But he always denied having sent any subversive telegrams to Washington. The question of intrigue is beside the point: George Thomas had successfully held out until he was ready to give John Bell Hood battle.[24]

Thomas would enjoy immense numerical superiority. By Wiley Sword's count, his veteran infantry totaled 33,000 (14,000 in the IV Corps [now under Thomas J. Wood, with General Stanley still recovering from his Franklin wound], 10,000 in XXIII, and 9,000 in A. J.

---

[23] Hay, *Hood's Tennessee Campaign*, 142; Grant to Thomas, December 2 (11 A.M. and 1:30 P.M.); Thomas to Grant, December 2 (10 P.M.); Grant to Thomas, December 5 (8 P.M.); Thomas to Halleck, December 5 (10 P.M.); Thomas to Grant, December 6 (8 P.M.); Grant to Thomas, December 6 (4 P.M. and 8:30 P.M.); Grant to Stanton, December 7 (10:20 A.M.); Grant to Halleck, December 8 (4 P.M.); Halleck to Thomas, December 9 (10:30 A.M.); Thomas to Grant, December 9 (1 P.M.); Grant to Thomas, December 9 (7:30 P.M.) and December 11 (4 P.M.); Thomas to Halleck, December 11 (9:30 P.M.), 12 (10:30 P.M.), 13 (9 P.M.) and 14 (8 P.M.), *OR*, vol. 45, pt. 2, 17–18, 55, 70, 84, 96–7, 114–15, 143, 155, 168, 180; Longacre, *Union Stars to Top Hat*, 183; Sword, *Embrace an Angry Wind*, 310–11; Hay, *Hood's Tennessee Campaign*, 144–45.

As for the "frayed relationship" between Grant and Thomas before the battle of Nashville, Brooks Simpson judges that "both of the principals bore some responsibility for the near-collapse of trust between the two men" ("A Failure to Communicate," 119).

[24] Thomas, *Indomitable Warrior*, 556, 564–65; McDonough, "Schofield," 14; Bobrick, *Nashville*, 96; Schofield, *Forty-Six Years*, 261; McDonough, *Nashville*, 152–53.

Smith's XVI). Maj. Gen. James B. Steedman had brought with him from the "District of Etowah" another 8,500 troops. Wilson's cavalry numbered 6,500, to which were added 4,000 men of the Nashville garrison, reserves, even soldiers assigned to the Quartermaster Department. This total of 52,000 parallels Livermore's calculation of 53,522 present for duty, which by his 93 percent effectives rule would reduce the number of Union soldiers in the battle to 49,773. This was still more than two times Hood's strength, which Livermore places at 23,207 men engaged; Horn echoes 23,053. The latter counts Thomas's number of troops " taking part in the fighting" at Nashville as 14,171 in Wood's corps; 10,207 in Schofield's; 10,461 in Smith's; 7,541 in Steedman's force; and 12,500 in Wilson's cavalry—54,881 in all.[25]

At least Hood knew it was coming. On December 10 he sent a circular stating that it was "highly probable that we will fight a battle before the close of the present year" and advising his corps commanders to see to their fortifications. In the meantime, quartered in John Overton's fine home, "Traveler's Rest," six miles south of the city, Hood had time on December 11 to complete a long report of the army's operations to date. He reviewed "our mortification" at Spring Hill and his accusation against Frank Cheatham for it; the battle of Franklin ("our entire loss was about 4,500"); the enemy's speedy retreat afterward; and his present position ("our line is strongly intrenched"). Hood sent his report to Secretary Seddon with copy to General Beauregard at Montgomery. By now Hood's insults to his superior officer, such as this refusal to communicate correctly through the chain of command, had become so standard that General Beauregard might well have shrugged this one off. But he did not. "General Hood does not seem to understand," Beauregard endorsed his subordinate's report, "that he is responsible directly to these Headquarters, and *not* to the War Department. Colonel Brent, Chief of Staff, will call his attention to that fact."

Despite the threat of the imminent enemy attack, Hood and his officers at Traveler's Rest enjoyed the visits of Nashville ladies. But Hood's thoughts were still on Buck Preston, who had begun to fret over his not writing. "Every insanity enters her head," recorded Mrs. Chesnut on November 17, "even that J.B. [Hood] may be tired of his engagement.... Buck wept," referring to Sally Buchanan Preston, to whom Hood con-

---

[25] Sword, *Embrace an Angry Wind*, 276–77; Livermore, *Numbers and Losses*, 132–33; McDonough, *Nashville*, 141 ("some twenty thousand" Confederates).

sidered himself engaged. But then, two weeks later, two letters from Hood arrived. "Buck is happy," Mary Chesnut wrote in her diary, "J.B. has written, and she says...he is coming in January to be *married.*"

In such sanguine spirit, Hood even seemed to exude confidence in the prospects of his army. "General Hood says that this campaign will change very greatly the movements of both armies," recorded Chaplain Quintard in his diary for Monday, December 12. "The enemy will have to seek our armies and fight them where he can find them...but that the losses will be on the side of the Federals." Hood's mood seemed to be infectious. On the 12th, at nearby Brentwood, one of Hood's staff officers, Maj. William Clare, married his fiancée, Mary Hadley, in a festive wedding ceremony. (The couple spent several days' honeymoon at the army's encampment.)

But the big battle loomed. O'Connor claims that Hood had spies in Nashville who slipped out and told him that the Yankees would attack when the weather cleared. Hood is said to have warned his three infantry corps commanders on the 14th that an attack was coming the next day. Hood regretted that the works on his left flank were not completed, but the break in the weather signaled to all that Thomas would soon stir.

On the afternoon of December 14, Thomas called his corps commanders together and discussed with them his plan of attack, which he had drawn up a few days earlier. After inclusion of a couple of changes suggested by his officers, Thomas's battle plan featured a diversionary attack by Steedman's force against the Rebel right to draw attention away from the main push, which would be against Hood's left. Wood's, Smith's, and Schofield's IV, XVI and XXIII infantry corps, with Wilson's cavalry on their right, were to attack the Confederates' west-most line and the five redoubts they had dug along the Hillsboro Pike. Given their numerical strength, the Federals' turning movement promised decisive results.

On December 15, after the morning fog cleared, Steedman launched his advance at 8 A.M. To their credit, Thomas and Steedman allowed two brigades of United States Colored Troops to participate in the attack. Steedman later wrote, "I was unable to discover that color made any difference in the fighting of my troops." But many Confederates were angered at the sight of Negroes in blue uniform. This was the first time, Pvt. Philip Stephenson recorded, that he and his comrades "had ever met our former slaves in battle." "It excited in our men the intensest indignation," and they poured a deadly, deliberate fire into the advancing enemy. Steedman's attack fell apart as both white and black regiments fell back. After midday his troops kept up a long-range musketry with the Rebels on the Confederate right, but without consequence. "General Hood was not deceived as to where the main Federal assault would come," writes McDonough. By mid-morning he had concluded that his left would be the enemy's main objective and moved his headquarters from Traveler's Rest to Judge John Lea's home, closer to that sector.[26]

Stewart's Corps held the Confederate left, from the Granny White Pike west to Redoubt 1, where the line angled sharply to the south. As early as 10 A.M. Brig. Gen. Claudius Sears, commanding French's Division (the latter absent, on sick leave from December 13) in the area of Redoubts 2 and 3, received report from his pickets that the enemy could be seen forming three lines of battle in their front. Sears forwarded the information: "the demonstration of the enemy on the left is increasing." Col. Robert Lowry, whose brigade in Loring's Division manned Redoubt 1, similarly reported, "there seems to be a movement of some magnitude on our left." This news was ominous enough; Stewart sent it on to Hood and to Lee to his right (Lee answered, "I feel secure in my line"). Making the situation worse was Stewart's knowledge that his line "was stretched to its utmost tension." He did not have enough troops even to extend it to Redoubts 4 and 5 on the army's extreme left. Hood therefore ordered reinforcements (Bate's Division) from Cheatham's sec-

---

[26] Mason circular, December 10, *OR*, vol. 45, pt. 2, 672; Sword, *Embrace an Angry Wind*, 286–87 ("Traveler's Rest"), 324 (fog); "Hood's Headquarters Near Nashville," 231 (illustration of house); Hood to Seddon, December 11, *OR*, vol. 45, pt. 1, 657–58; Roman, *Beauregard*, 2:612–14 (Hood's report of December 11 with Beauregard's endorsement of January 9, 1865, which I do not see in the *OR*); Woodward, *Mary Chesnut's Civil War*, 670, 683 (entries of November 17 and December 2); Sword, "The Battle of Nashville," 18–19; Elliott, *Doctor Quintard*, 195 (entry of December 12); J. J. Gaines, "Civilian Participants and Observers," 197 (the honeymooners); O'Connor, *Cavalier General*, 247; Hood, *A & R*, 302; McMurry, *Hood*, 178–79 ("on the thirteenth and fourteenth he correctly divined that Thomas would soon fight"); Einolf, *George Thomas*, 268–270 (including good map, "Nashville, Day One"); McKinney, *Education in Violence*, 406; Wesley and Romero, *Negro Americans in the Civil War*, 95; Stephenson, *Memoir*, 320; McDonough, *Nashville*, 166–68; A. J. Bailey, "The USCT in the Confederate Heartland, 1864," 235–37.

tor, no longer under threat on the right, and from Lee's in the center (Edward Johnson's division).

They did not arrive in time. The main Federal assault began at 1 P.M., when General Wood gave the order for his IV Corps to advance. The Federals did so in fine style, advancing up Montgomery Hill, which they found to be abandoned by the Rebels—it had been part of Hood's initial line, given up five days before as too close to the Federals'. Now it served only as a Southern picket line, with the main Rebel works another nine hundred yards to the south. Before Wood moved against them, the day's truly decisive fighting had opened to his right, when A. J. Smith's infantry and Wilson's cavalry moved against the Confederate Redoubts 4 and 5. Number four was held by Capt. Charles Lumsden's Alabama battery (four Napoleons) and a mere hundred infantry. After an artillery duel of an hour or more, the Federals charged Lumsden's position, sending the Confederates fleeing and capturing their four guns. Redoubt 5 also fell, with its two twelve-pounders. The Federals kept up their advance, crossing the Hillsboro Pike.

Ed Johnson's infantry came up to try to save the crumbling left flank. But they were just two brigades facing thousands of bluecoats and coming under artillery fire. They offered "but feeble resistance" (Stewart's words) before they turned and ran for the rear. The panic spread, and the arrival of Johnson's two other brigades could not stop it. Stewart had no choice but to order a retreat. Federals moved against the Rebels' remaining redoubts and took them, too. Just then, about 4:30, Wood launched his IV Corps attack against Loring's line, overrunning those Confederates who had not begun to fall back. "A. P. Stewart's entire corps was now in utter flight," writes Sword. Brigadier General Sears was struck by an enemy shell that passed through his horse and ruined his lower right leg. Toward sundown, when General Bate finally brought his division up, he saw "streams of stragglers, and artillerists, and horses without guns or caissons," all fleeing. Fortunately for the Southerners, early nightfall that December day prevented the Federals, themselves fatigued and in disarray, from further pursuit after they reached the Granny White Pike around 6 P.M. Hood ordered a withdrawal of his army to a line more than a mile farther south, which he ordered his engineers to lay out that night.[27]

---

[27] Sword, *Embrace an Angry Wind*, 329 ("the pulled back line, established on December 10"), 333 ("Map 13 Assault on the Confederate Left"), 334–42,

General Thomas was thrilled that he could report good news to Washington that night. "I attacked the enemy's left this morning and drove it from the river, below the city, nearly to the Franklin pike, a distance about eight miles," he wired General Halleck at 9 P.M. "The troops behaved splendidly"; he reported their having taken some 800 to 1,000 prisoners and sixteen pieces of artillery. Adding to his satisfaction, according to recent scholar Michael Thomas Smith, was "the thought of how the news would be received in Washington, where Stanton's and particularly Grant's doubts about his dispositions would now be put to rest once and for all." About a half-hour before Thomas's message was received by the War Department, a telegraph operator in Nashville, J. C. Van Duzer, had sent his own message to Washington giving news of the day's victory. From Van Duzer's dispatch Grant thus learned of Thomas's "splendid success," but in his telegram to Thomas that night he could not resist mentioning that he was just about to leave for Nashville (presumably to relieve Thomas of command). A quarter-hour later, having read Thomas's wire, Grant added, "I congratulate you and the army under your command for to-day's operations." Secretary Stanton was more effusive: "I rejoice in tendering to you…the thanks of this Department for the brilliant achievements of this day," he wrote. "We shall give you a hundred guns in the morning."[28]

"That last and bravest champion of a desperate cause in the west appears to have decided to remain and invite the total destruction of his army," was General Schofield's assessment of Hood's decision to stay and fight after the pummeling Stewart's Corps had received on December 15. (By 9 P.M. Forrest had heard of the defeat and issued orders for his command to head out on the 16th to rejoin the army.) Hood's troops took up their new line, some three miles long, in the Brentwood hills. "Hood's position," writes Hay, "was none too favorable." The Confeder-

---

349 ("Map 14 Situation Early Morning, December 16"); McDonough, *Nashville*, 170–72, 181–82, 192–97; Sears endorsement, December 15 (10 A.M.) ("Brigadier-General, Commanding Division") of Estes to Sanders and Sanders to Gale, December 15 (10 A.M.), Lowry to Loring, December 15, *OR*, vol. 45, pt. 2, 692; Lee to Stewart, December 15 (10:10 A.M.) *OR*, vol. 45, pt. 2, 691; Stewart report, January 20, 1865, *OR*, vol. 45, pt. 1, 709; Sword, "Nashville," 43.

[28] Thomas to Halleck, December 15 (9 P.M.), Van Duzer to Eckert, December 15 (10:30 P.M.), Grant to Thomas, December 15 (11:30 and 11:45 P.M.), Stanton to Thomas, December 15 (midnight), *OR*, vol. 45, pt. 2, 1194–96; M. T. Smith, *The 1864 Franklin-Nashville Campaign*, 82.

ate right rested on Overton Hill; the left on an eminence that came to be called Shy's Hill. Throughout the night the tired men dug in and set up barricades against the enemy attack expected the next morning. The infantry corps aligned from left to right, Cheatham, Stewart, Lee. The army's wagon train was sent farther south to Franklin. An inauspicious omen came from Hood's headquarters at 8 A.M. on the 16th: "Should any disaster happen to us to-day," Colonel Mason wrote Peter Stewart, his corps should retreat by the Franklin pike; "Cheatham would move by the Granny White pike."

A little after 5 A.M. Schofield sent word to Thomas that the Rebels were fortifying ahead of him. Thomas did not order his corps commanders to press a quick assault, but to advance cautiously, feeling for the enemy position. They aligned in the same order as on the previous day: Steedman on the left, then toward the west, Wood, Smith, Schofield, and Wilson. The latter two showed a lack of drive that morning. Schofield said he needed reinforcements as he ordered his artillery to shell the Rebel position. Wilson claimed that the ground on his flank was too boggy for cavalry operations. Wood and Steedman were also cautious, using the morning to bring their troops up in front of the Rebel lines while their artillery loosed a heavy bombardment. Once in position, though, Wood decided that he could take the key position on the Confederate right, Overton Hill. Wood and Steedman coordinated their assault, which involved four brigades and began around 2:45 P.M. against Clayton's and Stevenson's divisions (Lee's Corps). "On they came in splendid order," recalled a Confederate artilleryman. General Hood worried that his right needed strengthening and ordered two brigades from his left (which proved to be a big mistake). "My men reserved their fire until they were within easy range," General Lee reported, "and then delivered it with terrible effect." Under heavy musket and cannon fire, Steedman's USCT again demonstrated conspicuous bravery. Three regiments took part in the charge, advancing closer to the Rebel works than their white comrades and taking losses at least as heavy. "They fell like wheat before a mowing machine," according to one Confederate. Under such fire, Wood's and Steedman's troops faltered and broke, sustaining more than a thousand casualties in the failed assault.[29]

---

[29] M. T. Smith, *Franklin-Nashville Campaign*, 82–83; Henry, *"First with the Most,"* 410; Anderson to Jackson, December 15 (9 P.M.), *OR*, vol. 45, pt. 2, 693; McKinney, *Education in Violence*, 410–11; Mason to Stewart, December 16

So the Confederate right held. Not so the left, which was anchored on the hill (Compton's), soon to be named for Lt. Col. William Shy of Tennessee. The cards were stacked against the Southerners for three dramatic reasons, as outlined by McDonough. First, they were heavily outnumbered. On the second day of the battle, General Thomas wanted a repeat of the plan that had worked so well on the first: turning the enemy left. To that end he had again committed the bulk of his forces: Schofield's and the part of A.J. Smith's corps facing the hill probably numbered 15,000 men. Against them on Shy's Hill were ten Confederate brigades; unfortunately, Hood had ordered several other brigades off to other sectors of his line. Those ten likely had fewer than 5,000 troops on the morning of December 16. What was worse, the Southerners could clearly see the enemy massed against them. "They confronted us everywhere," General Stewart later noted, "with a force double or treble our own." Second, the Federals had more artillery with more ammunition. Against almost one hundred Northern guns, Cheatham's Corps had thirty-four, McDonough counts, "but many of them were bogged down in mud and could not be maneuvered into acceptable locations." The Federals shelled the hill heavily all day, inflicting casualties among infantry, artillerymen, and battery horses. "It seemed as if the Federals were shooting crooked cannons from their forts and batteries in Nashville," remembered Robert A. Jarman of the 27th Mississippi, "for they could throw their shells right in our midst, and from killed and wounded men and horses, a small branch in our rear ran red with blood." Third, during the night the Confederates had laid out their line and their works poorly, on the hill's topographical crest versus its "military" crest—the same mistake that had hurt them so badly at Missionary Ridge. At daylight General Bate, whose three brigades held the center of the hill-line, recognized its flaws. The works were "too far back from the brow of the hill" and in some places "flimsy." But he could do little to improve his line. "It was

---

(8 A.M.), *OR*, vol. 45, pt. 2, 696; Schofield to Thomas, December 16 (5:15 A.M.), *OR*, vol. 45, pt. 2, 214; Einolf, *George Thomas*, 276 ("telling his commanders to make a cautious advance"); McDonough, *Nashville*, 223–35, 239 (Hood's "big mistake"); Hay, *Hood's Tennessee Campaign*, 156; Trudeau, *Like Men of War*, 345–47; E. W. Smith, "Douglas's Battery in Battle of Nashville," 531; Lee report, January 30, 1865, *OR*, vol. 45, pt. 1, 688.

impossible to remedy it," he reported. "The constant fire of sharpshooters from the neighboring hills made it fatal."[30]

John Schofield's battle-blood was not up that day. "I have not attempted to advance my main line," he wrote General Thomas at 1:30 P.M., "and do not think I am strong enough to do so." Union Brig. Gen John McArthur, commanding an infantry division north of Shy's Hill, perceived the weakness of the Rebels' line, noting that their artillery would not be able to be depressed enough against assaulting troops, and that their infantry were too far back on the hill crest to fire effectively. He decided to wait no longer and began planning an attack. He asked for support from Maj. Gen. Darius Couch, commanding the division to his right, but Couch declined, saying he had no orders for an assault. Unfazed, McArthur sent message to Major General Smith announcing that he would send his troops forward unless he received orders to the contrary in five minutes. He did not, and at 4 P.M. McArthur's infantry stepped forth. Sure enough, the Confederate artillery could not stop the Yankees, and the infantry could get off only one volley before the Federals stormed into the works held by Thomas B. Smith's brigade. In hand-to-hand fighting, Colonel Shy, holding a musket, was shot in the face and killed. To Smith's right, Finley's Florida Brigade began to break and run. The panic spread and in minutes the whole of Bate's Division had surrendered or fled. That exposed Walthall's brigades on the eastern side of the hill. "To save any part of my command," Walthall recalled, he had to order a withdrawal. The Federals who took Colonel Shy's hill eventually counted as captives 8 cannon, 4 battle flags, 85 officers (including Brigadier General Smith), and 1,533 men.

Panic on a battlefield is hard to stem. "Bate's division, French's two brigades, Walthall's and Loring's divisions, now the remnant of Frank Cheatham's Corps, all had been routed within a few minutes," Sword

---

[30] Marshall, "Shy's Hill—Whence the Name?," 522; McDonough, *Nashville*, 222 ("Rebel artillery…conserving ammunition, which was very scarce"); 238–44; Roth, "That Cruel Hole in His Brow," 49; Wills, *True as Steel*, 322 ("turning the Confederate left"); Sword, *Embrace an Angry Wind*, 371 ("Map 16: Assault on the Confederate Left," showing the brigades of Quarles, Shelley, Reynolds, Sears, Jackson, Finley, Smith, Ector, Strahl, and Vaughan on the hill); abstract from Army of Tennessee field return, *OR*, vol. 45, pt. 1, 678–79 (showing the ten brigades' strength on December 10 as roughly 5,600–5,700—and that was before the first day's fight at Nashville); Jarman, "A Mississippian at Nashville," 14; McDonough, *Chattanooga*, 183.

writes. General Cheatham tried to rally his men. He managed to stop one soldier, who just hunkered down under the general's horse and kept running. Even S. D. Lee's troops on the right got caught up in the chaos and took to flight, losing sixteen cannon in the mayhem. General Lee sought to stem the flight, riding among his troops, waving a battle flag. "Rally, men, rally!" he cried, "For God's sake, rally!" Maj. Gen. Edward Johnson, one of Lee's division leaders, fell prisoner (he had just been released from a Northern prison four months before, after having been captured at Spotsylvania).

The sight of Southern men running for their lives, throwing away their weapons, appalled the nearby citizens. One, a Mary Bradford, ran out even under the Yankees' fire and pleaded to the men to stop running. Henry Clayton's division refused to buckle and fell back in somewhat good order against Steedman's and Wood's advance. "I withdrew the brigade and attempted to arrest the enemy," reported Randall Gibson, one of Clayton's brigadiers, "but at this time confusion prevailed over every thing." "When soldiers are in a stampede, they are without control, and this was our condition," remembered a private in the 18th Alabama. Clayton's troops fell back as rearguard for the rest of Hood's retreating army.

General Hood had not seen it coming. He was near his headquarters at the Lea house when a staff officer brought news of the break on the left. He rode off and into the flood of his fleeing men. He was seen by one "with other commanding officers, trying to rally the men, but in vain." Hood had to give up and join the sorrowful retreat southward. "I beheld for the first and only time a Confederate Army abandon the field in confusion," Hood recorded in his memoirs. Confederate Pvt. Philip Stephenson caught a glimpse of the army commander. "He seemed overwhelmed by his defeat and sat on his horse, a picture of despair." A cold rain had begun to fall, adding to everyone's dejection. With "his head sunk on his breast, reins dangling loosely, an oil cloth blanket draping him," Stephenson recalled, "if anyone looked 'whipped' that day, our unfortunate leader certainly did."[31]

[31] Schofield to Thomas, December 16 (1:30 P.M.), *OR*, vol. 45, pt. 2, 215; McDonough, *Nashville*, 246–48; Bailey, *Chessboard*, 161; "The cover illustration," *Civil War Times Illustrated* 3/8 (December 1964): 3 (Ralph McDonald's painting of Colonel Shy's death); Sword, *Embrace an Angry Wind*, 373–79, 382–85, 391–92; Walthall report, January 14, 1865, *OR*, vol. 45, pt. 1, 723; C. B.

"Our men were flying to the rear in the wildest confusion," General Lee reported, "and the enemy following with enthusiastic cheers." Hood's staff officers tried to halt the rout. "It was sad to see men disgrace themselves so," wrote Halsey Wigfall a few days later. The Federals scooped up many prisoners (including Brig. Gen. Henry R. Jackson) before Clayton's Division could anchor a temporary line a mile or so south of the battlefield to slow the enemy's pursuit. Even darkness and rain did not bring a halt to Yankee cavalry, which was resisted by Chalmers's horsemen. In a sharp little night-fight 2½ miles south of the overrun Rebel line, Wilson's troopers pressed Col. Edward Rucker's cavalry brigade; Rucker was wounded (he would lose his arm) and captured. The Yankees eventually allowed the Southerners to withdraw into the night. Hood's troops kept moving in the darkness. "Night closed down," wrote a disconsolate officer in Tyler's Brigade of Bate's Division, "upon a thoroughly demoralized and routed mob making all possible haste to get into the Franklin Pike before their only way of retreat should be blocked."

General Hood informed Richmond of the calamity in a telegram sent December 17: "Early on the 16th they made a general attack on our entire line, and all their assaults were handsomely repulsed, with heavy loss, till 3.30 P.M., when a portion of our line to the left of the center suddenly gave way, causing, in a few minutes, our line to give way at all points, out troops retreating rapidly down the Franklin pike." He acknowledged the loss of fifty cannon, plus generals Johnson, Smith, and Jackson captured. Otherwise "our loss in killed and wounded is very small." In this he would not be proven quite correct. Acknowledging the absence of Confederate officers' reports, Hay estimates 1,500 Confederate killed and wounded at Nashville. Sword's "educated guess" is higher, about 2,300—a tenth of the army's strength in the battle. Hood was

---

Martin, "Jackson's Brigade in Battle of Nashville," 13 (Cheatham's horse, and Hood "trying to rally the men"); Gale, "Confederate Disaster at Nashville," 46; Gibson, "Report of Brigadier-General R. L. Gibson of Operations in Vicinity of Nashville," 141 (*OR*, vol. 45, pt. 1, 702–703); Dozier, "Concerning Hood's Tennessee Campaign," 192 ("in a stampede"); Hood, *A & R*, 303; Stephenson, *Memoir*, 334.

Sam Watkins's claim that after the battle he saw General Hood at his headquarters, "pulling his hair with his one hand (he had but one), and crying like his heart would break" is well known, but I believe not credible, if only for the implausibility of a private soldier entering a commanding general's headquarters to request a furlough (Watkins, *Co. "Aytch,"* 269).

much further off the mark when he claimed, "Our loss in prisoners is not yet fully ascertained, but I think it comparatively small." (Walthall's Division alone lost 844 officers and men missing, as opposed to 17 killed and 127 wounded.) Thomas claimed 4,462 prisoners, including 287 officers from the rank of major general (Edward Johnson) and brigadier general (T. B. Smith and H. R. Jackson) down the scale. A subsequent Federal provost marshal report put the number of enemy officers and men captured during December 1864 at 8,635. Thomas proudly claimed fifty-three pieces of artillery captured, and for once a Civil War officer did not exaggerate: Hood reported fifty-four guns taken by the enemy. Maj. Daniel Truehart's entire battalion, three Alabama batteries supporting Stewart's Corps, had been swept over by the Federal onslaught. Lee's Corps' artillery chief reported sixteen cannon lost in his command. Among them were those of Douglas's Texas battery. Douglas had managed to bring his guns off the battlefield, but hard-charging Federal cavalry overwhelmed them. Further on in the retreat evidently more were lost. In the month from the army's departure from Florence to December 21, the number of guns with the army dropped from 124 to 59—65 pieces of artillery.

Assuming 1,500 killed and wounded plus 4,500 captured and missing, Hood's casualties December 15–16 totaled approximately 6,000. Thomas's A. A. G. later submitted a report of Federal casualties: 363 killed, 2,289 wounded, 1,197 missing for a total of 3,849 (which did not include Steedman's but did include Wilson's at Franklin).[32]

---

[32] Lee report, January 30, *OR*, vol. 45, pt. 1, 689; McDonough, *Nashville*, 257–65; Sword, *Embrace an Angry Wind*, 365 (Halsey Wigfall), 425 (Hood's 2,500 K & W); Brennan, "Last Stand in the Heartland," 44 (Rucker's wound); Garrard, "Gen. S. D. Lee's Part in Checking the Rout," 350; Tyler, "Forrest Covers Hood's Retreat," 437; Logsdon, *Eyewitnesses at the Battle of Nashville*, 109; Hood to Seddon, December 17, *OR*, vol. 45, pt. 2, 699; Hood report, February 15, *OR*, vol. 45, pt. 1, 655; Hay, *Hood's Tennessee Campaign*, 180 (Hood's killed and wounded "probably less than 1,500"); R. D. Smith, "Record of Walthall's Division at Franklin and Nashville Late in 1864," 265; Ritter, "Sketch of the Third Maryland Artillery," 170 (stating that the army lost 72 pieces of artillery at Nashville, but without substantiation); "Abstract for inspection report of artillery, December 21," *OR*, vol. 45, pt. 1, 682; Thomas report, January 20, *OR*, vol. 45, pt. 1, 40; Scaife, *Hood's Campaign for Tennessee*, 87–88; report of Lt. Col. Llewellyn Hoxton, February 1, *OR*, vol. 45, pt. 1, 692; Shands, "Dr. Lea Williamson," 370 (Truehart's battalion "without horses or guns"); L. R. Doug-

las, *Douglas's Texas Battery*, 151; "report of prisoners of war captured," February 4, 1865, *OR*, vol. 45, pt. 1, 47; "Report of casualties of the Army of the Cumberland," *OR*, vol. 45, pt. 1, 47.

The Army of Tennessee's loss of fifty-four pieces of artillery thus makes the Civil War's Top Dozen List of Artillery Captures, which begins with the astonishing number of cannon captured by Virginia forces when they overran the U.S. Navy's shipyard at Norfolk, Virginia, in April 1861: 1,195 guns of all different calibers (*Message from the Executive of the Commonwealth*). Commissioner Peters's report is summarized in Scharf, *History of the Confederate States Navy*, 133–34. Fifty-two of the guns were prized nine-inch Dahlgrens (West, *Mr. Lincoln's Navy*, 31). After that, the list continues with Vicksburg, 172 pieces of artillery captured by the enemy; at Island No. 10, 109; Fort Jackson, 74; Harpers Ferry, 73; Chickamauga, 66; Fort Donelson, 65; Nashville, 54; Yorktown, 53; Fort St. Philip, 52; Port Hudson, 51; 50 left behind by Johnston, unspiked, at Yorktown; Fort Sumter, 48; Fort Pulaski, 48; and Missionary Ridge, 42. Most of these capture-events involved garrison surrender. Only three were open field battles, and, sadly, two of them ended in the humiliating rout of the Army of Tennessee (Winschel, *Triumph & Defeat*, 185; L. J. Daniel and Bock, *Island No. 10: Struggle for the Mississippi Valley*, 145; "The Opposing Forces in the Operations at New Orleans, La.," 2:75; Frye, "Drama between the Rivers," 33; O. T. Gibbes's report ["Statement of captured stores taken at battle of Chickamauga"], *OR*, vol. 30, pt. 1, 40–42; Gott, *Where the South Lost the War*, 263; Hood report, cited above; J. G. Barnard, "Journal of the Siege," *OR*, vol. 11, pt. 1, 337; Hearn, *The Capture of New Orleans 1862*, 274; N. P. Banks to H. W. Halleck, July 10, 1863, *OR*, vol. 26, pt. 1, 55; Hewitt, *Port Hudson*, 185–87; Bryan, "The Siege of Yorktown Part II," 28; Swanberg, *First Blood*, 307; Schiller, *Sumter is Avenged!*, 64; Downey, *Storming of the Gateway*, 1951).

After that, the "Top List" falls down to such lower figures as 28 pieces of artillery surrendered by the Federal garrison at Plymouth, North Carolina (Carbone, *The Civil War in Coastal North Carolina*, 115); 27 cannon of various calibers lost by McDowell's army at Manassas (E. P. Alexander, "Return of captures and abstract of prisoners taken," October 12, 1862, *OR*, vol. 2, 571); 24 guns captured by Federals at Fort McAllister (Durham, *Guardian of Savannah*, 165); 24 cannon lost by Early at Cedar Creek (Mahr, *The Battle of Cedar Creek*, 334); 23 guns lost by Milroy at Second Winchester (Maier, *Gateway to Gettysburg*, 304); 22 guns lost by Federals at Gaines' Mill (Sears, *To the Gates of Richmond*, 249); 21–23 guns lost by Pope's army at Second Manassas (Naisawald, *Grape and Canister*, 455 n.52); 18–20 cannon captured by the Federals in their attack at Spotsylvania, May 12, 1864 (Matter, *If It Takes All Summer*, 267; William N. Pendleton report, February 25, 1865, *OR*, vol. 36, pt. 1, 1044 ["the enemy could thus boast of getting twenty guns"]); 18 cannon captured by Confederates at Glendale (Sears, *To the Gates of Richmond*, 307); 16 guns taken by Forrest at

In flight from the battlefield the Confederates lost so many cannon and wagons that their retreat was aided by the absence of this *impedimenta.* The army's route south was the same as it had been to Nashville: down the pike to Franklin, Spring Hill, and on to Columbia. A Columbia resident described the Confederates as "the worst looking and most broken down looking set I ever laid eyes on." Yankee cavalry dogged them along much of the way. General Wilson was up at 3 A.M. on December 17, promising Thomas that he would "move forward at once." The Union cavalry advanced and ran once more into the Rebel rearguard, of which General Lee had taken charge. As Hood's despondent troops marched to and then through Franklin, Lee formed a battle line on the north bank of the Harpeth while engineers laid their pontoon bridge for the crossing of the army's trains and wounded. In afternoon fighting the lieutenant general was struck in the foot by a piece of shell; hours later he handed command over to Carter Stevenson. That day Forrest's command began to rejoin the army to help Chalmers's cavalry with the rearguard. The army bivouacked around Spring Hill, where it had spent that momentous night two and a half weeks before. On the 18th, with Wilson's cavalry continuing to harass them, the Confederates managed to get across rain-swollen Rutherford Creek, burning the bridge behind. Wilson had to spend a day or more rebuilding it and waiting for rations and forage to catch up with his column. This bought some time for Hood's troops to push southward to Columbia. When General Forrest arrived there, Hood put him in charge of forces guarding the army's rear, adding eight brigades of infantry under General Walthall in support. One soldier from Georgia recalled that General Walthall remarked that he had never sought a hard place for glory nor a soft one for comfort

---

Brice's Cross Roads (Bearsss, *Forrest at Brice's Cross Roads*); 14 guns lost by Hooker's army at Chancellorsville (report of Brig. Gen Henry J. Hunt, commanding artillery, August 1, 1863, *OR*, vol. 25, pt. 1, 253); "about fourteen guns" seized at Gaines' Mill (Krick, "Decisive Charge of Whiting's Division," 203) and the dozen guns taken by Hood's soldiers in their attacking battle of July 22 (S. Davis, *What the Yankees Did to Us*, 124 n.26).

Larry Daniel states that on the first day of Shiloh, the Confederates captured 33 pieces of artillery, but the Federals retook many of these on the second day of the battle. In their retreat, the Southerners could not haul off some guns for want of horses, but probably came off with a net of four cannon to their advantage (Daniel, *Cannoneers in Gray*, 44).

but took his chances as they came. Forrest and Walthall would find their hard place in protecting the army's rear in the next stages of its retreat.[33]

Hood had a decision to make. He needed a river between his army and Thomas, and there were only two under consideration: the Duck at Columbia, and the Tennessee more than sixty miles to its south. On the night of December 18 he discussed the question with Charles Quintard. The chaplain urged that Hood should try to hold a line at the Duck; falling back across the Tennessee would admit the defeat of the campaign, give the North even more to holler about, and provoke desertion among Tennesseans in the army. "The General is not decided in his future movements," Quintard concluded in his diary entry that night. The next morning Forrest arrived at Hood's headquarters and offered the opinion that the army should retreat across the Tennessee. "If we are unable to hold the state," as Quintard recorded the thinking, "we should at once evacuate it." Hood accepted the advice and made up his mind to continue the army's retreat. One Confederate remembered that at some point General Hood, riding along with his marching soldiers, was asked when he would grant furloughs. "After we cross the Tennessee," he answered, adding "the cards have been fairly dealt, for I cut them and dealt them myself, and the Yankees have beat us in the game." This caused one veteran to call out, "Yes, General, but they were badly shuffled."

Hood's chief chore after the battle of Nashville was to get the army southward to safety. This he achieved in good measure, thanks to the tenacity of Forrest's rearguard, Wilson's fagging pursuit, some favorable luck (getting a pontoon bridge across the Tennessee and driving off two snooping Yankee gunboats), and the Yankees' halt of their pursuit at the river. The army's itinerary was recorded by Col. Edwin J. Harvie, adjutant inspector general.

> *December 16.*—Yankees charged our lines and were repulsed. About 3 o'clock the whole line gave way in confusion, losing guns and property, retreating in great disorder to Franklin.

---

[33] Hay, *Hood's Tennessee Campaign*, 171; McDonough, *Nashville*, 271 ("worst looking set"); Wilson to Whipple, December 17 (3 A.M.), *OR*, vol. 45, pt. 2, 237; Hattaway, *Stephen D. Lee*, 145–46; Lindsay, "The Retreat from Nashville," 311; Henry, *"First With the Most,"* 411–12; Longacre, *Union Stars to Top Hat*, 189–91; Sword, *Embrace an Angry Wind*, 407 (Walthall); Bohannon, "'He exhibited the highest soldierly qualities,'" 217–18, 236; McWhirter, "Soldier in the Western Army," 623.

December 17.—Army falling back. Headquarters to-night near Spring Hill.

December 18.—Headquarters to-night at Columbia.

December 19.—Headquarters still at Columbia.

December 20.—Moved to Pulaski.

December 23.—Left Pulaski for Bainbridge.

December 24.—Some of the army reached Bainbridge to-day; the rest on the route.

December 25.—Army at and around Bainbridge.

December 26.—Pontoon bridge was completed to-day and the army commenced crossing. Headquarters at Bainbridge.

December 27.—Headquarters moved to Tuscumbia to-day.

December 28.—Headquarters at Tuscumbia.

December 29.—Moving on toward Corinth.

December 30.—Headquarters at Burnsville, Miss.

December 31.—Headquarters at Burnsville.

January 1, 1865.—Still at Burnsville.

January 2.—Moved to Corinth in the cars.

This laconic recounting, however, does not tell the tragic story of the army's retreat. "As the army trudged south through intermittent rain and falling temperatures," Anne Bailey writes,

> Men with tattered clothing suffered, and those without shoes covered their bare feet as best they could. Food was scarce, and as the precipitation froze, the ice covered the trees and made it impossible to build fires to dry clothing or warm cold bodies. Then came the snow, and a blustery north wind froze wet clothing. The intense cold was almost a blessing, however, for before the roads froze, the earth was a sea of mud that exhausted the animals as they tried to draw supply wagons and artillery through the mire. Weary horses and mules collapsed for lack of forage.

"Through the bleak and chilling blasts of December 31 our ill-clad, barefooted, hungry soldiers marched in slush and mud, and at nightfall drew their foot-sore and weary bodies into a tentless bivouac," recalled Bromfield L. Ridley. When the emaciated animals could no longer pull their wagons out of a rut or up a hill, the men had to help. In these efforts their lack of shoes was most pitiable. Once General Cheatham ordered an officer, Maj. James D. Porter, to find a hundred well-shod men to push some wagons. In all of Cheatham's Corps, after scouring the ranks, Porter found only two dozen men who could show whole shoes on both of their feet.

The supply problem for Hood's suffering men was made worse by a Union cavalry raid conducted by Brig. Gen. Benjamin Grierson. Riding out from Memphis on December 21, the Federal raiders struck the Memphis & Ohio Railroad, running south from Corinth, the only rail line able to bring supplies from Meridian and other points. The Yankees wrecked track and bridges, knocking the road out of operation from Christmas to January 6. They also burned train cars and wagons loaded with commissary and quartermaster stores headed for Hood's troops. Shipments resumed after that. "By the 6th the line was declared open for traffic," writes Ed Bearss, "and the locomotives, though they coughed like consumptives, were again hauling supplies to Hood's decimated legions."

Given the suffering of Hood's soldiers in their retreat from Nashville, I think we can fairly dismiss one of the war's more durable legends: that they sang merrily on the long march home. Stanley Horn tells the tale. "Hood's soldiers, as they tramped down through the muddy pike through the sleet and rain," he writes, "were singing a song—an improvised parody of 'The Yellow Rose of Texas'—which just about summed up the military situation: "But now I'm going to leave you; / My heart is full of woe. / I'm going back to Georgia to see my Uncle Joe. / You may talk about your Beauregard and sing of General Lee / But the gallant Hood of Texas played hell in Tennessee."

It is not hard to dismiss this legend. First of all, in the hard, cold retreat from Nashville there was little for the men to be merry about. They knew that their army had been routed and badly cut up, which was hardly Hood "playing hell in Tennessee." Then, too, in the last days of December 1864, men of the Army of Tennessee most certainly could not have suspected, much less known, that they would somehow go back to Georgia and fight under Joe Johnston. After reviewing the sources, Stephen Hood concludes that the "played hell in Tennessee" lyric "appears nowhere in the many memoirs, letters, or diaries" of Hood's veterans.

Those who had endured the hardships of the long retreat later expressed pride in their grim accomplishment. "You talk of hardships," wrote W. D. Gale on January 15. "You should have been with us during the march into and out of Ten. The mud & water, the sleet & snow & ice & rain. The sleeping on the ground, the bare & bleeding feet of the men as they limped along." Amidst these travails hundreds of men, especially Tennesseans, deserted, slipped away for home, or just fell out of the line of march. Carroll H. Clark of the 16th Tennessee recorded, "Several of the boys in our regiment said it was 'no use going any further.'" Clark saw them walk off "and never thought hard of them for leaving." Hay calculates that about 18,000 men took part in Hood's retreat, but that when it reached the Tennessee River on Christmas Day, "the army probably numbered probably less than 15,000 men and had lost all coherence and morale." Given the number of troops who had set out for Tennessee the month before, one commentator concludes "the army had been almost carved in half." "The demoralized army became disorganized and was rapidly degenerating into a rabble," wrote one Southerner. "The days of the Confederacy were numbered and it was easy to foresee that its extinguishment was near."[34]

From Corinth the army moved southward to Tupelo; its arrival there ended Hood's retreat from Tennessee. To all observers, after the battle of Nashville the Confederate Army of Tennessee was finished as a cohesive fighting force. "General Forrest expressed himself to this effect many times," recalled the cavalry wizard's chief of artillery, John W.

---

[34] Elliott, *Doctor Quintard*, 200–201; Finlay, "Another Report on Hood's Campaign," 406 ("badly shuffled"); Wills, *Battle from the Start*, 289–90; "Itinerary of the Army of Tennessee," *OR*, vol. 45, pt. 1, 675; Bailey, *Chessboard*, 164–65; Ridley, "The Last Night of Sixty-Four," 539; Horn, *Decisive Battle of Nashville*, 154–55 (Porter); Bearss, "Grierson's Winter Raid on the Mobile and Ohio Railroad," 37 (my thanks to Greg Biggs, of Clarksville, TN, for a copy of this article); Horn, "Nashville During the Civil War," 22 (the song); Hood, *Rise, Fall, and Resurrection*, 283; McMurry, *Hood*, 181–82 (including Gale letter); McDonough, *Nashville*, 271 (Clark); Losson, *Tennessee's Forgotten Warriors*, 242–43; Stockdale, *The Death of An Army*, f.p. 111 (map, "Hood's retreat December 16, 1864 to January 13, 1865"); Hay, *Hood's Tennessee Campaign*, 176 (halted Federal pursuit), 180; D. Smith, *In the Lion's Mouth*, 258 ("carved in half"); Gayarre, "Military Operations of General Beauregard," 441 ("demoralized army"). Recently, O.C. Hood has counted Confederate casualties in Hood's retreat from Nashville, which he dates from December 17 to December 26: 9 dead (killed, mortally wounded, or drowned), 1 wounded, 103 captured (O.C. Hood, *Army of Tennessee Retreat*, 199-213).

Morton. Few would disagree with Surgeon George W. Peddy of the 56th Georgia, who wrote his wife two days after Christmas: "Hood is a compleet failure.... We have nothing left but a remnant of a demoralized army." "Our Campaign has been the most disastrous of the war," he added. "It seems that they are not going to get over it Soon," commented another soldier, W. H. Reynolds, to his wife on January 15, "especially if Gen. Hood remains in Command." "Genl Hood's army is...the worst whipped army you ever saw," wrote L. B. Welch to Confederate brigadier Basil Duke on January 17. Mobile newspaper editor John Forsyth was relating what he had probably only heard when he wrote General Bragg in Richmond on January 17 that Hood's "is a shattered debris of an army now." Other papers blasted Hood for his failure. "He has been tried and found palpably wanting"; "the army has no confidence in Hood—they had none at first"; "the campaign was above Hood's capacity to manage"; "Hood stepped out of the way, hat in hand, and asked Sherman to walk through Georgia," is a sampling of the invective from just two Richmond papers, the *Whig* and *Examiner.* Several decades after the war, former major D. W. Sanders framed the events as many of Hood's survivors doubtless would have: "General Hood, in a campaign of thirty-eight days on the north side of the Tennessee River, fought the battle of Franklin and Nashville and wrecked his army."

Historians have echoed the verdict. "The Army of Tennessee suffered so severely in the Nashville campaign that it no longer was an effective fighting force" (Hattaway); Hood "had all but wrecked his army" (Horn); "the Army of Tennessee, one of the Confederacy's two major armies, had essentially ceased to exit" (McDonough); "the entire army was on the verge of collapse" (Sword). "The disaster at Nashville had cost it sixty-seven hundred casualties and had destroyed it as a viable combat force," writes John D. Fowler, who adds that "the Army of Tennessee was finished, and the war in the western theater was essentially over." "Thomas' campaign broke the back of the Army of Tennessee," write Gallagher and his co-authors (peculiarly giving the Union commander's name to a campaign that usually bears Hood's); afterward, "it no longer existed as an army in the true sense." "After the Battle of Nashville," Emory Thomas concludes, "the Army of Tennessee offered little in the way of resistance to further Federal conquest." Destruction of the Confederacy's second most important army leads Stanley Horn to call Nashville the battle that "decided the fate of the Confederate States of America." Charles P. Roland agrees: "the battle of Nashville was the most decisive of the war." But it was not just the outcome of the battle fought on December 15–16. Leading his army into Tennessee, Hood conducted what Richard McMurry has called "probably the most poorly planned and executed major campaign of the war."[35]

A man courageous on the battlefield, John Bell Hood should have owned up to his responsibility for this disaster, but he did not. On December 17 Hood had wired General Beauregard and Secretary Seddon from Spring Hill that the day before his line had been broken at Nashville, his troops were retreating toward Franklin, and that he was "moving to the south of Duck River." The problem was, though, that Hood's telegrams took so long to get to his superiors. His request of December 7 that Beauregard visit the army before Nashville did not reach the Creole for at least a week; it was routed through Mobile on December 12, picked up by Colonel Brent, Beauregard's A. A. G. at Montgomery, and from there transmitted to the commanding general, who was then at Macon. (Telegrams to Richmond took even longer; Hood's wire to Seddon sent December 7 was received in the capital on the 18th). Almost a week after Hood sent his telegram from Spring Hill, Beauregard had still

---

[35] Morton, *Forrest's Artillery*, 300; Cutino, ed., *Saddle Bag and Spinning Wheel: Being the Civil War Letters of George W. Peddy, M.D.*, 204; McMurry, *Hood*, 182 (Reynolds and Welch); Forsyth to Bragg, January 17, 1865, *OR*, vol. 52, pt. 2, 808; Bergeron, *Confederate Mobile*, 109 (Forsyth as editor, *Mobile Advertiser and Register*); Dyer, *Gallant Hood*, 306 (quoting the *Richmond Whig* and *Examiner*); D. W. Sanders, "Hood's Tennessee Campaign," 404; Hattaway, *Stephen D. Lee*, 146; Horn, *Army of Tennessee*, 421; McDonough, *Nashville*, 274; Sword, *Embrace an Angry Wind*, 426; Fowler, *Mountaineers in Gray*, 174; Gallagher et al., *American Civil War*, 285; Thomas, *Confederate Nation*, 281–82; Horn, *Decisive Battle of Nashville*, 166; Roland, *An American Iliad*, 234; McMurry, *Two Great Rebel Armies*, 130–31.

Stanley Horn was not alone in emphasizing the decisiveness of the battle of Nashville. He mentions that the British military historian J. F. C. Fuller argued that "it was Thomas's victory at Nashville and not Sherman's march through Georgia and the Carolinas that settled the war in the West, and thereby decided the result of the war" (Horn, "Most Decisive Battle of the War," 4). After the war, both Thomas and Schofield resented the acclaim being accorded to Sherman's march through Georgia, when their battles at Franklin and Nashville "were a great deal more important," Wesley Moody writes. The two Union generals bristled that Sherman's march "caught the public's attention while Franklin and Nashville were pushed to the periphery" (Moody, *Demon of the Lost Cause*, 72).

not gotten it, as evidenced by the Creole's wire of December 23 to Brent, "Should General H. be unable to gain any material advantage in Tennessee with his present means, he must retire at once behind the Tennessee River." Unaware of Hood's message from Spring Hill about the reverse at Nashville, Brent confessed to Beauregard, who had returned to Charleston, "Have nothing official from Hood." The colonel dutifully obeyed Beauregard's instruction and over his name telegraphed Hood: "Should you be unable to gain any material advantage in Tennessee with your present means you must retire at once behind the Tennessee River, and come with or send to Augusta, by best and quickest routes, all forces not absolutely required to hold defensive line referred to."[36]

Hood could claim that he had communicated to Richmond on December 17 about his defeat at Nashville, even if he had tried to minimize its consequences (killed and wounded "very small"; loss in prisoners "comparatively small"). What he had not done was to send any message as to the condition of the army or announce that as of December 18 he had decided to retreat across the Tennessee River. First clue for Beauregard was when Colonel Brent received a message from S. D. Lee, dated December 23 from Florence, that he would soon be in Okolona and would like to discuss with the department commander "recent events in Tennessee." Brent wired Beauregard on Christmas Day that he did not understand Lee's message, but added, "am apprehensive that some reverse may have occurred." Lee's biographer Hattaway asserts, "Hood had asked Lee to see the Creole and try to explain the events in Tennessee." If Hattaway is right, then Hood was doing more than hiding behind a slow-moving telegraphic system; he was refusing to inform his superiors of the calamity at Nashville and that his campaign had ended in failure. Worse, he was asking a junior officer to break the unpleasant news.

On December 25 Hood wired both Seddon and Beauregard that he was at Bainbridge, Alabama, about to lay a pontoon bridge across the Tennessee. But enemy possession of the north Alabama telegraph lines meant that the message had to be transmitted via Corinth; the *Official Records* suggest it was not sent from there until January 3. To Beauregard

---

[36] Hood to Seddon, December 17, *OR*, vol. 45, pt. 2, 699; Hood to Beauregard, December 12 (*sic:* December 7, "via Burton 12th & Mobile 12th") and Beauregard to Brent, December 23, Roman, *Beauregard*, 2:626; Brent to Beauregard, December 24, *OR*, vol. 45, pt. 2, 729; Beauregard to Hood, December 23, *OR*, vol. 45, pt. 2, 726.

he added a request that his superior "please come to Tuscumbia or Bainbridge." Again he avoided adding any detail as to the plight of his army. "That was all it said," notes Shelby Foote.

Even more self-serving and duplicitous was the letter, dated December 25 from Tuscumbia, written by Governor Isham Harris of Tennessee to President Davis. The governor was going to entrust the epistle to Col. J. P. Johnson, one of Adjutant General Cooper's staff officers then with Hood's army ("I will not detain Colonel Johnson"). The conclusion is unmistakable that Harris was writing the president at Hood's request, and that the content of his letter was at least approved by the general if not dictated by him. Harris mentioned "the unfortunate result of the battle at Nashville, and immediate retreat of the army," but offered no details. Again, the army commander (assuming that Hood was behind the letter) could not brace himself to tell his government directly what he had done or what had befallen his officers and men.

Both Hood and Harris knew that Colonel Johnson's train trip to Richmond would take at least three or four days. By then they could expect that news of Nashville and the retreat and failure of the campaign would have reached Richmond not just from rumor or word of mouth, but from Northern newspapers, which customarily reached the Confederate capital within a day or two after printing. The Yankee press was hailing Thomas for his victory. As early as December 19, John Beauchamp Jones, the war clerk, entered into his diary, "Northern papers say Hood has been utterly routed, losing all his guns!" To be sure, the Southern press regularly discounted Yankees' gloating in their war reports, which they dutifully passed on to readers with appropriate caveats. "The news from Tennessee is through United States sources," observed the *Richmond Sentinel* on December 20, "and requires to be received with what the lawyers call 'abundant caution.'" Still, after "making all reasonable allowance," the paper had to conclude that the news "is still unpleasant"—that Thomas had attacked at Nashville, broken the Confederate line on its left, and forced Hood's army to retreat at least several miles. The next day the *Sentinel* quoted a Washington paper, citing its report that the enemy in Tennessee boasted of capturing 5,000 prisoners and forty-nine artillery pieces in the recent battle. All of this swirl led the *Richmond Examiner* on December 21 to declare, "The news from Tennessee is bad.... The army in Tennessee has been terribly misused and has suffered great injury."

Amidst this swirl of negative news reaching Richmond before Christmas, Governor Harris's gratuitous compliment to Hood's generalship—written to Jefferson Davis on December 25—was as unctuous in nature as it is obvious in intent:

> I have been with General Hood from the beginning of this campaign, and beg to say, disastrous as it has ended, I am not able to see anything that General Hood has done that he should not, or neglected anything that he should have done which it was possible to do; indeed, the more that I have seen and known of him and his policy the more I have been pleased with him; and regret to say that if all had performed their parts as well as he, the results would have been very different.

Was this last statement another slap at Frank Cheatham for Spring Hill? Probably. Then, as if G. T. Beauregard needed another slap, Harris closed his letter by advising the president, "if General Hood is to command this army he should, by all means, be permitted to organize the army according to his own views of the necessities of the case."[37]

News of the Army of Tennessee's disastrous campaign did indeed begin to leak out. On December 25, Andrew J. Watt, Richard Taylor's A. A. G. in Meridian, wired an officer, "Rumors from above, and the official reports of the enemy, seem to establish the fact that General Hood has sustained a serious reverse," with the consequence that "Hood [could] cross the Tennessee." General Taylor had Watt telegraph this

---

[37] Brent to Beauregard, December 25, *OR*, vol. 45, pt. 2, 731; Hattaway, *Stephen D. Lee*, 148; Hood to Seddon and Beauregard, December 25 ("Via Corinth," January 3, 1865), *OR*, vol. 45, pt. 2, 731; Foote, *Civil War*, vol. 3, 757; Harris to Davis, December 25, *OR*, vol. 45, pt. 2, 732; J. B. Jones, *A Rebel War Clerk's Diary* (2015), 2:324; "FROM TENNESSEE," *Richmond Sentinel*, December 20; "ONE DAY LATER FROM THE UNITED STATES," *Richmond Sentinel*, December 21; J. M. Daniel, *The Richmond Examiner during the War*, 214; Elliott, *Dr. Quintard*, 107 n.26 (on Johnson, but missing that he was with the Army of Tennessee till at least December 25); S. D. Elliott, *Isham G. Harris of Tennessee*, 176.

Governor Harris's letter to the president is reprinted in full in Crist, et al., *Papers of Jefferson Davis*, 11:248–49. In it the governor apparently underlined *all*: "if *all* had performed their parts as well as he…." Hood included Harris's entire text in his memoir, also emphasizing "if *all* had performed their parts" (*A & R*, 306). I do not see in Crist, et al., reference to when Harris's letter reached the Confederate White House.

news to other officers; Colonel Brent would have been among them, allowing Beauregard to share apprehension that "some reverse may have occurred." Still, the military division commander had received no full and definitive report from Hood. "No intelligence yet from General Hood," Brent wrote on December 27; he informed another officer that his last message received from Hood was dated December 15 (about the assignment of an officer of engineers, one Capt. W. G. Hazlehurst). Nonetheless, by December 27 some word from Hood, probably his wire of the 17th, must have reached Colonel Brent in Montgomery, as in a telegram to Hood he referred on the 27th to "your retrograde march." Brent warned that enemy cavalry in north Mississippi could threaten Hood's line of supply. "Be good enough to let us know your plans," Brent in effect pleaded, "so that we may provide for you accordingly." Later that day Brent sent another message, "Let us know by courier, via both Columbus and Talladega, of your plans." Also on the 27th Beauregard complained about Hood to General Cooper in Richmond: "I have no advices whatever from that quarter." To get the facts he therefore announced his intent to travel to Hood's army at Tupelo and see for himself.

Even in the winter of his discontent, Hood still had his supporters in Richmond society. Isabella Martin had all along been a "Hoodite," as Mrs. Chesnut characterized her. When Louis Wigfall, the Texas senator, happened to pay a compliment Joe Johnston, Isabella snapped, "Well, if the end has come, try Joseph the Xenophontic. He will retreat the last thousand with grace and dignity, down into the Gulf of Mexico (and misery)."[38]

---

[38] A. J. Watt to Maj. Gen. Franklin Gardner and Watt to Maury, December 25, *OR*, vol. 45, pt. 2, 733–34; Brent to Beauregard, Brent to Mackall, and Beauregard to Cooper, December 27, *OR*, vol. 45, pt. 2, 738–39; Brent to Hood, December 27, *OR*, vol. 45, pt. 2, 739–40 ("we have no advices, since telegram of the morning of the 15th instant, about Captain Hazlehurst"); Hood to Beauregard, December 15, *OR*, vol. 45, pt. 2, 690 (Hazlehurst); Woodward, *Mary Chesnut's Civil War*, 627 (entry of July 26, 1864), 698 (Wigfall-Martin conversation, under date of December 27, 1864).

McMurry scolds Hood for withholding news of his defeat in Tennessee. "Even after he reached Tupelo," he writes, "Hood's reports were skimpy at best, misleading at worst" (*Hood*, 183).

"Joseph the Xenophontic" is an arcane allusion to Joseph E. Johnston. Christopher J. Tuplin's column for Xenophon (c. 430–350 BCE) in the *Encyclopedia Britannica* offers a clue as to why Miss Martin compared Joe Johnston to

Lt. Gen. Richard Taylor wrote one of his officers on December 27 that "information from Federal sources, and confirmatory intelligence on our part, satisfies me that General Hood has met a serious reverse and is retreating from Tennessee." "No tidings yet from General Hood," Brent telegraphed Beauregard two days after that. Nevertheless, aswirl all the rumor, headquarters assumed that the Army of Tennessee had retreated across the Tennessee. Brent wired an officer in Corinth that "if General Hood has not recrossed the Tennessee, all stores at Corinth should be moved toward Tuscumbia to supply his army." Hood had already recrossed his army, which he announced in telegrams from Bainbridge on December 25 to both Seddon and Beauregard. Again, however, the circuitous path of transmission meant that neither the military division commander nor the secretary of war knew what Hood was doing. Finally, from Taylor on December 31 came report "on authority of an officer from General Hood's army" that it had crossed the Tennessee at Bainbridge on the 26th and 27th.

Beauregard may have had this information when on December 31 he telegraphed President Davis, "on reaching the Army of Tennessee, am I authorized to appoint General Taylor to its command, should I find its condition such as to require a change of commander?" Either Beauregard had gotten so exasperated with Hood's noncommunication and virtual insubordination, or Taylor's confirmation that the Tennessee campaign had ended in failure brought Beauregard to the conclusion that he should relieve Hood of command. As if to set the stage for this act, Colonel Brent sent another telegram to Hood, repeating that General Beauregard wanted to hear from him; that he should "advise by telegraph" and "write fully the condition of the army." On January 2 Beauregard left Charleston for Montgomery, wiring Cooper. His message—"No report from General Hood since December 3. I leave to-day for the West"—seems to be a further, careful effort to set the stage for relieving Hood from army command. On the day Beauregard left for "the West," Davis answered

---

an ancient Greek soldier, historian, and philosopher. Speaking of a "typical Xenophontic puzzle," Tuplin declares, "his Socratic world often resembles a sanitized version of reality; Xenophon created a fictive history in which propositions about the pursuit of virtue—though they derive authority from being rooted in the past—acquire either a mythical aura or an intriguing piquancy through the use of a deviant version of the past." The keys appear to be in "sanitized version of reality," "fictive history," and "deviant version of the past," proclivities toward which critics of General Johnston have long accused him.

his question: "If you find it necessary to make the change suggested, you are authorized to employ General Taylor as proposed." (It is likely that Governor Harris's letter, borne by Colonel Johnson, had reached Richmond by then. If Johnson discussed the pathetic state of the army when he had left it around Christmas, the president would have had sound reason to grant Beauregard's request so quickly.) Beauregard left Charleston at 11 A.M. on the 2d; he may not have seen Davis's telegram, which was sent to Brent's office at Montgomery. But there it would be waiting for him on his arrival. Just as he was about to see Hood, the departmental commander had been given authority to relieve him from command.[39]

---

[39] Taylor to D. W. Adams, December 27, *OR*, vol. 45, pt. 2, 741; Brent to Beauregard, December 29, *OR*, vol. 45, pt. 2, 745; Brent to Col. William R. Miles, December 29, *OR*, vol. 45, pt. 2, 746; Hood to Seddon and Beauregard, December 25, *OR*, vol. 45, pt. 2, 731; Brent to Beauregard and Beauregard to Davis, December 31, *OR*, vol. 45, pt. 2, 749; Brent to Hood, January 1, 1865, *OR*, vol. 45, pt. 2, 751; Beauregard to Cooper and Davis to Beauregard, January 2, *OR*, vol. 45, pt. 2, 753; Brent to Taylor, January 2, *OR*, vol. 45, pt. 2, 756 ("left Charleston to-day at 11 o'clock"); Roman, *Beauregard*, 2:332 ("the power with which he had been clothed" [to relieve Hood]).

Chapter 5

Final Months of the War

Once at Tupelo Hood was able to resume regular correspondence with Montgomery and Richmond, but his efforts to do so were less than energetic. For instance, by January 3 he had gotten Beauregard's/Brent's telegram instructing him, once on the south side of the Tennessee River, to send troops eastward to Augusta—they would be needed to contest Sherman's army about to raid through South Carolina. He must have seen it at least a day earlier, as General Forrest mentioned to General Taylor on the 2nd, "I have just had an interview with General Hood, and am informed by him that the Army of Tennessee is ordered to Augusta." Hood did not acknowledge receipt of these instructions, which essentially would take his army away from him. Instead, in telegrams to Seddon and Beauregard on January 3, he wrote, "the army has recrossed the Tennessee River, without material loss since the battle in front of Nashville"—declining to explain what "material loss" meant. He stated that the army would be assembled at Tupelo and resupplied with shoes and clothing. In a separate message to Beauregard, he stated that the army would need some time to rest and refit, during which time he asked the military division commander to come and visit it. Addressing Beauregard's order of January 1 that he "write fully [on] the condition of the army," Hood became defensive, offering only that he considered his telegram of December 17 to be an adequate report on the battle of Nashville and stating only that he considered withdrawal from Tennessee to be his best course. Also on the 3rd in a telegram to the president, Hood continued to be evasive. "Lieut. Col. J. P. Johnson will explain to you the campaign in Tennessee"—further indication that Hood considered Governor Harris's letter en route to Richmond to be what Hood wanted the government to know—and turned instead to the matter of furloughing Trans-Mississippi troops in his army.[1]

General Beauregard's patience was nearing an end. He was at Macon by January 7, when he instructed Colonel Brent, "Order General Hood, in writing, to make report of his operations from Tuscumbia to Nashville, and back to Tupelo." Meanwhile Richard Taylor was traveling from Meridian to meet Hood at Tupelo. He was there and back by the 9th, when he reported to President Davis on the condition of the Army of Tennessee. Not only did it need "rest, consolidation, and reorganization": "if moved in its present condition it will prove utterly worthless." Thus when Beauregard informed Hood that by presidential order he was to send troops to help Hardee contest Sherman's advance, Hood explained that he could not do so before Beauregard's arrival in Tupelo. The military division commander was still on his way there on the 13th, when he met with General Taylor at Meridian. Davis had earlier asked Taylor to meet the Creole: "General Beauregard went to Army of Tennessee with large discretionary power. Your name was referred to, and it would be well for you to see him." Thus Taylor learned, if he did not already know, that Beauregard intended for him to replace Hood in command of the army. In this context Taylor would have told Beauregard even more about the sad state of Hood's command. Based on Taylor's report of the "disorganization and demoralization of the Army of Tennessee," Beauregard wired Davis, no troops from it could likely be sent to Hardee in South Carolina. Besides, he added, Hood at the time had no more than 15,000 infantry—a number he could only have gotten from Taylor after the latter's visit to Tupelo.

Having tried for the past month and a half, beginning with Franklin, to shield from his superiors the magnitude of his failures as army commander in Tennessee, and with Beauregard now coming to inspect the army personally, Hood had run out of room for denying and dissembling. On January 13 he telegraphed Secretary Seddon: "I respectfully request to be relieved from the command of this army." It is not known whether or how Hood may have sensed that Beauregard was coming to meet him armed with the authority to relieve him of duty. Dyer implies

[1] Beauregard to Hood, December 23, *OR*, vol. 45, pt. 2, 726 ("Received January 3, 1865"); Forrest to Taylor, January 2, *OR*, vol. 45, pt. 2, 756; Hood to Seddon, Beauregard and Davis, January 3, *OR*, vol. 45, pt. 2, 757–58.

that Hood's request of January 13 was a preemptive act to save himself from the humiliation of being fired.[2]

The speed with which the government granted Hood's request says as much about its eagerness to relieve him as it does about his utter failure as army commander. Within two days, Secretary Seddon in Richmond replied, "Your request is complied with," adding that General Beauregard was instructed to formally relieve him. Once that formality was performed, Hood was to report to the capital for further orders.

In the meantime, Hood could not tell his troops, but he had to continue issuing orders as commanding general. Food was in short supply; he asked that rations be brought up quickly from the rear. Soldiers were roaming about the countryside committing "depredations daily"; these men had to be reined in and disciplined. Officers were instructed to look out for deserters and bring them back in for punishment. Returns dated January 20, 1865, list 1,795 officers and 16,913 men present for duty, for a total of 18,708. "Over the span of little more than a month Hood's army had been wrecked," Wiley Sword writes. Between September 7, 1864, and January 20, 1865, he tallies Hood's losses as 13,189 POWs, 2,000 deserters, and 72 field pieces. These figures, together with 6,300 K/W at Franklin and 2,300 at Nashville, allow Sword to stake Hood's casualties for his Tennessee Campaign at more than 23,500. But numbers alone did not tell the full story. When Beauregard arrived at Tupelo on the 15th he could readily see how the army was in danger of coming apart. "Very little—if anything—remained of its former cohesive strength," writes Roman. "If not, in the strictest sense of the word, a disorganized mob, it was no longer an army." Sergeant Chambers of the 46th Mississippi confirmed this dismal view in his diary entry of January 16:

---

[2] Beauregard to Brent, January 7, *OR*, vol. 45, pt. 2, 769; Taylor to Hood, January 3, *OR*, vol. 45, pt.2, 759; Taylor to Davis and Beauregard to Hood, January 9, *OR*, vol. 45, pt. 2, 772; Hood to Beauregard, January 10, *OR*, vol. 45, pt. 2, 773; Beauregard to Davis, January 13, *OR*, vol. 45, pt. 2, 780; Davis to Taylor, January 12, *OR*, vol. 45, pt. 2, 778; Hood to Seddon, January 13, *OR*, vol. 45, pt. 2, 781; Dyer, *Gallant Hood*, 304.

Stephen Hood affirms, "many modern-day scholars accuse Hood of concealing the army's depleted condition from Richmond, both after Franklin and even after the retreat from Nashville." I tend to disagree with his contention that "these accusations directly conflict with the written historical evidence" (Hood, *Rise, Fall, and Resurrection*, 196).

> The regiment numbers about one hundred and fifty men, about half of whom are barefoot. All are ragged, dirty and covered with vermin, some not having sufficient clothing to hide their bodies. There are perhaps twenty guns, but not a single cartridge box in the regiment. The men are jovial enough in regard to their condition and their environment, but when it comes to discussing the prosecution of the war they are entirely despondent, being fully convinced that the Confederacy is *gone.*

Earlier opposed to Hood's proposal for mass furloughs of his soldiers, Beauregard now endorsed it as a means of raising the men's broken spirits. Troops from across the Mississippi were sent home, as were men from Tennessee regiments. The rest of the army was ordered to head by rail eastward, to help stop Sherman's forces, which were then about to head out from Savannah for South Carolina. As for the issue of removing Hood, Beauregard was relieved to learn that Hood himself had made the request. The Creole, according to Roman, "had not the heart to virtually disgrace him," so he kept the president's instructions quietly in his pocket. Seddon telegraphed Beauregard that he was to place Richard Taylor in command. Taylor did not want the job, however, and on the 18th asked the president to reappoint Joe Johnston instead. "The army and the public unanimously desire General Johnston," he wrote; "this alone will restore confidence.... Johnston should be sent at once." Jefferson Davis would do no such thing, but he did not say so. After Taylor arrived at Tupelo on the night of the 18th, he and Hood spent several days arranging the change of command.

Meanwhile word leaked out that day among the officers and men. Henry Watterson, the *Mobile Advertiser & Register*'s correspondent with the army, writing as "Shadow," sent on January 18 a wire from Tupelo announcing that Hood had been relieved and would be succeeded by Dick Taylor.

The next day Watterson wrote a long column reporting the mood of the troops. "The army, rank and file," he observed, "seem to be perfectly indifferent as to any change which may be made, that does not bring back to them their pride and idol, Joseph E. Johnston." (Indeed, the *Jackson Mississippian* a few days earlier had commented, "the feeling among the soldiers is unanimous for the return of Gen. Johnston.")

On the evening of January 18, a throng of soldiers gathered outside Hood's headquarters and called him outside. The band from Randall Gibson's Louisiana brigade serenaded the general with a few operatic airs, then the men requested a speech. "Shadow" described Hood's several points: 1) "he had been placed in command…by the orders of the government and that the position had been entirely unsought by him"; 2) "since that time he had never had the support of his country; but on the contrary, he had been from first to last attacked by the croakers and scribblers in the rear"; and 3) "he entreated them to yield a hearty support to his successor."

"Shadow"/Watterson then noted that Hood retired "without making any allusion to the campaign in Tennessee."

Watterson offered his opinion that Hood was being egotistical in thinking that the "scribblers" had been out to get him. No, the journalist explained, the press and the people were not opposed to Hood; they simply wanted Johnston to be kept in command of the army. "The many arrows which Gen. Hood imagined were poised at him," "Shadow" explained, "were actually aimed at the powers in Richmond."

The reporter made a few more points in his column. He had heard a rumor that since the end of the campaign, "General Hood is said to have remarked…that he had not been mistaken in the campaign, but that he had in the valor of his troops." Watterson termed this "strange language," especially given the army's courageous charge at Franklin and its consequent casualties. The reporter then got his own dig in. "If men who charged so brilliantly at Franklin, gave way before enemy at Nashville," he concluded, "we can but think that there was a lack of confidence on their part in Hood. They had reasoned themselves into the butchery at Franklin was uncalled for, and they were probably not willing to test another experiment."

General Taylor had asked Richmond on January 18 that Johnston, not he, should replace Hood. Finally, on the 23rd, "without waiting longer for a response from the President," as he wrote, Taylor assumed command of the Army of Tennessee. That day Hood issued his farewell to the troops, thanking them for their endurance of so much hardship during the recent campaign. "I am alone responsible for its conception," he magnanimously acknowledged, "and strived hard to do my duty in its execution." Then, urging his men to give their "entire support to the distinguished soldier who now assumes command," Hood left Tupelo for Richmond.

Hood, however, did not consider himself washed up. "I am ready to command a corps or division," he wired the president on January 16. Two days later he sent Davis another telegram: "Please consider the propriety of my going west of the Mississippi River." "I would like it," he added; "I think I can be of more service there than east of the river." Then, several days after that, he sent another message to the president: "I wish to cross the Mississippi River to bring to your aid 25,000 troops. I know this can be accomplished."

But Hood's Confederate military career was over. By the time he reached Richmond on February 8, the Confederate Congress had passed a resolution calling upon the president to reappoint Joe Johnston to command the Army of Tennessee, a move which, the measure asserted, would "be hailed with joy by the army and receive the approval of the country." Five weeks later Hood received a further insult from Congress. The previous summer he had been appointed to the rank of full general only temporarily, as the Congress, not in session at the time, could not confirm the president's appointment. Congress never did, which allowed a senator (probably Wigfall) on March 8, 1865, to propose the following resolution: "Resolved, That General J. B. Hood, having been appointed General, with temporary rank and command, and having been relieved from duty as Commander of the Army of Tennessee, and not having been reappointed to any other command appropriate to the rank of General, he has lost the rank of General, and therefore cannot be confirmed as such."[3]

---

[3] Seddon to Hood, January 15; Harvie to Fleming, January 14; Mason Circular, January 14; Mason to Stewart, January 14, *OR*, vol. 45, pt. 2, 782–83, 785; "Consolidated abstract from returns of the Confederate Army on or about December 31, 1864," *OR*, ser. 4, vol. 3, 989; Sword, *Embrace an Angry Wind*, 424–25; Beauregard to Cooper, January 15, *OR*, vol. 45, pt. 2, 785; Roman, *Beauregard*, 2:331–32; Chambers, *Blood & Sacrifice*, 197 (journal entry of January 16); Grimsley, "Learning to Say 'Enough,'" 41–42; Seddon to Beauregard, January 15; Hood to Davis, January 16 ("Taylor will arrive to-morrow"), *OR*, vol. 45, pt. 2, 784–86; "TELEGRAPHIC," *Augusta Constitutionalist*, January 22 (quoting *Mobile Advertiser & Register* of the 18th); "Shadow," "From the Army of Tennessee. Correspondence of the *Mobile Register*," *Augusta Constitutionalist*, February 1; Andrews, *South Reports*, 489; "GEN. JOHNSTON," *Augusta Constitutionalist*, January 20 (quoting *Mississippian*); Parrish, *Taylor*, 429; Taylor to Brent, January 23, *OR*, vol. 49, pt. 1, 929; "SOLDIERS," January 23, *OR*, vol. 45, pt. 2, 805; McMurry, *Hood*, 183; Hood to Davis, January 16, 18 and 23, *OR*, vol. 45, pt. 2, 786, 792, 804; King, *Wigfall*, 211–14; Wright, *General Officers*, 13.

Alvy King, Wigfall's biographer, suspects that the Texas senator's relationship with Hood, which had been quite friendly for the first part of the war,

John Bell Hood was not yet through. As unbecoming as it was understandable, he began to employ all the means at his disposal to justify his record and salvage what was left of his reputation. Even before he left Tupelo on January 23, he wrote S. D. Lee, asking if he would send his campaign report as an aid to the official account Hood was about to write. "I am anxious to make my own," he affirmed, requesting that Lee comment on "the morale of the army after the fall of Atlanta as we all regarded a change necessary"—apparent reference to Hood's decision to march the army into north Alabama and Tennessee. It is evident that in his report Hood would thus seek to justify his decision to launch a campaign that had ended so disastrously.

Thus en route to Richmond, and maybe before that, Hood had begun to compose his report to the government, which would cover the entire period from his assumption of army command in mid-July 1864. He apparently had an outline or early draft of it when his train stopped in Augusta on February 3 (a crowd serenaded him, and he gave a little speech). There he met his distant cousin G. W. Smith. Either at Hood's request or with his knowing consent, Smith wrote a long defense of his kinsman and arranged for it to be printed in the local *Constitutionalist.* The Augusta paper had already printed a column absolving Hood of blame for the failure of the Tennessee Campaign. "The people may see that the campaign was not so 'ill-starred' and 'fruitless' as has been general imagined," it advised. First of all, Franklin had been a "dearly bought victory," which had led to the enemy being driven from the field. Then, the collapse of the Confederate line at Nashville had been "one of those inexplicable affairs that so mystify and darken the pages of history…. It was wholly unaccountable, and we fear must ever remain so." Hood was simply the victim of circumstance. "That he was not wholly successful and that his name is not now hailed as one of the chiefest of our military

---

changed when he began to suspect that Hood was complicit in Joe Johnston's removal from army command (King, *Wigfall*, 210). The senator's daughter, Louise Wigfall Wright, states that the strain in his relationship with President Davis began earlier, in fall 1862, when "dissatisfaction was felt in many quarters with President Davis's policy and conduct of affairs." Things got so bad that General Longstreet wrote Wigfall on November 7, 1862, expressing concern ("I heard yesterday that you and the President had had an unpleasant interview") and hope that "you not to allow anything to bring about any difference between you" (Wright, *Southern Girl*, 87, 89–90).

leaders," the *Constitutionalist* concluded, "is attributable to the accident at Nashville against which no human intelligence could have provided."

Thus the *Constitutionalist* had already emerged as a friend of John B. Hood's. Smith's article for it, titled "Atlanta and Nashville," published on February 5, made a number of points. "The spirit of the army" had been "almost ruined" by Johnston's constant retreating. Seldom had a young officer been promoted to take charge of an army, the article declared, "under more difficult and trying circumstances." Still, Hood had held Atlanta for more than a month and a half. His battles at Peachtree Creek, east of Atlanta, and at Jonesboro would have succeeded had his orders been properly executed. Indeed, he had done everything he could to save the city. Then, after its fall, he had sought to revive the Confederate fortunes with an offensive into Tennessee. He had attacked Schofield's forces at Franklin before they could unite with Thomas at Nashville. In the end, Hood had had to retreat from there, much as Lee had done in his two raids after Sharpsburg and Gettysburg. Indeed, Hood's Tennessee campaign "will compare favorably with either of Lee's." With these bold assertions, "Atlanta and Nashville" offered such a stout defense of Hood's performance as army commander that it should have carried the name of its author. Instead, the column carried the initials "G. W. Y." Two days later, the Augusta paper printed a correction: the author's initials should have been printed as G. W. S. McMurry has shrewdly deduced that General Smith was the author, and his conclusion is now generally accepted.[4]

Hood's train route to the capital from Augusta took him through Columbia and Chester, South Carolina, and via Charlotte, Greensboro, and Danville. At Chester he stopped to visit friends, staying for a night with the Prestons (Buck was away). Mrs. Chesnut was also there, having taken some rooms in town. "How plainly he spoke out these dreadful words," Mary wrote in her diary on February 5. "'My defeat and discomfiture'—'My army is destroyed'—'My losses'—&c&c. He said he had

[4] Hood, *Lost Papers*, 101–102 (Hood to Lee, January 22, 1865); Snow, *Lee and His Generals*, 417 (Hood serenaded February 3); Hudson, *Odyssey*, 166 (Smith as "distant cousin"); "Hood's Tennessee Campaign," *Augusta Constitutionalist*, January 17, 1865; "G. W. Y.," "Atlanta and Nashville," *Augusta Constitutionalist*, February 5; "Correction," *Constitutionalist*, February 7; McMurry, *Hood*, 183–85; Symonds, *Johnston*, 352 ("Johnston suspected that the author was Gustavus Woodson Smith," but without elaboration).

nobody to blame but himself." Jack Preston, Buck's brother, was there and noticed one evening how Hood sat staring into the fire, ignoring all conversation. "Yes, he is going over some bitter hour," Mrs. Chesnut recorded Jack as saying. "He feels the panic at Nashville and its shame.... And the dead on the battlefield at Franklin.... And that agony in his face comes again and again. I can't keep him out of those absent fits," Jack lamented. "When he looks in the fire and forgets me and seems going through in his own mind the torture of the damned—I get up and come out."

Hood arrived in the Confederate capital on February 8 and took rooms at the Spotswood Hotel. A week later he had his campaign report drafted and ready to submit to the War Department. The speed with which it was written—it runs for sixteen dense pages in the *Official Records*, and twenty-one as an appendix in *Advance and Retreat*—suggests that he had been working on it for quite a while, and that he may have had help with its composition. There is reason to suspect that President Davis may have lent a hand in its content, for its two opening pages are harshly critical of General Johnston's handling of the first phase of the Georgia campaign. (In February 1865 Davis faced criticism from such politicians as Louis Wigfall for having relieved Johnston; Hood's report could serve as the president's rejoinder.) In his report Hood argued that Johnston had so badly weakened the army during his time of command ("lost 22,750 men, nearly one-third of its strength") that Hood's failure in the ensuing campaigns was virtually ordained. "If, as is untruly charged," he asserted, "the Army of Tennessee ceased to exist under my command, it is also true that it received its mortal wound when it turned its back in retreat in the mountains of Georgia." Yet Johnston was not the only culprit. Hardee, as we have seen, received blame regarding the battles around Atlanta, as did Frank Cheatham in Tennessee. "In his report," Vann Woodward wryly observes, "Hood blamed everyone but himself for the plight of the Army of Tenn." Just as bad were Hood's efforts to whitewash the failure of his campaigns. He claimed that "from Palmetto to Spring Hill the campaign was all that I could have desired." Had it not been for the surprising collapse of his left flank at Nashville, "I think we would have gained a complete victory." Astonishingly, he asserted, "Notwithstanding that disaster I left the army in better spirits and with more confidence in itself than it had at the opening of the campaign." Through it all, he claimed that his total loss would not "exceed 10,000 men."[5]

Jefferson Davis was pleased with Hood's report. Governor Harris of Tennessee, in Richmond at the time, got a chance to hear Hood read his draft. Harris guessed that Hood "had been put up to write this by some one else & that it had been seen & approved by either the Pres't or the Secy of War." Through these channels Hood also would have been able to read Johnston's campaign report, which he had submitted in October 1864, and respond to his statements. After the House and Senate in late February called for the administration's correspondence regarding the recent campaign, Davis sent Hood's report to both chambers. Louis Wigfall, probably Joe Johnston's closest friend in the Confederate Congress, did not like it at all. Wigfall at the time was speaking out in the Senate against President Davis, calling him at one point "an amalgam of malice and mediocrity." Clement Clay told Mrs. Chesnut that "the Sen-

---

[5] Woodward, *Mary Chesnut's Civil War*, 708,774–75; Muhlenfield, *Chesnut*, 127 ("her husband…had found three vacant rooms in Chester"); McMurry, *Hood*, 183; Hood report, February 15, 1865, *OR*, vol. 38, pt. 3, 628–36 (re: July 18–September 2); vol. 39, pt. 1, 801–803 (re: September 2–November 21); vol. 45, pt. 1, 652–56 (re: November 21, 1864–January 23, 1865); Hood, *A & R*, 317–37.

In the closing months of the war, in addition to having to defend his decision to relieve Joe Johnston, President Davis also faced pressure from Congress to create the position of general in chief, and to appoint Robert E. Lee to it. "We are not afraid of being conquered by the enemy so much as being defeated by Mr. Davis," bellowed the anti-administration *Richmond Examiner*. "He despises all warnings and snubs Congress. The press of the country, and the army backs it, call on him to keep his hands off and let General Lee be generalissimo of the forces." Davis gave in and on February 6 named General Lee to the new post (Basso, *Beauregard the Great Creole*, 270 [quoting the *Examiner*]; Freeman, *Lee*, abridged by Richard Harwell, 445).

Davis held out as long as he could before bowing to political pressure—and Robert E. Lee's own suggestion—that he reappoint Joe Johnston to lead the Army of Tennessee. Special Orders No. 3, dated February 22, announced that Johnston would take over departments embracing Georgia, Tennessee, and South Carolina. The same order announced that General Beauregard would report for duty to Johnston. Beauregard readily complied (Symonds, *Joseph E. Johnston*, 341–42; Special Orders No. 3, February 22, 1865, and Beauregard to Lee, February 22, *OR*, vol. 47, pt. 2, 1248).

ate were inclined to pity Hood after his failure—that is, until his report came out. Then Wigfall fell foul of him savagely."

But Hood did not stop there. He arranged for G. W. Y.'s apologia also to be printed in the *Richmond Sentinel*, the unofficial organ of the Davis administration. General Hardee, incensed at having been maligned in the G. W. Y. article, wrote Hood, complaining about it and Hood's dissemination of it. "I know that the same article appeared in Richmond immediately after your arrival in that city," he protested. "I know indeed that you procured its publication in the 'Sentinel.'"

After submitting his report to Adjutant General Cooper, Hood spent the next month and a half in Richmond and talked his narrative up to his friends. Mrs. Chesnut recorded in her diary that her husband had been given by Hood "a most interesting account of his western campaign…that Hood left 18,000 muskets in the western army and had not lost more than 8,000 all told—dead, wounded, and prisoners!" During that time he once again saw Buck Preston, to whom he still believed he was engaged to be married. But her parents objected, and the couple's relationship effectively ended, despite Buck's later confession to Mary Chesnut that she would have married John Hood "if he had been persistent." Dyer's comment, "the romance just played out," is worth quoting. Meanwhile, Hood's presence in the Confederate capital, in which just a bit more of a year earlier he had been the celebrated hero, was a dismal reflection of how the general's fortunes had changed. The War Department clerk Jones recorded in his diary, "Gen. Hood is here, on crutches, attracting no attention, for he was not successful." "Hood stock is going down," is the way Mrs. Chesnut put it.

In early March Hood suggested to General Cooper that he be sent to the Trans-Mississippi Department on an inspection tour, reporting back on the feasibility of sending troops across the river to help the cause's failing fortunes. Cooper liked the idea and issued the order for Hood to set out west. Before leaving, though, Hood wrote President Davis—his last letter of the war. "Before leaving for Texas allow me to say that I more than appreciate all of your kindness to me," Hood affirmed, adding, "I am more content & satisfied with my own work whilst in command of the Army of Tenn, than all my military career in life." After he left Richmond in the first days of April Hood learned that Joe Johnston was even angrier than Hardee at his campaign report. After reading it, Johnston informed him that he would press charges through the Adjutant General's office. Hood replied that he looked forward to a court of inquiry, but Cooper declined to grant it, instructing him to continue heading to Texas as ordered.[6]

Hood tarried at Chester, again seeing his friends. On March 5 he paid a visit to Mrs. Chesnut. "He was so well dressed, so cordial, not depressed in the slightest," she observed, which was a considerable improvement from the last time she had seen him, two months before. One may guess that Hood's spirits had lifted with the submission of his campaign report and the administration's eager acceptance of his criticism of Johnston. "He calls his report self-defense," Mary recorded, "says Joe Johnston attacked him, so he was obliged to state things from his point of view." She noted that Hood had gotten others to advocate for him: Francis Lawley, the English correspondent in Richmond writing for the *London Times*; Humphrey Marshall, former brigadier, now congressman from Kentucky (and a Davis supporter); T. W. MacMahon, whose pro-Confederate polemic printed in 1862 had been dedicated to "His Excellency Jefferson Davis, chosen chief of united Southern patriotism"; plus "a New Orleans editor whose name…escapes me."

Hood stayed in Chester more than a week and saw the Confederate government's "treasure train." As they evacuated Richmond, government officials had packed three-quarters of a million dollars in gold, silver, and other metal currency onto a train. The loaded cars rumbled through Danville and Charlotte and arrived at Chester on April 13. There the loot was placed on wagons, as the railroad was out of operation farther south. Aboard the train were the president's wife, Varina, and their children. Upon her arrival, Mrs. Davis was greeted by Hood, Brig. Gen. John Preston, and Brig. Gen. James Chesnut. The officers offered to help her in any way, even as they exchanged sorrowful admissions that

---

[6] McMurry, *Hood*, 185–89; Johnston report, October 20, 1864, *OR*, vol. 38, pt. 3, 612–21; Symonds, *Johnston*, 353; J. D. Richardson, *A Compilation of the Messages and Papers of the Confederacy*, 1:534–35, 538–39; Pollard, *Southern History*, 2:458 ("malice and mediocrity"); Woodward, *Mary Chesnut's Civil War*, 710, 774–75 (including Clement Clay's remark, "Hood did not write that report"), 804; Andrews, *South Reports*, 33; Crist, et al., *Papers of Jefferson Davis*, 11:489 (Hood to Davis, March 31); Hood to Cooper, March 8, 1865, *OR*, vol. 53, 1047; Dyer, *Gallant Hood*, 308; J. B. Jones, *A Rebel War Clerk's Diary* (1866), 518 (entry of March 15); Williams, *Diary from Dixie*, 510 (entry of March 27, "Hood stock going down"); Johnston to Hood, Hood to Cooper, and Cooper to Hood, April 4–5, 7, 1865, *OR*, vol. 38, pt. 3, 637–38 (also in Hood, *A & R*, 159).

their cause was collapsing. Mrs. Davis remembered Hood as saying, "If I have lost my leg and also my freedom, I am miserable indeed."

Hood left Chester on April 15, making his way through Georgia, Alabama and Mississippi. Unable to secure passage across the river, he decided to surrender to Federal troops. He did so, on May 31, riding into Natchez and offering his sword to Brig. Gen. John W. Davidson, who courteously declined it and issued papers of parole.

John Bell Hood's Confederate service was over, yet he adhered to the plan he had developed with General Cooper's permission: heading across the Mississippi. In his war memoir he states simply that after he and his accompanying staff officers had been handed their parole papers, General Davidson "allowed us to proceed without delay to Texas, via New Orleans."[7]

---

[7] Hood, *A & R*, 159 (Hood to Johnston from Chester, April 4); Woodward, *Mary Chesnut's Civil War*, 779, 785 ("Hood has gone"); Furgurson, *Ashes of Glory*, 319 (Lawley in Richmond, April 2); Warner and Yearns, *Biographical Register of the Confederate Congress*, 168 (Marshall elected February 1864); McMahon, *Cause and Contrast*, v; W. C. Davis, *An Honorable Defeat*, 85; W. H. Parker, "The Gold and Silver in the Confederate States Treasury," 305–306; Hanna, *Flight into Oblivion*, 33; V. Davis, *Jefferson Davis, Ex-President of the Confederate States of America*, 2:610–11.

O'Connor claims that "Hood had gained the distinction of being the last of the leading Confederate generals to surrender," but this is not so. Edmund Kirby Smith, for instance, gave himself up near Galveston on June 2 (O'Connor, *Cavalier General*, 265; A. Bailey, "Edmund Kirby Smith," 5:167).

Epilogue

Hood after the War

Within a week of surrendering himself, Hood was in New Orleans, still on his way to Texas. He was in Houston by June 24. He spent several months in his adopted state, often in the company of old friends and admirers. "Though the cause for which General Hood has battled so nobly and suffered so manfully has failed, still he has a home in the hearts and under the roof of every brave and generous Texan," opined the *Dallas Herald* in July. "He returns to his home in our State poor in purse but rich in the rewards coveted by a brave and chivalrous soldier."

From Galveston in early October 1865, Hood wrote to President Andrew Johnson, stating that he had taken an oath of loyalty to the United States and asking for pardon. Johnson demanded that ex-Rebels of high rank appeal directly to him, and word was spreading among Southerners that it would be hard to find an occupation without a presidential pardon. Writing in the third person, Hood asserted that "he had nothing to do with bringing on the war, but being a soldier by education, he deemed it necessary and right that he should take part in the struggle.... He submits he can now be as earnest in his efforts to restore the Union as he was in his attempts to tear it asunder."[1]

The hostilities were not even over when Hood began seeking evidence to defend his war record. In March 1865 he asked General Shoup, his former chief of staff, for affirmation of how he had strengthened the army during the semi-siege of Atlanta by calling in the detailed men *et al.* Hood applied himself even more after the peace. Three months later in Houston, he secured an officer's statement as to the Army of Tennessee's strength and losses under Johnston. In October 1865, while

[1] Dyer, *Gallant Hood*, 309–310; McMurry, *Hood*, 192–93; Perman, *Reunion without Compromise*, 122; T. H. Williams, *Beauregard*, 258 (loyalty oath and employment); S. Davis, "Citizenship for General Lee," 94–95 (on Johnson's requirements for Confederate officers); Stephen M. Hood, "Once a Confederate, Twice an American," 1–2.

Jefferson Davis was being confined at Fort Monroe, Hood sought permission from Federal authorities to visit and converse with him. He asked General Phil Sheridan, in charge at New Orleans, for his help. Sheridan relayed Hood's request to President Johnson, who refused. With James Longstreet he then steamed upriver to visit his mother in Kentucky; on the way, he told a newspaperman that he was planning to write his memoirs. "In my shattered condition I may before I die write my own memoir of this great struggle just ended," he confided to S. D. Lee on November 29. In this letter, however, is an unbecoming tinge of self-pity. "Injustice has been done me," he confided to Lee, but he added, "I expect to die more proud of my defense of Atlanta & my Tn Campaign than all my career as a soldier—considering the small & dispirited capital I had to work with." It was evident that Hood was pondering his campaigns and his justifications of them. The *Louisville Daily Journal* in mid-November reported a correspondent's report of a conversation with the general about his expedition into Tennessee: "it was necessary for him to go on, he said" (meaning, to take the offensive), "for the men were losing spirit by failure, and only some bold effort could reanimate them."[2]

Further correspondence ensued as Hood sought documents to help him tell his story. Bragg in December 1865 explained how the government was prepared to send Johnston thousands of reinforcements if he would commit to an offensive movement in spring 1864. "I have always considered it a great misfortune," Bragg wrote, "that the generous offer of President Davis was not promptly accepted, and the campaign energetically undertaken." Hood even wrote a former enemy staff officer, J. S. Fullerton, about the strength of the Federal forces at Spring Hill on November 29. Hood continued his work into 1866, writing Joe Wheeler

---

[2] Hood, *Lost Papers*, 127–28 (Shoup to Hood, March 15, 1865, and Smith to Hood, June 21); McMurry, *Hood*, 192–93 (including citation of *Washington Chronicle*, November 3, 1865, leading McMurry to assume that Hood traveled to Washington despite Johnson's refusal); Sheridan to Johnson and Stanton to Sheridan, October 13 and 18, 1865, *OR*, ser. 2, vol. 8, 766–67; Sanger and Hay, *James Longstreet*, 321 (Hood traveled with Longstreet from New Orleans only to Kentucky, and not with him on to Washington); "Gen. Hood…on Board Steamer Clinton, at Sea, October 9, 1865," *New York Times*, November 2, 1865 (on Hood writing memoirs); 196; Hood to Lee, November 29, 1865, Lee Papers, Chapel Hill (typescript courtesy of Keith Bohannon); Hood, *Rise, Fall, and Resurrection*, 77 (*Louisville Daily Journal*, November 19, 1865).

and Red Jackson about the strength of their cavalry forces during the Atlanta campaign. Meanwhile, S. D. Lee was cautioning him not to proceed too quickly in publishing his reminiscences; wartime passions still ran high. From Louisville on January 9, Hood wrote that he agreed with his friend, that on war matters he would "act on the defensive" only. "The injustice done me by my countrymen has ceased to trouble me," he claimed (not convincingly). Hood put his discomfiture in perspective: "Look at Mr Davis today in chains—for doing what he knew & believed to be right—the greatest man of America—the martyr of modern times."[3]

Hood was evidently much troubled by Jefferson Davis's confinement by Federal authorities after the war. The president had been captured by Union cavalry in south Georgia on May 10, 1865, confined at Fort Monroe in Virginia, indicted for treason but never tried, and finally released on May 13, 1867. Even during Davis's imprisonment, Dr. John J. Craven, the surgeon assigned to his medical care from May to December 1865, published a book in 1866 on Davis in captivity. Davis later acquired a copy of Craven's book and did not like what he read. He marked the margins of more than 180 pages with comments, criticisms, and corrections. After his death, among her late husband's papers, books, and other material, Mrs. Davis found the book and gave it to the Louisiana Historical Association. It reposes today at Tulane University. Of interest is Craven's one reference to John B. Hood and the Georgia Campaign. If Joseph E. Johnston "had vigorously attacked Sherman at Atlanta when directed, the fortunes of the war would have changed," Craven wrote, "and Sherman hurled back to Nashville, over a sterile and wasted country—his retreat little less disastrous than Napoleon's from Moscow. He did not do so, and was relieved—General Hood, a true and spirited soldier, taking his place—but the opportunity was then gone." Craven harshly judged Davis's delay in relieving Johnston, adding, "and to this delay, more than to any other cause, the Southern people will attribute

---

[3] Hood, *Lost Papers*, 55–56 (Bragg to Hood, December 17, 1865, also in *A & R*, 90–91); 114 (Fullerton to Hood, October 20, 1865, also in *A & R*, 288); 129–30 (Jackson to Hood, July 15, 1866); Hood, *A & R*, 75–76 (Wheeler to Hood, June 1, 1866); Hood to Lee, January 9, 1866, Lee Papers, Chapel Hill.

their overthrow, whenever history comes to be truly written." In the margin of this page, Jefferson Davis simply wrote, "misconception."[4]

During winter 1865–1866, Hood settled in New Orleans, the city he would henceforth call his home. The Crescent City attracted a number of other former Confederate officers. Beauregard arrived in May 1865. Simon B. Buckner, with whom Hood had fought at Chickamauga, joined the staff of the *New Orleans Crescent* in October 1865 and soon thereafter found work with a firm of commission merchants and cotton factors. Joe Wheeler came to manage a hardware and carriage business. Longstreet came, too, partnering with William Miller Owen, formerly of the Washington Artillery, in a cotton-factoring venture. Braxton Bragg came to New Orleans in early 1867 to take a position connected with a railroad, then with the city waterworks, before moving a few years later to Mobile. Randall Gibson, who had commanded a brigade in Hood's army, arrived to practice law. Richard Taylor and Harry Hays, who had been prominent citizens of New Orleans before the war, returned to their home after surrender.

In January 1867, an event brought many of these former officers together: the disinterment of Gen. Albert Sidney Johnston. After his death at Shiloh, Johnston had been buried in New Orleans's St. Louis Cemetery, but his brother-in-law remembered that Johnston had expressed a wish to be buried in Texas. The Lone Star State legislature appropriated funds in fall 1866, and on January 23, 1867, Johnston's remains were removed from his New Orleans tomb, with Beauregard, Bragg, Buckner, Gibson, Hays, Hood, Longstreet, and Taylor all serving as pallbearers. The event was described by the *New Orleans Picayune* the next day; the paper did not say whether Hood, with his physical infirmities, could have participated in the heavy lifting.[5]

---

[4] Eckert, *"Fiction Distorting Fact,"* xvii, xxi, xl, xlviii (Davis got the book in August 1866), 1, 47 ("misconception"); McElroy, *Jefferson Davis the Unreal and the Real*, 558; Craven, *Prison Life of Jefferson Davis*, 117–18.

[5] Gallagher, "Pierre Gustave Toutant Beauregard," 1:90; Stickles, *Buckner*, 278–79, 282; Dyer, *"Fightin' Joe" Wheeler*, 244; Bergeron, "Randall Lee Gibson," 2:188; Parrish, "Richard Taylor," 6:29–30; T. L. Jones, "Harry Thompson Hays," 3:78, 80; Piston, *Lee's Tarnished Lieutenant*, 95; Hallock, *Braxton Bragg*, 261–62; Roland, *Albert Sidney Johnston*, 352–53; W. P. Johnston, *The Life of Gen. Albert Sidney Johnston*, 694, 700–701.

Hood's activities were never far from the war. Early on he became involved with the Southern Hospital Association for Disabled Soldiers, helping to raise money for it and accepting its presidency. Most important were his efforts to gather material for his memoir. In early November 1866, Hood issued a circular calling upon former Confederates to send him information about their war service. The circular was essentially an open letter to all officers who had fought under Hood to send him wartime after-action reports they may still have possessed, along with a statement of particular facts regarding specific battles. Hood did not propose to mold the information he received into a narrative of his army's history but, as the *Nashville Banner* put it, "to collate these facts in such a form that the future historians may render justice to his gallant comrades." The "future historians" would want facts, so in his circular Hood listed the kinds of information he sought. For a battle he suggested that the officer provide data in a prescribed form, e.g., "Hours of action," "Position of field," "Troops to the right," and "Troops on left." Hood even asked that officers send "with his report, his photograph, and the photograph of any one of his commanders killed in action, together with such facts relating to his career and death as may be interesting."

Hood apparently sent his circular to newspapers across the South, such as the *Banner*, which urged other papers to publicize Hood's appeal. "As the greater portion of the troops composing the army of the West, at the time of the surrender, were Tennesseans," the *Banner* declared, "the General's request will, we trust meet a hearty response in this and other sections of the State."

Hood was asking a lot. Col. Orlando S. Holland of the 37th Mississippi explained to Hood that his papers, left in spring 1865 at his mother's home in Columbus, Georgia, had been destroyed by Wilson's raiders in the last weeks of the war. But Holland wrote a long letter, setting forth some facts as well as he could remember them. Similarly, W. S. Featherston, who had led a brigade in Loring's Division, wrote a "brief account"—eighteen handwritten pages—of his unit's service under Hood from Atlanta to Tupelo, "without the aid of official reports."

If Hood intended his circular to elicit some flattery from his former officers, it worked. In his letter of February 14, 1867, former lieutenant of engineers William W. Fergusson explained that he had no photograph of himself ("I am too homely in appearance"), but that he hoped Hood would share a likeness of himself: "the resemblance of one whose fortunes I shared, whose conduct & courage I applaud, & whose association from the time of taking charge of part of the line of works around Atlanta to Tupelo, after the ineffectual attempt to redeem Tennessee & revive the lost spirit of the Army of Tennessee."

In his letter of February 24, 1867, Lt. Col. Bushrod W. Frobel offered even more praise. "The world will never know the full extent of the difficulties and trials," Frobel wrote, that Hood had faced in assuming command of the army and guiding it in his campaigns. "With the skeleton of an army, dispirited and broken down, filled with politicians in Military garb, who were striving solely for their own advancement, with the press in reality arrayed against you tho endeavoring to maintain an appearance of friendship," he waxed, "you stood alone in the breech, staking upon the issue a reputation second to none in the Confederate army."[6]

Frobel did not stop there. After the war, employed as superintendent of public works by the state of Georgia, he contributed articles to an early postbellum journal published in Atlanta, *Scott's Monthly Magazine.* In its monthly issues from September 1866 to December 1867, Frobel published a series of reminiscences about his war service in Virginia in 1861–1862, part of which he had served on General Hood's staff. In its issues from February 1868 to March 1869, Frobel also contributed to *Scott's* a series of eleven chapters of an historical narrative titled "The Georgia Campaign; or, a South-side View of Sherman's March to the Sea." In them he unabashedly praises Hood's command record in Virginia and Georgia. Hood was "the idol of his own division, and the pride of Lee's army.... It is true Gettysburg had claimed an arm, and Chickamauga had taken a leg, but it was still Hood—unconquered and undaunted as ever."

Needless to say, Frobel takes Hood's side against J. E. Johnston. "He had abandoned the mountains because he did not believe he could longer hold them, and had been forced back step by step, to the flat lands

---

[6] Miller, *Hood and Memory*, 189 (Southern Hospital Association); L. G. Marshall to Hood, December 4, 1866 ("in response to your circular of the 1st ult."); O. S. Holland to Hood, November 22, 1866, and W. S. Featherston to Hood, December 15, 1866, all in Hood Papers, National Archives, photocopy courtesy of Stephen M. Hood; "Forward Your Reports," *Pulaski (TN) Citizen*, November 23, 1866 (transcription courtesy of Jamie Gillum, Franklin, TN, and Stephen M. Hood); Hood, *Lost Papers*, 102 (William W. Fergusson to Hood, February 14, 1867), 104 (B. W. Frobel to Hood, February 24, 1867).

of Georgia to a point where no strong nature barrier presented itself between the invading army and the sea. Georgia was thoroughly alarmed." When Johnston, at the gates of the prize city, told the president that "he intended to garrison Atlanta with the Georgia militia, so that the army might be more free in its movements," Johnston had to be removed from command. Hood had consequently taken charge of the army. At Peachtree Creek "he had met the enemy in open fight, had foiled his movement, and had for the time saved Atlanta." Frobel also clearly takes Hood's side of the debate with Hardee. Concerning the battle of July 22, he asserts, "this partial success is but stronger evidence of the victory that would have resulted to our arms had Hood's order been fully executed" (meaning, the order for Hardee to swing far to the rear of McPherson's flank). Even Ezra Church, S. D. Lee's bloody repulse, is depicted as something of a victory by Frobel: "the road remained in our possession." As a result of Hood's battles in July, Southerners thus "began once more to take heart," even to the point that they "confidently expected that Sherman would not only be met, but hurled back in one grand and overwhelming victory." After Wheeler and Iverson had defeated McCook's and Stoneman's raiders south of Atlanta, Hood was justified in sending forth Wheeler with 4,500 horsemen to fall upon Sherman's rear and railroad. Even here Frobel offers defense of his former commander. "The raid, though a failure so far as its main object was concerned, had the effect of causing Sherman considerable uneasiness—compelling him to send large bodies of troops to the rear for the protection of his communications." Frobel's most gallant—read, *puzzling*—defense of Hood occurs in his recounting of the critical events of August 31. Discerning the wide sweep of Sherman with much of his army marching toward Jonesboro, Hood, Frobel asserts, sent Hardee's and Lee's corps to attack Howard's army there because he hoped the action "would draw the attention of Sherman in that direction, and that he would abandon his works on the left [meaning the XX Corps, still at the Chattahoochee], giving Hood a favorable opportunity of attacking him in flank."

In the end, Frobel absolves Hood for his failure to hold Atlanta. "I do not think that anybody," he reasons, "supposed it possible to hold Atlanta, and the expectation was general that it would be abandoned so soon as Sherman was prepared to make one of his flank movements." The writer concludes, "the defense of Atlanta, under the innumerable disadvantages with which it was conducted, reflected the highest credit upon both the army and the General."[7]

Two years after the war Hood was sounding the same notes he had in his campaign report. When he took command of the Army of Tennessee, he wrote Sarah Dorsey in March 1867, it had already lost more than twenty thousand men and, demoralized by Johnston's retreats, "was not in condition for pitched battle." After the loss of Atlanta—"this untenable city"—he had launched his offensive into Tennessee "as the only hope of improving & increasing the army." "Accidents however, perhaps beyond human control," he concluded, "caused the campaign to fail at a time when the fruits of victory were seemingly within our grasp." Only John Bell Hood could have cast his disastrous final campaign in such a benign light.

After 1867, however, Hood largely turned away from historical pursuits. For the first time in his adulthood, Hood was without a military posting and had to earn a living. For a while he engaged in Texas land speculation. Occasionally business would lead to travel. In a Louisville hotel in November 1866 he ran into his former enemy, George Thomas. "Gen. Thomas rose from his seat to meet the maimed and gallant Hood," reported a local paper, and the two sat down for lunch. "The two Generals dined together like old friends."

Hood then became a commission merchant, buying and selling cotton. But after a few years he turned to the insurance business, taking over in 1869 the insurance company James Longstreet had previously run. The work allowed him at least a comfortable income ($5,000 a year) through most of the 1870s. In April 1868, Hood also married Anna Marie Hennen, the well-educated daughter of a fine Louisiana family (and six years younger than Hood). The couple moved into a comfortable home on Third Street in the city's fashionable Garden District and started their family, which over the years grew to eleven children. The first of these was a daughter, Lydia, born in May 1879. Hood asked his friend and former comrade Randall Gibson to serve as her godfather; Gibson gladly agreed. As his family grew in size, the number of the Hoods' chil-

---

[7] Bonner, "Bushrod Washington Frobel," 2:328–29; Atchison, "*Scott's Monthly Magazine*," 301; [Frobel], "The Georgia Campaign; or, a South-side View of Sherman's March to the Sea," (March 1868): 85; (October 1868): 696–701; (December 1868): 843; (January 1869): 6; S. Davis, "Full-Throated Defender of Hood," 16–17.

dren became the talk of New Orleans. After the birth of the third set of twins in October 1876, a resident of the city wrote the former general, expressing his appreciation of "your most laudable and successful endeavors to increase the population, and thereby indirectly to add to the general prosperity of the State."[8]

Hood wisely chose not to express his opinion on political matters, keeping his pledge to President Johnson that he would be as earnest in restoring the Union as he had been in trying to tear it apart. In spring 1867, Congress passed a "military reconstruction" act, over Johnson's veto, which divided the former Confederate states into five military districts with a U.S. general in charge of each of them. The law sparked much heated discussion among Southerners. The *New Orleans Times* asked the ex-C. S. generals in the city to express their opinions for the newspaper. James Longstreet was the first to do so. "We are a conquered people," he counseled; "recognizing this fairly and squarely there is but one course left for wise men to pursue. Accept the terms that are offered us by the conquerors!" Beauregard also submitted an article, urging Southerners to submit to the government's policies, as resistance was futile. The two generals were taking significant risk in offering opinions sure to be criticized by many Southerners, but Longstreet went further still. He wrote another letter to the *Times*, expressing his thoughts with the "bluntness of a soldier." Afterward he was invited by a local businessman, John M. G. Parker, to set down his views still more fully. Longstreet did so in a letter to Parker, which he again intended to give to the New Orleans press. In it he asserted that because the war had been based on the Republican Party's political issues, the peace should be built upon them, too. "The military bill and amendments are peace offerings," he urged. "We should adopt them as such, and place ourselves upon them as the starting point from which to meet future political issues as

---

[8] Hood, *Lost Papers*, 106 (Hood to Sarah A. Dorsey, March 30, 1867), 198–99 (Hood to J. H. Brower, June 12, 1866 about land in Waldo County, Texas), 203 (Gibson to Hood, June 11, 1869), 210 (Thomas Hunton to Hood, October 25, 1876, "increase the population"), 240–41 (land speculation; wife and children); McMurry, *Hood*, 193–96; "The Amenities—How Soldiers Meet," *Atlanta Weekly Intelligencer*, November 28, 1866 (quoting the *Louisville Journal* on Hood and Thomas meeting); Hardin, *After the War*, 106 ("$5,000 a year"); Dyer, *Gallant Hood*, 315 ("Third and Camp Streets").

they arise." Longstreet showed a draft to his uncle, Augustus Longstreet. "It will ruin you," Uncle Augustus warned, "if you publish it."

Hood offered the same counsel to his friend. "General Hood met me and we talked about the editorial," Longstreet recalled years later: "He said in effect that it aimed at a condition that we all devoutly hoped for, but he added, 'If you declare yourself on those lines the Southern press and the Southern people will vilify and abuse you. It may be very patriotic and all that, but it will be very foolish.'" Hood was correct. He held views like Longstreet's, having recommended a few months after the war that Southerners adopt a policy of "submission with dignity," but he decided against going into print with them. Longstreet, however, did not take Hood's advice, and gave his letter to Parker to the *Times.* That paper, even as it printed the letter, called Longstreet a traitor, deserter, and collaborator. "Old comrades passed me on the streets without speaking," he recalled in his memoirs. His insurance business consequently began to drop. "General Hood (the only one of my old comrades who occasionally visited me)," Longstreet remembered, offered his help in saving the business. But, he concluded, "in a few weeks I found myself at leisure."[9]

As reflection of Hood's relative lack of interest in Confederate history during the late 1860s, he was absent when the Southern Historical Society was being formed in New Orleans. Dabney Maury, who had commanded at Mobile for much of the war, had come to New Orleans, like Hood and many others, to find work and income. In 1868, as he wrote in his memoir, "I determined to set on foot a plan for the systematic collection and preservation of the Southern archives relating to the war." Richard Taylor, Braxton Bragg, Harry Hays, and Beauregard all supported Maury in the effort. In April 1869 Maury, Bragg, Simon Buckner, and six other prominent men of New Orleans—Hood was not among them—met to discuss an organization dedicated to promoting the study of the war from the Southern perspective. Its purpose was suggest-

---

[9] R. W. Patrick, *The Reconstruction of the Nation*, 98–99; Dyer, *Gallant Hood*, 313–14 (quoting Longstreet in *New Orleans Times*, March 19, 1867, and in *New York Times*, June 8, 1890), 363 n.34 ("submission with dignity"); T. H. Williams, *Beauregard*, 266; Sanger and Hay, *Longstreet*, 331–34; Piston, *Tarnished Lieutenant*, 105; Longstreet, *Manassas to Appomattox*, 636–37 (Longstreet to Parker, June 3, but in a version different from that printed in the *New Orleans Times* of June 8, as noted by Sanger and Hay, *Longstreet*, 333 n.40).

ed later by Jubal Early, who complained that in its published reports and other documents, the federal government had "sought to make our cause and our conduct in the struggle...odious, and to present a one-sided view of the history of events preceding, attending and following the war." To counter this stream of falsehood, the society would compile and propagate historical evidence supporting Southerners' view of their war for independence. All agreed that such an organization would "secure the co-operation of intelligent and well-informed gentlemen residing in the different sections of the Southern States."

The planners decided to call a larger meeting within a week, which drew forty-five men of the city—again Hood was not present. The group agreed upon a name for their organization: the Southern Historical Society. Its founding members included former generals Bragg, Taylor, Buckner, Gibson, Hays, and Beauregard—but not Hood. Benjamin M. Palmer, Presbyterian minister in the city and a longtime Southern propagandist, was elected president. Braxton Bragg was named vice-president. Dr. J. William Jones, Virginia-born Baptist minister, wartime confederate chaplain, and postwar church pastor, was named as secretary-treasurer. Vice-presidents were appointed for each of the former Confederate states. Despite the appointment of such distinguished Southerners as Robert E. Lee as state vice presidents, and six thousand promotional flyers mailed out, the SHS did not thrive much outside of New Orleans. The society in a year had fewer than a hundred members, "and of this number," lamented Dr. Jones, "little more than one-fourth attend the regular meetings."

One organization Hood joined was the Louisiana Division of the Association of the Army of Tennessee; he delivered an address at one of its meetings in 1869. Hood was much more engaged with the Hood's Texas Brigade Association. When the group was being organized in May 1872, Hood made his way to Houston for the meeting. He was elected president by acclamation, but declined to serve as more than temporary chairman, telling the sixty-odd veterans assembled that traveling from New Orleans would make it hard for him to attend their reunions. Yet reuniting with his former soldiers was a moving experience. "Since coming into your midst," Hood told them, "recollections have rushed through my brain that almost stagger me." Hood missed the following two meetings but managed to attend the reunion in Waco in June 1877. During that meeting he delivered an address that was interrupted repeatedly by ovations; the Texas Brigade still loved their old commander. In their meeting of July 1879, the Texas veterans resolved to pay for Hood's railroad trips for the rest of his life if he could attend their reunions.

Hood's comparative lack of participation in historical societies did not harm his social standing in New Orleans. Friendship with prominent ex-Confederate generals in the city certainly helped. When Dick Taylor helped to reorganize the Boston Club, the exclusive meeting place of New Orleans gentlemen, Hood was invited to join. In August 1870, when Hood applied for an insurance policy, Braxton Bragg vouched for his "strong constitution," in spite of his physical disabilities. Amidst such camaraderie, every now and then Hood became involved in activities pertaining to the war. In early 1870 he was among a group of New Orleans citizens who sought to raise money to buy the large painting by Everett B. D. Julio, *The Last Meeting of Lee and Jackson* (1869), which showed the two generals conferring before Jackson's legendary attack at Chancellorsville. A half-year later, upon the death of General Lee, Hood was elected as vice president for Louisiana of the Lee Memorial Association. In August 1873, the Southern Historical Society held a meeting in western Virginia to discuss reorganizing itself and moving its headquarters to Richmond. Hood did not attend, but he was nonetheless appointed vice president for Louisiana. In 1877, when the society began publishing its *Papers*, Hood was among the "prominent Confederates" asked by editor J. William Jones to submit a paper about Gettysburg. As we have seen, Hood responded by sending Jones his letter to Longstreet written two years earlier when the latter had asked for a campaign report on the battle. But beyond this, there is no evidence that Hood even joined the historical society until probably spring 1879. At that time ex-Confederates in New Orleans organized a "Louisiana Branch" of the SHS. Among the officers elected were eight vice-presidents; Hood was one of them.[10]

---

[10] Maury, *Recollections*, 243–44, 251; G. M. Foster, *Ghosts of the Confederacy*, 50–51, 54; Maury, "The Southern Historical Society," 349–52, 356–62; "Origin and History of our Society," 244–45; Osborne, *Jubal*, 433; Parrish, *Richard Taylor*, 460–62; Rable, *God's Almost Chosen People*, 63 (Palmer as Confederate intellectual "heavyweight"); Moore, *Apostle of the Lost Cause: J. William Jones*, 75; Hess, *Bragg*, 258 (Hood's insurance application); Miller, *Hood and Memory*, 193–94 (Julio), 196–97; Pennington, *The Last Meeting's Lost Cause*, 17; "Causes of the Defeat of Gen. Lee's Army at the Battle of Gettysburg," 49; "Leading Confederates on Gettysburg," 145; C. R. Wilson, *Baptized in Blood*, 125 (on Jones; "he devoted seven issues of the *Southern Historical Society Papers* to examining the reasons for Confederate defeat at Gettysburg"); "A Louisiana

Branch of the Southern Historical Society," 253; McMurry, *Hood*, 197 ("Army of Tennessee Association"); Simpson, *Reunion and Memory*, 16–17, 158–59, 161; "The Southern Historical Society," 43–44 (Hood among "Vice-Presidents of States" for Louisiana).

Brian Craig Miller is off the mark in asserting that Hood was the target of "critical and fierce…attacks from the SHS," and that he was eventually among those blamed by Jubal Early, J. William Jones, and other Virginia leaders of the society for Confederate defeat. James Longstreet, not John B. Hood, was the S.H.S. target, as Gaines Foster demonstrates with his phrase, "Longstreet-lost-it-at-Gettysburg excuse" (*Ghosts of the Confederacy*, 58, 119). Miller seems peculiarly unaware of the "Lost Cause *versus* Longstreet" literature. More than thirty years ago, Connelly and Bellows wrote, "Longstreet's fate as the scapegoat of the Lost Cause is so well known that no full explanation is required here" (Connelly and Bellows, *God and General Longstreet*, 32). Miller makes such statements as "the society, while blaming Hood" and "the SHS remained derisive toward Hood." In doing so he seems to confuse the use by such ex-Confederates as B. F. Cheatham and T. B. Roy of the *Southern Historical Society Papers*, a prestigious publication (as vehicle for their articles arguing with Hood about war events), with animus toward Hood by the *society itself* and its leadership (Miller, *Hood and Memory*, 200, 209, 213–14). Thomas J. Brown makes the same mistake when he claims that "Southern Historical Society writers assailed Hood after death for contributing to Confederate defeat" ("John Bell Hood and the Lost Cause," foreword to Hood, *Rise, Fall, and Resurrection*, xx).

Gaines Foster suggests that the "Virginia coalition" of Jubal Early, J. William Jones, Fitzhugh Lee and others "seemed to be searching for some factor that would have changed the results of the war: If only they had not faced overwhelming numbers, if only Longstreet had been on time."

> They appeared captivated by a dream of victory, a dream of a return to an undefeated Confederacy. This aspect of their historical vision does not appear very different from another revitalization movement of the late nineteenth century, the Ghost Dance among the Plains Indians. The Ghost Dancers preached that white culture would disappear if only Indians believed and danced. Similarly, the Virginians seemed to believe that if they wrote their articles and kept southerners from deserting after the war, the Yankees and all that had occurred after Appomattox would simply disappear. One can easily imagine a gathering of Early and company, wearing their gray ghost shirts and clutching volumes of the *SHSP* that Jones promised would be impervious to the slings and arrows of northern slander. Formed in a circle about a statue of a recumbent Lee, the true believers dance in and back, chanting, on one foot, "overwhelmed by numbers," and on the

Tending to his own business and raising his family did not mean, however, that Hood did not keep up with what was being said about him. A few years after the war, W. T. Sherman gave a speech to Union veterans in Cleveland and included in it an episode he had learned from Joe Johnston when the two had met on a Mississippi steamboat. There the two generals talked about the war and Johnston related how Hood, on the evening of May 19, had urged retreat from the Cassville lines when they came under enfilading artillery fire. Sherman mentioned Johnston's claim about Hood and the retreat in his Cleveland speech, which was reprinted in newspapers around the country. One of them must have been the *Macon Telegraph*. In December 1870, while he was in Macon on insurance business, Hood wrote the paper's editors and cheekily challenged Sherman's implication that at Cassville Hood had "saved" the Army of Tennessee by urging the Johnston to retreat. In his letter Hood offered no explanation of events at Cassville; he could not, because Johnston and Sherman *were* correct: he and General Polk *had* urged withdrawal from the exposed Cassville position. Instead, Hood rehashed all of his triumphs in Virginia and at Chickamauga, and he closed by reminding the *Telegraph* that he had "a fair portion of body buried beneath Ga's soil."

In June 1870, Jefferson Davis wrote Hood, asking if he were progressing in the preparation of his war memoirs ("the material which we should leave for a true history of the great war"); Hood could not say that he was. A few letters of 1872–1873 show that Hood was at least corresponding with some former officers about the war. Stephen M. Hood has found in minutes of a South Carolina Confederate veterans' reunion, held in December 1872, the text of a speech Hood delivered to them. (In it, he praised Lee and Jackson for their offensive spirit, observing how

---

> other, "betrayed by Longstreet"—waiting for an undefeated, marble Lee to rise and lead them to victory. (*Ghosts of the Confederacy*, 55, 60)

Foster is having fun, but the "Virginia coalition" had a long and pervasive influence. Chris Mackowski sees it at today's Fredericksburg National Battlefield Park. There, because of "his fallout with the most ardent of the Lost Cause writers," Longstreet's name tends not to be mentioned in metal tablets marking the Confederate line (e.g., "CONFEDERATE LINE—Hood's Division"); in contrast, Stonewall Jackson's name *is* used ("CONFEDERATE TRENCHES: Jackson's Line") (Mackowski, "Fredericksburg in Memory," appendix E in Mackowski and White, *Simply Murder*, 136–37).

keeping soldiers in defensive trenches, "even for a short time, was to dampen the ardor of even the bravest.") A year later, Hood asked Francis Shoup what he remembered about his time as the army's chief of staff. But as we have seen, it was the publication of Johnston's *Narrative* in April 1874 that stirred Hood to resume his historical self-defense. A "prospectus," allowing for advance reviews, was circulated several months before the book's publication. One, written by "A.," a former Confederate who had "served prominently" under both Johnston and Hood, was published in the *Picayune* in January. "A." was extremely critical of Johnston's book and its purpose of "blackening the military reputation of the young and noble Hood." The author was evidently a resident of New Orleans and thus in touch with the ex-general, for he quoted fully documents he could only have gotten from Hood himself: Stonewall Jackson's letter of September 1862 recommending Hood for promotion to major general after Sharpsburg, and Longstreet's letter of a year later recommending Hood for promotion to lieutenant general (with Bragg's, Seddon's, and Davis's endorsements of the recommendation). Also printed was General Lee's letter of May 21, 1863, to Hood ("I have always had you in my eye & thoughts"). "A." concluded with a warm tribute to Hood: "His aspirations and services were to render services to his country; he never sought place or position; he was an example of great modesty, united to undaunted courage and valor, and could say, with the great Admiral de Ruter, 'I willingly dispense with all praises, if I only satisfy my conscience and follow the commands sent me.'" "A." may have been John E. Austin, who as major had commanded a battalion of Louisiana sharpshooters in the Army of Tennessee. Austin was living in New Orleans in 1874 and was definitely on Hood's side in his arguments with Johnston; as we have seen, he offered a letter in May of that year supporting Hood's version of events at Cassville.[11]

If "A."/Major Austin had seen the "prospectus" for Johnston's *Narrative*, Hood likely saw it too and could read some of Johnston's statements. In January 1874 Hood began writing other officers on a number

[11] Sherman, *Memoirs*, 2:39–42; Hood, *Lost Papers*, 205–207 (Hood to *Telegraph;* Davis to Hood, June 3, 1870; and Shoup to Hood, December 30, 1873), 133 (T. B. Roy to J. E. Austin, January 12, 1874, in New Orleans); Hood, *Rise, Fall, and Resurrection*, 78 (Hood oration); "A.," "Gen. Johnston's Book," *New Orleans Picayune*, January 8, 1874; Hood, *A & R*, 101–102 (Austin to Hood on Cassville).

of points. In a letter to S. D. Lee, he sought opinion of "the general effect of entrenchments upon an army." He offered his own: soldiers' experience in defensive earthworks weakened their ability to successfully attack and carry enemy fortified positions. "An army cannot at one time fight behind breastworks," he wrote, "and, at another time…forget its own experience in the use of entrenchments, and charge the enemy's works, and force him to fly from the field, as was so often the case in Virginia." Hood remembered that Gaines' Mill, Second Manassas, Sharpsburg, Fredericksburg, and Chancellorsville had all been fought without entrenchments, implying that Lee's greatest victories had come when he avoided digging in. Once Lee's forces had been forced into the entrenchments of Petersburg, "it had a depressing effect even upon the stern veterans who made up that grand old Army." S. D. Lee agreed with Hood: "to attack entrenchments, give me troops who have never served behind them." In his memoir Hood used his and Lee's correspondence to reinforce his general premise that "the longer an army retreats, entrenches, and fights at the same time, the more numerous the desertions, and the more thorough the demoralization." With this contention, however, John B. Hood demonstrated sadly that during the war his views had not matured as the tactics on the battlefield had evolved. By 1864, armies of both sides had come to rely upon earthworks and other fortifications as protection against enemy musketry and artillery. For Hood to argue against their use justifies McMurry's conclusion, "Like a knight riding forth with a lance in the late nineteenth century, Hood was out of place at the head of an army" in mid-1864. Yet, in fairness, John Bell Hood may not have been alone. In his letter to Hood of January 16, James Longstreet sounded Napoleonic himself in replying to Hood's query about earthworks: "I am clearly of the opinion that a Field Marshal should rely upon his own skill, and the valor of his troops, as his great and only Fort."

Hood also sought to marshal evidence as to Johnston's numbers. He wrote Wheeler's A. A. G., Elisha S. Burford, asking the strength of the cavalry during the retreat from Dalton. Major Austin apparently offered his help in the letter-writing, as when he inquired of T. B. Roy the strength of Hardee's Corps at the start of the campaign. Other officers offered their comments. Gustavus W. Smith opined on Johnston's astonishing statement to the administration in mid-July 1864—assuming that Johnston was serious, and not just penning his "death-wish"—that he was about to leave the Georgia militia in the defensive works of the city so that the movements of his three infantry corps could be "freer and wider." "Atlanta would, in all probability, have been taken within twenty-four hours after its defence was entrusted to the Georgia militia," Smith wrote Hood, "because in number this force was entirely inadequate, under the circumstances."[12]

Within weeks of the publication of Johnston's *Narrative of Military Operations*, Hood had the first installment of his "Reply to Gen. Johnston's Book" printed in the pages of the *New Orleans Times*, suggesting further that Hood had begun drafting his articles from the advance-released prospectus of Johnston's *Narrative.* The length of the articles and their detailed counterarguments indicate that Hood had spent a good deal of time and work on the six essays (chs. 4–9 in *A & R*). He was understandably proud of his "Reply." After the first article appeared, he sent it to friends, such as Charles M. Conrad of Washington, DC. Leonidas Polk's son William wrote Hood to thank him for correcting Johnston's version of events at Cassville. A surgeon who had served in the Kentucky mounted infantry was so impressed by Hood's ability in his *Times* articles to demonstrate "the inability and inefficiency of Genl Johnston in managing an army" that he wrote to ask if Hood planned to publish the articles together in a booklet. Matthew C. Butler, who had been a brigadier of cavalry in Lee's army, wrote Hood in July, asking if he would send "your articles on Genl Johnston's narrative." When he used them as the central portion of *Advance and Retreat*, however, there is reason to believe that Hood chose to hide the fact that at least a couple of the six chapters had previously been printed, albeit in a New Orleans newspaper; he pointedly makes no mention of this fact in his memoir. When he included the text of his letter to C. M. Conrad in *Advance and Retreat*, he deleted its last sentence, "I send you a copy of the Times."

As one might expect, Hood's "Reply to Gen. Johnston's Book" provoked criticism and objection from Joe Johnston and his partisans. Johnston wrote Kinloch Falconer on June 4, 1874, "General Hood is publishing a series of attacks," adding that he would gather numerical statistics to refute Hood's claims. Charles G. Johnsen, living in New Orleans, and formerly of the Washington Artillery serving in the Army of Tennessee,

---

[12] Hood, *A & R*, 135–39 (Hood to Lee, January 17, 1874, and Lee to Hood, January 26), 147 (Smith to Hood, January 23, 1874); McMurry, *Hood*, 191; Hood, *Lost Papers*, 133–34 (Roy to Austin, January 12, 1874, and Burford to Hood, January 20, 1874), 208 (Longstreet to Hood, January 16, 1874).

wrote the *Picayune* about Hood's first column of May 10. "Gen. Hood seeks to convey the idea that the army was demoralized by Gen. Johnston's policy and its *morale* restored by his own," he wrote. "*The contrary is the fact.*" Johnsen stressed that "Gen. Johnston had the unlimited confidence of the army," which he claimed was stronger in mid-July, at the time of his relief, than it had been at the start of the campaign at Dalton. He criticized Hood for having failed to deliver Johnston's attack at Cassville, and for whining about his enfiladed line on the evening of May 19. Moreover, in the two generals' war of words, Johnsen asserted, "I have carefully read Gen. Johnston's narrative of the campaign from Dalton to Atlanta, and believe in every particular it is strictly correct." He closed by inviting Jefferson Davis and J. B. Hood to offer their own version of events "and give the future historian material that he will never otherwise find."

Johnsen did, however, help Hood on a matter of fact. In his "Reply" article of May 10, Hood stated that in the army's march into north Georgia in October 1864, his troops had been unable to capture an enemy blockhouse guarding Mill Creek Gap, and that the army consequently had to make a twenty-mile detour to get past it. Quite to the contrary, Johnsen stated, "the 5th Company Washington Artillery, Lieuts. Adolphe Chalaron and Abe Leverich, captured and took possession of the fort in Mill Creek Gap." Johnsen was right. On October 13, while his infantry tore up the railroad, Major General Bate directed Lieutenant Chalaron and his artillery to reduce the blockhouse. They did so, firing accurately at a side of the little fort until it was "completely torn to pieces." Inside, a dozen of the garrison members were killed or wounded before they surrendered. Besides the fifty prisoners taken, however, were their valuable stores, including a lot of coffee. "Gen. Billy Bate," Johnsen concluded, "claimed a division of the spoils" from the encounter, which reflected well on all involved: "the Battery got the glory; Bate the coffee." The result was that in *Advance and Retreat* Hood revised his earlier error in the *Times*, so that his memoir reads, "this block-house was surrendered to a detachment of our troops."

Hood was not afraid to include in his "Reply" statements that he knew could not be corroborated or confirmed by former comrades. He had written that before leaving New Hope Church, General Johnston had called the three corps commanders "assembled alone, at night, in his quarters" and had told them he was thinking of eventually giving up Atlanta and retreating all the way back to Macon. To C. M. Conrad on

July 1, 1874, he acknowledged, "I have publicly stated Genl. Johnston foreshadowed to his corps commanders at New Hope Church, during his campaign of '64 from Dalton, his intention to retreat to Macon Georgia," adding, "I am the only living witness of this historic truth." Which was true: Lieutenant General Polk had been killed less than a month later, and former Lieutenant General Hardee had died just months before Hood's article had appeared.[13]

General Sherman's published reminiscences appeared in print a year after Johnston's. He had begun drafting his war memories in the late 1860s, and by early 1872 had a draft. He took his time securing a publisher, which ended up being D. Appleton of New York. Sherman's *Memoirs* then appeared in May 1875. Reading it, Hood took careful note of how he was treated in its pages. Sherman was not particularly complimentary of Hood's generalship, but at least the two former enemies had no cause to quarrel over the facts of what had happened during the war. Still, in August 1875 Hood wrote Steve Lee, "since the appearance of Sherman's book I have determined to give a full account of the siege of Atlanta and the Tennessee Campaign."

One particular passage of Sherman's book brought up the old Johnston-Hood argument over Cassville. Sherman wrote of his Cleveland speech, c. 1868, and how Johnston had told him that on the night of May 19 Hood and Polk had urged retreat. Sherman further writes of Hood's reaction at the time. In spring 1870, when Sherman was in New Orleans, Hood looked him up at the St. Charles Hotel and brought up the Cleveland speech. "He stated that he had argued against fighting the battle purely on the defensive," Sherman recalled in his memoir, and elaborated on a claim Hood made to him: how at Cassville he had asked Johnston for permission to attack Schofield when the latter had become

[13] "Hood's Reply, No. 1," *New Orleans Times*, May 10, 1874; Hood, *Lost Papers*, 63, 65 (Hood to Conrad, May 19 and July 1, 1874), 66 (W. J. Byrne to Hood, June 26, 1874), 67 (Butler to Hood, July 18, 1874), 88–89 (Polk to Hood, June 17, 1874, which is also in *A & R*, 108); Miller, *Hood and Memory*, 204 (Johnston to Falconer); Chas. G. Johnsen, "Gen. J. B. Hood's Letter," *New Orleans Picayune*, June 14, 1874 (copy courtesy of Anthony Delrosario of the Tulane Library); Bate report, January 26, 1865, *OR*, vol. 38, pt. 5, 826–27; Stephenson, *Memoir*, 225 (quoting Johnsen's article in the *Picayune*, but without editor Hughes noting that it was in reply to Hood's letter of May 10 in the *Times*); Hood, *A & R*, 85 (correction on blockhouse), 124, 154; Warner, *Generals in Gray*, 125 (Hardee's death, November 6, 1873).

separated from the rest of the army group, and how "General Johnston had taken offense, and that it was for this reason he had ordered the retreat that night." If the Northern general's recollection of the St. Charles conversation is correct, one must conclude at least in this case, as Sherman sometimes did, that Hood was prevaricating through Georgia, too. Cump charitably concludes, "as subsequent events estranged these two officers, it is very natural they should now differ on this point."

There is no record of Hood having written Sherman to contest points in his book, for the two men had become friends. In December 1873 Hood had written him, asking how many Confederate soldiers had been captured or received as deserters by his troops from Dalton to the Chattahoochee; Hood was clearly building his numerical case against Johnston. When Hood was in St. Louis on business, in February 1875, Sherman paid him a call and invited him to dinner, saying (as Hood wrote to Anna) "that our names share association in history and he wanted to know me &c." Hood frankly wanted to dodge the dinner, but felt he could not risk further offending Sherman, who evidently remembered having earlier been in New Orleans and that he had not been invited to the Hood home. "He was very kind and courteous," Hood recalled. Several years later the two friends got together again. During winter 1878–1879, Sherman and his daughters were in New Orleans, and this time Hood did indeed invite them to his home for a visit. "I cannot help thinking of that wonderful and beautiful group of children you paraded before us last winter at your home in New Orleans," Sherman wrote Hood afterward.

"The General has written a book, I think, and I want to get it," Hood informed his wife after meeting with Sherman in St. Louis in February 1875. When it came out, Braxton Bragg thought Sherman's book was better than Johnston's, insofar as it would shed "much light on the invisible darkness of Johnston's narrative." Part of that "invisible darkness," he explained, involved Johnston's use of Dr. Foard's casualty numbers, which only counted killed and wounded during his part of the campaign, and not prisoners, deserters, and sick. The number of sick soldiers alone, based on Bragg's memory of a conversation with Dr. Foard in 1874, was 11,000. Add this to the killed and wounded during Johnston's stage of the campaign, he explained, "and you have losses (with prisoners & deserters) of 25,000."

As we have seen, in the early 70s, well before Sherman's *Memoirs* appeared, Hood had begun writing other officers about the Georgia

Campaign, as shown in *Advance and Retreat* (e.g., Harry Sellers, former A. A. G. to Hood, October 20, 1872, as to the strength of his corps at Dalton). Recent publication of Hood's "lost letters" confirms his interest in matters of the Tennessee Campaign as well. In March 1875 Hood wrote Jefferson Davis, asking his recollections of their meeting at Palmetto in September 1864, when they planned the campaign to get into the rear of Sherman's forces in north Georgia. On the other hand, it cannot be denied that the publication of Johnston's *Narrative* spurred Hood's renewed interest in his war memories. Of the two dozen letters quoted in *Advance and Retreat*, all but nine were written by or to Hood in 1874, the year he wrote his "Reply" articles.[14]

After that, for the rest of the decade until his death, Hood maintained intermittent correspondence with ex-Confederates as he was writing *Advance and Retreat*. As we have seen, he had written his *Times* arti-

---

[14] Kennett, *Soldier's Life*, 317; Lewis, *Fighting Prophet*, 615; Hirshson, *White Tecumseh*, 355; Hood to Lee, August 19, 1875, and E. P. Alexander to Lee, December 23, 1890 (addressed to "My very dear old Steve"), Lee Papers, Chapel Hill (typescripts courtesy of Keith Bohannon); Sherman, *Memoirs*, 2:41; McMurry, *Hood*, 197 (Hood to Sherman, December 26, 1873); Hood, *Lost Papers*, 70–71 (Bragg to Hood, August 24, 1875), 107 (Hood to Davis, March 27, 1875), 168–70 (Hood to Anna, February 21 and 22, 1875), 220 (Sherman to Hood, August 26, 1879); Hood, *A & R*, 80 (excerpt of Sellers to Hood, October 20, 1872, fully printed in *Lost Papers*, 130–31); McMurry, "Disappointment in History: The Papers of John Bell Hood," 162 n.4.

Writing in 1974 of the disappointing thinness of the Hood Papers in the National Archives, McMurry was aware that John Dyer in the late 1940s, writing his biography, had found a "limited" collection of the general's papers in the possession of his daughter Ida but correctly surmised there must have been more ("he must have collected a large file of important correspondence"). Neither biographer knew that another daughter, Lillian, had another cache, which is the basis of Sam Hood's *Lost Papers*. McMurry makes two statements about Hood's papers: "historians have speculated that the documents were destroyed in the general house cleaning that followed his death" and "since there was no one in New Orleans who would have been interested in preserving the papers they were simply thrown away." Today, these statements read as not so much incorrect deductions as simply the product of what Richard knew (and did not know) forty years ago ("Disappointment," 162; Dyer, *Gallant Hood*, 367; McMurry, foreword to Hood, *Lost Papers*, xviii ["future historians of Hood's military campaigns and his next biographer will have the use of crucial documents that were not available to earlier writers," such as himself]).

cles by summer 1874; the "Reply" chapters would form a third of his eventual book. After that he wrote four chapters on his efforts to save Atlanta, and what eventually amounted to be almost seventy pages on his Tennessee campaign. In August 1875 he was writing former brigadier George B. Hodge, who had himself just finished an article on Johnston's retreat from Dalton; Hodge noted that when the two had earlier met, he "concurred with you [Hood] of the campaign." The *Official Records* were yet to appear; in 1878 and 1879 Hood was seeking copies of Bragg's and Longstreet's reports on Chickamauga as well as the reports of his own army's general officers.

In January 1879 the reminiscences were not yet complete. At that time he wrote Anna, "I have decided to add another Chapter to the book, to be the first chapter, which will embrace a brief sketch of my life & career up to joining Johnston." "Gibson says I must by all means do this," he added, "and as it will take me but little time I shall do so." As for book title, Hood was thinking merely of "Hood's Memoirs." "Gibson says the young generation must have this minder to know who I am &c.," he related to his wife. By spring 1879 he was far enough along in his research and writing that comrades looked forward to his eventual book. Henry Clayton wrote Hood, "please put me down as a subscriber to your Book when published and send me a copy."

Hood tailored his memoir in large part as a critique of Joseph E. Johnston's leadership in the Atlanta Campaign. He may not have coined the phrase "Lee and Jackson school," but he used it repeatedly to contrast himself to what could have been termed "the Joe Johnston school." The "Lee and Jackson manner of handling troops" committed them to bold, open-field battles such as Gaines' Mill, Sharpsburg, and Gettysburg, and refrained from putting them into entrenchments except as a last resort in the closing months of the war. Hood claimed that the Confederate soldiers' elan and morale were strengthened by Lee's and Jackson's aggressive tactics. The "Joe Johnston mode of warfare," with its reliance upon earthworks and endless retreats, on the other hand, tended to demoralize the men and dull the effectiveness of the army as a whole. "The one school elevates and inspirits," Hood concludes, "whilst the other depresses, paralyzes, and, in time, brings destruction."

Hood seems to have almost enjoyed lampooning Johnston's claim, made in his *Narrative*, that if his alleged plan of attacking the enemy before Atlanta failed, his army could "secure place of refuge in Atlanta, which it could hold forever." Just short of a dozen times in *Advance and*

Retreat does Hood quote "forever," asking such questions as why Johnston did not make this claim to Jefferson Davis when his job was on the line and the president was searching for any sign of optimism from his general in Atlanta. Hood went so far as to write Gustavus Smith about Johnston's claim that he could have held Atlanta "forever," particularly given his stated intention of leaving the city guarded by the Georgia militia while he led his three army corps in maneuver against Sherman. Smith bluntly answered that Atlanta probably would have fallen in twenty-four hours; Hood included the full text of Smith's letter in *Advance and Retreat.*

Hood's "Reply to General Johnston" chapters, some of which are based on his articles in the *New Orleans Times* during 1874, form the narrative core of *Advance and Retreat*, almost a hundred pages. In contrast, his description of the rest of the Atlanta Campaign, from his assumption of army command to the fall of the city, fills four chapters, some sixty-seven pages.

Hood concluded his book with a chapter titled "Rashness—Johnston—Fabius—Scipio." In it he added more criticism of Joe Johnston, rehearsing Roman military history to show that Johnston's supporters were wrong. He was no Fabius Maximus, Hood argued. Fabian policy meant holding onto mountainous terrain; Johnston gave up the high ground of north Georgia. And Johnston was certainly no Scipio Africanus. That Roman commander carried the war to Africa, defeated Hannibal and conquered Carthage; Johnston lacked such "spirit and boldness."

Hood was in turn sensitive to charges of "rashness" during his time as army commander. For these accusations he blamed Sherman, Johnston, and "Pollard." Edward A. Pollard, wartime editor of the *Richmond Examiner*, was a political enemy of Jefferson Davis, and hence no friend of John B. Hood. In his postwar history of the war, *The Lost Cause* (1867), Pollard had charged Hood with having ordered "fanciful attacks" that brought "useless sacrifice of life." The Confederate assault of July 22 he termed "one of the most reckless, massive, and headlong charges of the war." Against such an "accusation of rashness," Hood's self-defense in his memoir is restrained. "I shall merely state that the confidence reposed in me…by Generals Lee, Jackson, and Longstreet," he wrote in his closing chapter, "expressive of satisfaction with my course, is a sufficient refutation of the charge." The reader will notice that Hood is summoning his record as Virginia division commander to refute putative blunders committed during his time two years later as army commander in Georgia and Tennessee.

One reason Hood could devote time to his historical writing was that his insurance company was prospering. In 1874 he did well, earning fully $18,000 that year alone. "He engaged in various enterprises that at one time enriched him," noted the *New Orleans Picayune* in August 1879, a year after Hood's good fortune ended. In summer 1878 yellow fever struck New Orleans. The Hood family was spared serious illness, but the epidemic shook the city, closed the Cotton Exchange, and ruined a number of companies, among them Hood's insurance firm. The Hoods also lost money through failed investments in gold and state bonds. Thrown onto hard times, the ex-general thought of ways to raise money. One of them was to sell to the federal government certain documents he had retained or acquired pertaining to the war. The War Department was assembling and arranging documents in its possession (including captured Confederate papers) for publication in the *Official Records.* Former C. S. Brig. Gen. Marcus Wright was in charge of soliciting papers from important officers. Sometimes Wright asked for their donation or loan for transcription; sometimes he settled for a typed copy. Because the federal government also bought important documents, Hood believed he had a good case to ask for a princely $20,000 for his most important papers. In January 1879 Hood went to Washington to make his sales pitch. To help him he called on a friend of notable stature: William Tecumseh Sherman.[15]

---

[15] Hood, *Lost Papers*, xxii ("heavy losses in investments"), 69 (Hodge to Hood, August 21, 1875), 137–38 (Hood to Marcus Wright, February 11, 1878, and Bate to Hood, January 25, 1879), 121–22 (Clayton to Hood, February 24, 1879), 167 (Hood to Anna, January 18, 1875, "my earnings for the year reach full 18,000"), 180 (Hood to Anna, January 11, 1879), 241 (yellow fever in 1878); Hood, *A & R*,130–36, 141–53, 313–16; Johnston, *Narrative*, 350; Pollard, *The Lost Cause*, 577–78; Dyer, *Gallant Hood*, 315 (quoting *Picayune*); McMurry, *Hood*, 201–202; McMurry, "*Official Records*," 369; S. Davis, *What the Yankees Did to Us*, 318 n.41 (on the *OR* staff receiving from an unidentified donor only a typed copy of Confederate major William Clare's list of Atlantans expelled from Atlanta in September 1864); McMurry, "Disappointment," 164.

One scholar has stated that Hood's financial problems in the late 1870s occurred "probably from ignoring his insurance business to concentrate on *Advance and Retreat*" (Moody, *Demon of the Lost Cause*, 63). Moody's supposition is not supported by Hood's correspondence. In 1878, the year that his company fell

At the start of 1879 Sherman was living in Washington, serving as general-in-chief of the United States Army. He had offered his assistance to Hood in his money-making project, having suggested that he send his papers to Washington for examination. Hood did more than that; he took his papers to the capital himself. Arriving there on the night of January 10, 1879, he visited Sherman the next morning. The general-in-chief took him to see the secretary of war, George W. McCrary. Hood was pleased with his warm reception. "The Secretary has appointed a committee to examine my papers at once," "Bell" (as he signed himself) wrote to Anna.

Sherman did not stop there. On the evening of the 12th, Sherman took Hood to pay a call on his brother, Secretary of the Treasury John Sherman, and no less than President Hayes at the White House. There he was introduced to Chief Justice Morrison Waite, Vice President William Wheeler, and other dignitaries. "So you perceive that I am going all around about," Hood proudly informed his wife. Even better were prospects for a good price tag for his war papers. "Have just returned from seeing Gibson," Hood wrote, meaning Randall Gibson, his longtime friend. Gibson at the time was in Washington serving in the House of Representatives. "He saw Sherman & the Sec. of War this morning. The former is trying to have my papers valued at 20,000. Trust he will succeed." Hood was delighted that Sherman, his former enemy, was showing him such courtesy. "His attention & kindness has been marked," he wrote Anna, along with this treat: "he told Gibson this A.M. that I had done in my Campaign the best thing that could be done."

Hood was writing on Monday, January 13, and closed his letter, "I hope to be able to leave here Wednesday or Thursday." When he did so, he left his "carefully packed valise" of documents in the care of Sherman's daughter Elizabeth, "Lizzie," who with her father had visited the Hoods in New Orleans the previous winter, and to whom Hood "seemed to [have] taken a fancy," as Sherman later wrote. Eventually the War Department assessed Hood's papers as worth $12,500, but Sherman and Congressman Gibson lobbied Congressional appropriations committees

---

through (according to the *Lost Papers*, which is the best reflection of his letter writing), Hood wrote all of one epistle to a former comrade (to Marcus Wright, asking for Bragg's and Longstreet's Chickamauga reports). One may figure that if Hood in 1878 was in the midst of writing his memoir, he would have been sending a lot more letters out asking comrades for factual assistance.

for the requested $20,000. They would be unsuccessful. The papers remained with Sherman after Hood's death later that year. They made their way to the National Archives in 1938; the Hood papers are still there.[16]

Hood was writing and working hard on his memoir in winter and spring 1879. On February 19, Jefferson Davis wrote W.T. Walthall that Hood "is enlarging the scope of his work, and says he is nearly ready to publish." It is not known whether Hood had contacted a publisher, but he was thinking of one, for in a letter to S. D. Lee in April 1879 he mentioned that he had copied his official campaign report "for the publisher." Just a few days before he had delivered "brief remarks" at an Army of Tennessee veterans' reunion; he may have mentioned the book he was finishing. On April 18 he wrote Lee, "I have just examined my manuscript"; the author was evidently looking for mistakes and double-checking details. To Lee, for example, he related that the five o'clock sunset at Franklin had been a factor limiting the success of his attack, adding that the fact was "acknowledged by Van Horne"—author of an early history of the Army of the Cumberland.

John Bell Hood would not see his book published; he died of yellow fever later that summer. The epidemic struck New Orleans in late July

[16] Kennett, *Soldier's Life*, 303; Hood, *Lost Papers*, 180–81 (Hood to Anna, January 11 and 13, 1879), 220–21 (Sherman to Hood, August 26, 1879 and to Mrs. E. R. Hennon, December 24, 1879; Bergeron, "Gibson," 188; McMurry, "Disappointment," 164; McMurry, *Hood*, 202; Povall, *Rebels in Repose*, 92.

McMurry and Hirshson write about Hood's visit to see Sherman in Washington either with uncertainty or inaccuracy: "Hood journeyed to Washington in either late 1878 or early 1879" (McMurry, *Hood*, 202); Hood "came to St. Louis to see Sherman about selling his Civil War papers to the War Department" (Hirshson, *White Tecumseh*, 335). Cause derives from these previous writers' lack of awareness of the letters discovered by Stephen Hood in 2012, including Hood's letter to Anna, cited above, which specifies date and place of Hood's meeting with Sherman. Moreover, McMurry has termed Hood's leaving his valise with Sherman's daughter "a gesture typical of the impractical man he often was" ("Disappointment," 164), but Richard was writing in 1972 and did not have the "lost" Hood papers. Now we do, and we know from Hood's letter in 1879 that he and Sherman were close—close enough for Hood to entrust his papers with Sherman's daughter. That's not impracticality—that's *friendship.* I make these statements to remind future biographers of John Bell Hood that they must consult and study Sam Hood's *Lost Papers.*

1879. Dr. T. G. Richardson, Hood's longtime friend, urged the Hoods to leave the city, but they did not (Stephen Hood assumes they did not have the money for travel and lodging). On August 21, Anna came down with the fever and died in three days. On the 25th, Hood also began to show symptoms. On Friday, August 28, he told his physician, Dr. Samuel Bemiss, "Doctor, if you cannot overcome the enemy, do not try any experiments." Later that day there seemed to be some improvement in his condition. "Doctor," Hood told Bemiss, "I believe that we have routed the enemy." Hood was wrong. Shortly after his eldest daughter, Lydia, died of the fever at noon on the 29th, Hood became delirious. Like Jackson and others, his mind returned to the years of war. "I want those stores taken from my own commissary," he once called out. Then he regained lucidity enough to explain his hope that the Hood's Texas Brigade Association might see to the care of the ten children surviving so far. Then he slipped back into unconsciousness. At 3:30 in the morning of August 30 he died. Hood was 48.

The general was laid beside the grave of his wife in Lafayette Cemetery No. 1. The *Picayune* announced the death of "the accomplished gentleman and distinguished soldier" in its issue of August 31. The paper stated that in his last hours, delirious, the general was "giving orders to his columns engaged in the battle flame and smoke." Its long eulogy reviewed his military career admiringly, absolving him for his failures at Atlanta and in Tennessee. "The very soul of honor and knighthood lived in that shattered frame," the column affirmed. "Neither wounds, nor sickness, nor danger, nor poverty, nor the defeat of the cause he fought for so heroically, could shake that dauntless spirit or ruffle the composure of the polished and modest gentleman." "A host of mourning friends," it concluded, "will join with us in this sincere tribute to his memory."[17]

Anna and John Hood had not been buried long before people in New Orleans began to wonder what would become of their children.

---

[17] Rowland, *Jefferson Davis Constitutionalist*, 8:358 (Davis to Walthall, February 19, 1879, including, "Genl. Hood informs that he has satisfactory evidence that Genl. J. E. Johnston before the battle of Seven Pines, contemplated the abandonment of the Capital, Richmond"); Hood to Lee, April 10, April 14, and April 18, 1879, Lee Papers, Chapel Hill (typescripts courtesy of Keith Bohannon); Hood, *Lost Papers*, 242–43; McMurry, *Hood*, 202–203; Dyer, *Gallant Hood*, 318–19; "Death of Gen. J. B. Hood," *New Orleans Picayune*, August 31, 1879.

Anna's aged and feeble mother, Eleanora Hennen, took immediate charge of the ten children, who at the time were aged nine (Annabel and Ethel), seven (John, Jr.), six (Duncan), five (Marion and Lillian), two (Odile and Ida), one (Oswald) and four months (Gertrude). Dyer claims that Mrs. Hennen traveled to Texas to discuss with the officers of the Hood's Texas Brigade Association their role in possibly raising money for a house and housekeeper for the children, but Simpson finds no mention of such an event in the association minutes. Before she died in April 1880, Mrs. Hennen stipulated that her brother-in-law John Morris, also in New Orleans, would assume care of the children. Morris and his wife Cora adopted the twin sisters, Annabel and Ethel, but could take in no more.

Friends of the family began circulating appeals for Southern women to adopt and raise the children. Solicitations for financial assistance were distributed to newspapers. Mary Gibson, wife of the general, joined eleven other ladies signing a circular issued to the press, announcing that donations should be sent to Isaac Patton, mayor of New Orleans, or to several other former business associates of General Hood. "Let the women of the South adopt these orphans," read the appeal, "and at once raise a sum sufficient for their maintenance and education." The ladies suggested that "an association for this purpose be formed in every county and town."

After John's and Anna's deaths, their friend Beauregard stepped forth to help the children. Hood had completed his manuscript, including the chapter on his "career up to joining Johnston" that he had told Anna about just that January. Beauregard organized a Hood Memorial Association, under whose auspices the memoirs were published in July 1880 under title of *Advance and Retreat: Personal Experiences in the United States and Confederate States Armies.* All proceeds from sales went to the Hood Orphan Memorial Fund. Depending upon the type of binding chosen by the purchaser, price for the "elegant octavo" ranged from three to five dollars.

The Hood Texas Brigade Association stepped forth to do its part in fund-raising. By June of 1880 it had raised more than $3,000 to send to New Orleans for the care of the orphans and to build a tomb for General Hood. In the years after that it raised more than another thousand dollars.

As one more fund-raising tactic, Beauregard and others arranged for a touching photograph to be taken of the ten little ones. They are seen seated in a nicely furnished parlor: Odile and Ida, Duncan, John Jr., Oswald, Ethel and Annabel, Lillian and Marion. An empty chair in the room would have been for Lydia. Little Gertrude, who had been all of four months old when her parents died, lies on a tiny sofa in front of them all. A photograph of Hood, in wartime uniform, stands on an easel; that of their loving mother hangs on the wall behind. Titled *The Late General Hood's Family*, the photograph was taken in the Lilienthal studios of New Orleans. On the back of each print was the statement, "Every Picture Sold adds to the Permanent Fund for the Education and Maintenance of these 'wards of the South.'" The pitch worked at least modestly; by mid-1885 sale of the photographs had raised $425.

Others across the South conducted their own money-raising efforts. The *Atlanta Constitution* launched one such project. A daily column in the paper titled "The South's Trust" and "Our Babies" began to run in the paper, announcing updates on contributors. The *Constitution* called for every school in the state to conduct a subscription drive. "Let every soldier who reads this paragraph," one column affirmed, "put his hand on his pocket and bring out a dollar for Hood's children." Most of the pledged contributions were in small sums, but Governor Brown pledged $300 and former general Robert Toombs $100. Even an ex-Union soldier sent in $5 with a note: "all good men must sympathize with the family of a good and brave soldier overtaken by misfortune." The *Constitution*'s campaign was effective. By the end of the 1880s, Georgians had contributed $3,580.97 to the Hood Memorial Fund.

There never was enough money, however, and benefactors of the orphans concluded that the children had best be separated and offered for adoption singly or in twin-pairs. In late 1882 a lady as far away as Denver wrote the *Southern Bivouac* and asked what had become of the orphans. The editors offered an assuring response:

> The twin girls, twelve years of age, are being educated by Mr. Morris in Hanover; John B., the eldest boy, has been adopted by Mr. and Mrs. Russell, of Mississippi; Duncan N., nine years, is being deducted by Miss Furniss, a wealthy lady of New York; Lilian and Maria, twins of eight years, have been adopted by Mr. and Mrs. Thatcher Adams of New York; Odile and Ida, twins of six years, are under the care of Mr. and Mrs. McGhee, Woodville, Mississippi; Oswald, five years, has been adopted by Mr. and Mrs. Harney of New York; Anna Gertrude, the youngest died two years since from teething. They are,

each and every one, fine children, handsome, intelligent, and full of character.

The sad tale of little Gertrude's death in March 1880 includes a warm anecdote. Mrs. Bettie Joseph had happened to see the photograph of *The Late General Hood's Family* at the home of a friend and fell for the infant lying in the foreground. She asked for and was given the picture. The couple had given up on having a child themselves and had not pursued adoption. But Bettie asked her husband Moses if they could not adopt Gertrude. Moses assented; she went to New Orleans in February 1880 and won the right of adoption. Bettie took the infant to their home at Columbus, Georgia. The Josephs' happiness, however, lasted only until Gertrude's death a month later from causes doubtless more complex than "teething."

From the locales of their adoptive parents, it is clear that the Hood orphans touched as many Northern hearts as Southern. A breakdown by state of almost $20,000 in donations to the fund showed contributions coming from every ex-Confederate state, as well as New York, Pennsylvania, California, and Oregon. General Hood's efforts to raise money for the victims of the Franco-Prussian War of 1870–1871 led to donations coming from as far away as France ($40) and even Poland ($120). Eventually, the efforts of Beauregard and others, the several fund-raising tactics, the work of many collaborative associations (not to mention interest income from careful investment) raised almost $75,000 by June 1895. This money was divided among the children when they turned twenty-one years of age.

Many had been talking about a funereal monument for John Bell Hood and his wife, but inevitably lack of funds prevented more than an inscription in the Hennen family mausoleum in New Orleans's Metairie Cemetery:

GEN. JOHN BELL HOOD, C. S. A.
BORN IN OWENSVILLE, KY.
JUNE 29, 1831
DIED AUGUST 30, 1879

The inscription remains to this day. A passerby standing before it now, though, will see a plaque, dedicated August 30, 2003:

> John Bell Hood was born June 29, 1831, in Owingsville, Kentucky, and was reared in Mt. Sterling, Kentucky. After graduating from West

> Point in 1853, he served in the elite U.S. 2nd Cavalry Regiment on the Texas frontier. In 1861, he joined the Confederate Army. He was promoted to brigadier general in 1862 and commanded the renowned Hood's Texas Brigade under Gen. Robert E. Lee at the important Confederate victories at Gaines' Mill (Seven Days Battles) and 2nd Manassas (Battle of 2nd Bull Run). He held the critical Confederate left flank at Miller's Cornfield at Antietam, after which he was promoted to major general by Thomas J. ("Stonewall") Jackson. In July 1863, while serving as a division commander at Gettysburg, he was severely wounded and forever lost the use of his left arm. In September 1863, while leading a decisive Confederate victory at Chickamauga, Georgia, Hood was again severely wounded and lost his right leg. Promoted to lieutenant general by Gen. James Longstreet, he returned to duty in 1864 in north Georgia under Gen. Joseph Johnston, as corps commander. He succeeded Johnston as commander of the Army of Tennessee and was temporarily promoted to full general in July 1864. In November, in an unsuccessful attempt to draw Union Gen. William T. Sherman from his March to the Sea, Hood led the Army of Tennessee in an invasion of that state. After decisive defeats at Franklin and Nashville, he retreated to Tupelo, Mississippi. In January 1865, he resigned command. He surrendered to Union authorities at Natchez, Mississippi, on May 31, 1865. After the war, Hood set up residence in New Orleans, where on April 30, 1868, he married Anna Marie Hennen, with whom he fathered 11 children. He died of yellow fever on August 30, 1879, within days of his wife and eldest child. Seven families in 5 different states adopted the surviving orphans. Hood was buried in the Garden District's Lafayette Cemetery, but was moved to this location in 1927.

Eulogizers of John Bell Hood spoke of him as "polished and modest," "unpretentious," and a "high-toned gentleman" who was "true to himself." The text of the marker at his gravesite reflects these tributes; it is no more florid nor less descriptive of his life than the record he might himself have composed.[18]

---

[18] Hood, *Lost Papers*, xxii (Eleanora Hennen), 180, 241 (orphans and ages), 245–50 (including photograph, noting that it dates from 1879); Dyer, *Gallant Hood*, 319, 364 n.58; Simpson, *Reunion and Memory*, 162–73; "What Will Become of the Children[?]," 36–37; T. H. Williams, *Beauregard*, 320; "Just Out Hood's Great Book of the War," *Los Angeles Herald*, July 29, 1880; Miller, *Hood and Memory*, 234 (photograph sales), 235 (fund breakdown by states), 282–83 (Metairie marker text, which was written by Stephen M. Hood); "Those Little Orphans. What Georgia Is Doing," *New Orleans Picayune*, September 13, 1879;

"Editors, Southern Bivouac," 177; "Fund for the Children of Gen. J. B. Hood," 15 ("investment and reinvestment"); "Graves of Gen. John B. Hood and Wife," *Confederate Veteran*, 123; R. Owen and J. Owen, *Generals at Rest*, 96; "Army of Tennessee. An Eloquent Tribute to the Memory of the Late Gen. J. B. Hood," *New Orleans Picayune*, September 10, 1879; "Death of Gen. J. B. Hood," *New Orleans Picayune*, August 31, 1879; "The Death of General John B. Hood," *Southern Historical Society Papers*, 496; "Hood's Legacy. Letters about It," *New Orleans Picayune*, September 29, 1879.

Conclusion

Spurs without Greatness

> "No man is justly entitled to be considered a great General, unless he has won his spurs."[1]

Toward the end of his memoirs—toward the end of his life, really—John Bell Hood set down this soldierly axiom. At the time, he was writing about another Confederate officer, but he doubtless was thinking of himself as well. Hood did not specify how a soldier "won his spurs" because he did not need to: all knew it was by brave, conscientious performance of duty.

West Point, the 2nd Cavalry out west, and C. S. cavalry commission in April 1861 set Hood off in his Confederate soldier's career. He quickly won commendation on the battlefield and by October was colonel of the 4th Texas Infantry. Before the next spring's campaign began, Hood was promoted to brigadier general and placed in command of the "Texas Brigade." At Gaines' Mill during the Seven Days' Battles, Hood led his troops in a heroic charge that broke the enemy line. Both Stonewall Jackson and Robert E. Lee praised Hood's and his Texans' performance. At Second Manassas he was "magnificent," in the estimation of Douglas Southall Freeman.[2] Some started calling him "the gallant Hood."[3] But there is evidence that Hood helped to plant stories in the Richmond press hailing his battlefield exploits. More damaging, when he disobeyed a superior's order (out of loyalty to his troops), he was placed under arrest as Lee's army marched into Maryland. With his soldiers calling for Hood's return, Lee suspended his arrest on the eve of the battle of South Mountain. At Sharpsburg on the morning of September 17,

---

[1] Hood, *A & R*, 316.

[2] Freeman, *Lee's Lieutenants*, 3:138.

[3] "An Eyewitness," "The Texas Brigade," *Richmond Enquirer*, September 13, 1862.

1862, the counterattack of Hood's troops stopped Hooker's corps and stabilized the left of Lee's line. Afterwards Jackson recommended his advancement to major general, which Lee endorsed. His promotion, effective October 10, placed him in command of an infantry division in Lee's army.

Thus far Hood had won an enviable reputation as brave soldier and excellent battlefield leader. His scope of command had included cavalry company, infantry regiment and brigade. At Sharpsburg he had commanded a small division. But over the next year he did not strengthen his reputation as division commander. With General Lee's concurrence in November 1862, a staff officer chided Hood for inattentiveness to his men's equipment and their camps' orderliness. At Fredericksburg his troops were only slightly engaged. During the battle he did not follow Longstreet's instruction for a counterattack; afterward, Longstreet chose not to pursue the matter. Five months later Hood missed Chancellorsville altogether, with Longstreet at Suffolk. There, chafing at his inactivity, Hood inappropriately wrote General Lee, breaking chain of command (it would not be his only time doing so during the war). Some Confederates began to comment on Hood's ambitious angling for still higher rank. Significantly, however, in his reorganization of the Army of Northern Virginia after Jackson's death, Lee passed over Hood for promotion to lieutenant general, writing that the Texan was "improving" as an officer.[4]

But at Gettysburg Hood got no opportunity to shine on the battlefield, as he had done earlier. On July 2, his division was placed on the Confederates' far right, set to lead Longstreet's afternoon assault. When scouts reported that the end of the Union line at the Round Tops was vulnerable to a flank attack, Hood asked Longstreet several times for permission to alter General Lee's plan. When peremptorily told to attack as ordered, Hood cryptically mentioned his intent to make a "digression."[5] If so, Hood is subject to criticism for tinkering with his superior officer's orders for the battle—no small infraction indeed.

But a shrapnel-ball wound to his left arm knocked Hood out of the battle before it had even begun. The arm was not amputated, but Hood had little use of it for the rest of his life. Two months later, he had recovered in time to join his division in the train transport of Longstreet's

[4] Lee to Davis, May 20, 1863, *OR*, vol. 25, pt. 2, 811.

[5] Trudeau, *Gettysburg*, 323.

command to the Army of Tennessee in northwest Georgia. They arrived there just in time to help win the battle of Chickamauga. On its second day, September 20, Longstreet launched a massive infantry attack on the Federals' right just as a colossal command blunder created a gap in their line at that very point. Hood helped guide the Confederate infantry in arguably their most decisive battlefield breakthrough of the whole war. During the fighting Hood, riding in rear of his troops, was shot in the right upper thigh. Amputation was quickly performed, after which Hood received the army's best medical care. General Longstreet urged the government to promote Hood yet again to lieutenant general.

Convalescing in Richmond over the winter, Hood was lionized by capital society and befriended by Jefferson Davis himself. By February 1864 he was sufficiently recovered to return to field service. The president submitted Hood's nomination for promotion, and the Senate confirmed. When Hood told Mary Chesnut of this good news, she recorded in her diary, "all that ambition still—in spite of those terrible wounds." Some jealous officers grumbled that Hood was "going up too fast." Maj. Gen. Wade Hampton was one of them. "Sam was promoted too fast," Mary Chesnut recorded his having said to her. "He wanted experience," Hampton contended; Hood "was not long enough major general." But then, Mrs. Chesnut observed, Hampton was a Johnston man "out and out."[6]

Lt. Gen. John B. Hood left Richmond in mid-February and joined Gen. Joseph E. Johnston's Army of Tennessee as infantry corps commander. Shortly thereafter, Hood began private correspondence with the president, telling him what Hood doubtless knew Davis wanted to hear. During the winter the president had urged Johnston to prepare his army for an offensive into Tennessee. Johnston had repeatedly balked, frustrating Davis, Bragg, and Seddon. When Hood wrote the president about his desire for offensive operations, his letter can only be construed as a devious maneuver to undermine his superior officer in hopes of someday taking his job. Yet after the war Hood did not attempt to hide his correspondence with Davis and Bragg; he even printed his letters in *Advance and Retreat.* Moreover, Hood was not alone. Hood, Wheeler, Hardee, and other generals were writing Davis, Bragg, Seddon, and Wigfall in early 1864, as I state in my text, all complaining about Joe Johnston.

---

[6] Woodward, *Mary Chesnut's Civil War*, 551, 557 (Capt. John Chesnut), 710 (Maj. Gen. Wade Hampton).

From the time of Hood's arrival, Johnston seems to have relied upon him much more than he did on the army's senior corps commander, Hardee. Johnston and Hood were frequently seen riding together, and when William T. Sherman opened his campaign, Johnston placed Hood's Corps at the point in his Rocky Face line that he viewed as most likely to be attacked. During May–June Hood enjoyed good relations with Johnston, who in major battles put him in the places of key responsibility. At Resaca it was Hood to whom Johnston entrusted the only attacks he ordered. In the nasty postwar sparring conducted by the two generals, Johnston blamed Hood for the loss of Van Den Corput's battery on May 15, but Hood was not at fault for the enemy's capture of the four Napoleons. Similarly, after the war, the two officers fell into disagreement when Hood at Cassville on May 19 canceled a proposed flank attack. The evidence accumulated today and accepted by historians is that Johnston was wrong to have accused of Hood for crumbling before enemy "phantom troops."[7]

The loud proclamation issued by Johnston on the morning of May 19 that he was finally leading the army to attack, and his subsequent eating of those words later that day did much to poison his relationship with Hood. Then, south of Cassville, when Hood and Polk urged a retreat from their enfiladed lines, Johnston became more bitter still. The army commander seems not to have known that Hood was communicating with President Davis, as when Hood sent a staff officer, Col. Henry Brewster, to Richmond in late May with a personal message. "No plan—no strategy," Brewster later told Mary Chesnut; "overcautious." "So much retreating," she recorded, "is breaking Hood's heart."[8]

In late May Hood's troops repulsed the Federal attack at New Hope Church although Hood played no role in the Confederate victory. Despite his disappointment at Cassville, Johnston continued to rely on Hood to launch his attacking battles. On the day after Pickett's Mill, Hood was to assail the Federal left, but when he learned that the Yankees had withdrawn their lines, he canceled the assault and notified headquarters. But unlike at Cassville, Johnston held no grudge against Hood for calling off the attack of May 28. Indeed, General Hardee remarked in a letter to his wife of June 20 that Hood "is helping the General to do the strategy, and from what I can see is doing the most of it."

---

[7] Kurtz, "Johnston's Review of Sherman's *Memoirs*," 9.
[8] Woodward, *Mary Chesnut's Civil War*, 616.

At Kolb's Farm on June 22, Hood ordered a frontal assault without direct orders from Johnston and against an enemy position that had not been reconned. The Confederates' bloody repulse has led to criticism of Hood, and rightly so. Hood thus showed his pugnacity at Kolb's Farm and apparently expressed it to others. General Hardee informed Mary that "both Hood & myself expressed to Gen'l Johnston our desire to fight before reaching Atlanta, and that we would prefer to go there whipped rather than not to fight at all." Yet when Johnston put his army into General Shoup's redoubts with its back to the Chattahoochee, Shoup remembered that Hood "stated his position, and urged that the army be moved across the river." Hood was known as a gambler, but risking battle against a stronger enemy with a wide, deep river behind was one gamble that Hood prudently wished not to make.[9]

Hood's favored relationship with General Johnston was affirmed on July 1 at Marietta, when Georgia Senator Ben Hill met with them. Hill had agreed with Governor Brown to interview Johnston and confirm the need for Forrest's or Morgan's cavalry to be sent against Sherman's railroad communications. Hood was apparently the only corps commander attending the meeting; General Hardee was not asked to participate. In his report to President Davis and Secretary Seddon, Hill stated that Hood had largely listened throughout, but at the end had expressed worry that after the Confederates gave up their Kennesaw line, there was no good defensible ground south of it and, in Hill's words, "we should go back much more rapidly to the Chattahoochee River." Hood's prediction of a fast retreat back to the river coincided with his knowledge that Johnston was under criticism for all of these retreats. At some point in early July he wrote Buck Preston. According to Mrs. Chesnut, Hood told her "there was a talk of some new man to be put in command." When he added, "He hoped not" ("God help the new commander, whoever he should be, for he had a rough road to travel"), Hood was being disingenuous even to his fiancée. During the past year, his ambition for promotion to still higher rank had been repeatedly evidenced. Now, as Joe Johnston's fate appeared to be trembling, the ambitious lieutenant general seemed to be smacking his lips at forthcoming opportunity.[10]

---

[9] Hughes, *Hardee*, 211; Hess, *Kennesaw Mountain*, 18; Shoup, "Dalton Campaign," 264.

[10] Hill to Seddon, July 14, 1864, *OR*, vol. 52, pt. 2, 706; Woodward, *Mary Chesnut's Civil War*, 630 (entry of August 3).

In his fact-finding visit to Atlanta, July 13–15, General Bragg met with Johnston and his corps commanders, but he spent considerable time with Hood alone. Then, before he left Atlanta, Bragg had Hood write a long letter explaining how he had urged Johnston to assume the offensive and offering his own idea as to what the army should do ("we should attack"). Thus far Hood had written Richmond six times: 1) to President Davis March 7 ("I am eager for us to take the initiative"); 2) to General Bragg March 10 ("if we can concentrate and fall upon the enemy before he is ready we shall beat him badly"); 3) to Secretary Seddon March 10 ("we have a sufficient number of troops, if thrown together, to defeat his entire army"); 4) to Bragg April 3 ("you know how anxious I have been to advance"); 5) to Bragg April 13 ("I have done all in my power to induce General Johnston...to move forward"); and 6) to Davis May 21 (Colonel Brewster "can give you an account of the operations of this army"). But Hood's letter of July 14 to Bragg takes the cake. True, Hood in all likelihood wrote it at Bragg's request to cinch the case for relieving Johnston of command. Moreover, Bragg's accompanying letter, carried by Lieutenant Colonel Walter, played up Hood's abilities in a way Hess has termed "painfully flawed": "Do not understand me as proposing him as a man of genius or a great general, but as far better in the present emergency than anyone we have available." Bragg's assessment of Hood may have been painfully flawed, but at least Bragg recognized the shortcomings of his subject. At the same time, Hood's willingness to go along with Bragg in his letter-writing represents a bold display of his soldierly ambition, whose incessant goal had been always to seek ever higher rank and responsibility.[11]

Hood's letter did not reach Richmond until after President Davis, his Cabinet and the War Department had already decided to relieve Johnston of command. Because of his earlier declinature, Hardee was passed over; it was Hood to whom they now turned to lead the Army of Tennessee. Hood's alleged surprise when the offer came to him cannot

---

[11] Hood to Davis, March 7; Hood to Bragg, March 10; Hood to Seddon, March 10, *OR*, vol. 32, pt. 3, 606–608; Hood to Bragg, April 3, Bragg Papers, Western Reserve; Hood to Bragg, April 13, *OR*, vol. 32, pt. 3, 781; McMurry, *Hood*, 110 (re: Brewster, May 21); Hood to Bragg, July 14, *OR*, vol. 38, pt. 5, 880; Bragg to Davis, July 15, *OR*, vol. 39, pt. 2, 714; Hess, *Bragg*, 233 ("painfully flawed," in a photographic caption Hess has affirmed he wrote himself [email to author, August 31, 2016]).

be believed. Likewise, his joining Hardee and Stewart in wiring Richmond, asking for a postponement of Johnston's removal, must be taken as artificial show. Hood's acceptance on July 18 represents the ultimate fulfillment of his lifelong ambition.

Hood's promotion to full general was only temporary—the Senate was not in session, but the administration had authority to make temporary appointment in time of military emergency. Now in command of the Confederacy's second most important army, Hood did all he could to stop Sherman over the next month and a half. First, he ordered attacking battles when he saw the prospect for gain. Hood had been in command just two days when the enemy's advance gave him opportunity to plan one such engagement. Thomas's three corps were pressing down on the city from the north and would have to cross Peachtree Creek and its deep gorge. An assault on the Yankees in their act of crossing—or just after they had done so, but before they dug in—offered the chance of giving them the thrashing Hood had hoped for throughout the campaign. The plan for battle at Peachtree Creek was so good that Joe Johnston later claimed credit for it, and he even alleged that he had imparted it to Hood before leaving the army. (Historians are split on the issue.) In their attack of July 20, the Confederates enjoyed some partial, temporary successes, but were ultimately repulsed. For Hood, the most notable development was his criticism of Hardee and his allegation of the latter's failure to push the assault vigorously at Peachtree Creek.

Sherman's sending Garrard's cavalry off to wreck more railroad toward Augusta on July 21 left McPherson's left flank in the air and dangerously vulnerable to flank attack. Hood's seizure of the opportunity to plan a bold "Jacksonian" maneuver and attacking battle were demonstrations of genuinely good generalship. But his unrealistic expectations of what Hardee and his infantry could achieve on July 22 were not. A fifteen-mile night march over hot, dusty roads for a morning assault was simply not possible, and in his zeal Hood failed to realize that he was asking too much. His order for Wheeler to ride ahead and attack McPherson's wagon train at Decatur also exemplifies Hood's blundersome planning; the Confederate cavalry could have rendered far more important service by helping Hardee's divisions make their way to their assault jump-off positions. It is not too much to say that Hood's youthful inexperience as army commander manifested itself in planning his battle east of Atlanta. Hood should also be faulted for trying to make Hardee a scapegoat for the failure of the battle. His deliberate creation of animosity with the army's senior corps commander was not helpful to the Southerners' efforts to defeat Sherman in the short term and led to Hardee seeking transfer away from the army in the long term.

After the battle of July 22, Sherman swung the Army of the Tennessee, now O. O. Howard's, around the city to the west to get at the railroad running to East Point and Macon, Hood's last. Hood sent S. D. Lee's corps out to block the Federals' advance on July 28 and ordered Peter Stewart to follow and strike them in flank on the 29th. The plan came apart when the Yankees got to the Lickskillet road first and dug in. Against orders but believing he held an advantage, Lee ordered an attack which, once started, sucked in Stewart's troops as they arrived on the battlefield. The resulting battle of Ezra Church ended in a sharp Confederate repulse, with 600 killed and another 2,2,00 wounded or missing. Hood's statement in his memoirs that some of his troops showed a "lack of spirit" in their fighting reflected more poorly on Hood than it did on his allegedly spiritless soldiers. Blaming one's men after a battlefield defeat is not the mark of a magnanimous commander. Worse was Hood's effort to conceal from the authorities in Richmond details of the defeat. Telegraphing Secretary Seddon that the battle ended "with no decided advantage to either side," and that "we still occupy the Lick Skillet road" were outright lies, a further demonstration of conduct unbecoming an army commander.[12]

At the same time, Hood should not be criticized for the outcome of Ezra Church. The accusation that he ordered a futile frontal attack on July 28, so often seen in the literature fifty years ago, is now hopelessly outdated and thankfully on its way out. On the other hand, those same students have allowed William T. Sherman to get away with having ordered futile frontal attacks at Resaca, New Hope Church, Pickett's Mill, Kennesaw Mountain, and Utoy Creek. If Hood had ever written "I begin to regard the death & mangling of a couple of thousand men as a small affair, a kind of morning dash," historians would be lacerating him.[13]

As we have seen, Cump Sherman was inclined to believe that the enemy would do what he wanted them to do, and during the campaign he was just as often wrong as right. He was wrong when at the start of his semi-siege of Atlanta he had predicted to General Halleck, "I doubt if General Hood will stand a bombardment." Hood indeed withstood

---

[12] Hood, *A & R*, 195; Hood to Seddon, July 28, *OR*, vol. 38, pt. 5, 917.

[13] Simpson and Berlin, *Sherman's Civil War*, 660.

Sherman's shelling, himself coming under heavy fire on August 9. Throughout the Federal bombardment, he also sent through the lines no appeal for Sherman to cease his cannonade. One can only guess what Joe Johnston would have done in the same situation.[14]

After the three costly battles of late July, Hood's efforts to find replacements for his army did not make good his losses from those fights, but his well-publicized search for manpower bolstered morale both among his troops and the citizenry. As "Rover" put it, writing from Atlanta on August 17, "the Confederate private has come to the conclusion that Atlanta can be held, and the citizen is making up his mind it will be." In the city itself and the fortifications around it, more through the industry of his supply and railroad officers than through his own efforts, enough food and forage came in from Macon on the Southerners' sole railroad to hold the army together and feed the few thousand civilians still in Atlanta. At least Hood earns credit for not having panicked over his tight logistical situation. Indeed, as he wrote Bragg on August 4, "at present everything here looks well."[15]

In early August, Hood enjoyed the advantage of interior lines in the enemy's semi-siege of Atlanta. Sherman was stymied, unwilling to extend his lines farther to reach the Macon railroad. "I feel mortified that he [Hood] holds us in check by the aid of his militia," he acknowledged on August 11.[16] The Federal commander resorted to cavalry in three separate raids (McCook, Stoneman, Kilpatrick) to break the Rebels' road; all three failed. To break the deadlock Hood sent half of his cavalry into the enemy rear to cut the Federals' railroad supply line. By some he has been criticized for this alleged tactical blunder, but the critics miss three points: 1) Governor Brown, other politicians and much of the Southern press had been calling for such a raid; 2) Secretary Seddon had suggested it when Hood took command of the army; and 3) Davis and the administration gave the proposal their blessing as probably the Confederates' last resort to force Sherman to retreat. That the raid failed reflected Wheeler's poor judgment and Sherman's logistical preparations,

---

[14] Sherman to Halleck, July 21, *OR*, vol. 38, pt. 5, 211.

[15] "Rover," "Letter from the Georgia Front" (Atlanta, August 17), *Augusta Chronicle & Sentinel*, August 21; Hood to Bragg, August 4, *OR*, vol. 52, pt. 2, 718.

[16] Sherman to Schofield, August 11, *OR*, vol. 38, pt. 5, 461.

but more broadly it reflected the incapacity of Civil War cavalry to affect armies' strategic retreats by railroad-raiding.

During August Hood extended his fortifications southwest from the city in order to guard his railroad supply line, stretching his army from north of Atlanta to East Point and beyond. When Sherman led six infantry corps in a wide swing southward toward Quick Station, Morrow, and Jonesboro, Hood did not have the manpower to prevent the enemy from cutting the railroad at one of these points, which they did on the afternoon of August 31. The Confederates abandoned Atlanta on the night of September 1–2; the Federals marched in the next day.

During his retreat, Hood *should* receive credit from historians for pulling together his army, which was deployed from Atlanta to Jonesboro on September 1 with more than half of the enemy's forces in-between—except that Sherman was not the slightest bit interested in destroying the Rebel army. He allowed Hood to get out of Atlanta with Stewart's Corps and reunite with Hardee and Lee, preferring instead to march victoriously into the city.

If Hood does not receive high marks for this feat, he should get demerits for how he censured his soldiers for having "attacked without spirit" at Jonesboro. Hood believed that the troops' confinement for so long in the trenches around Atlanta had imbued in them a resolve "that they will not attack breast-works." In making this statement, and further offering up relatively low casualty figures for August 31 to show that "our effort was not a vigorous one," Hood was not just being callous toward his soldiers' courage and sacrifice. As an army commander, he was conceptually unrealistic and short-sighted, clinging to a view of battlefield tactics that heralded open-field charges such as the ones he had led at Gaines' Mill and Second Manassas. The war had changed since summer 1862; John Bell Hood had not. Another way of looking at it is to compare Hood with his much-admired mentor. On September 6, Hood wrote President Davis, offering the opinion that "we should have saved Atlanta had the officers and men of the army done what was expected of them." One can only conclude that Robert E. Lee would never have said anything like that.[17] The same goes with Hood's simmering resentment of Hardee, his senior and most "reliable" corps commander, which was

---

[17] Hood to Bragg, September 4, *OR*, vol. 52, pt. 2, 730; Hood to Bragg, September 3, *OR*, vol. 38, pt. 5, 1016; Hood to Davis, September 6, *OR*, vol. 38, pt. 5, 1023.

reciprocated. One cannot imagine General Lee picking a fight with, say, James Longstreet and asking for his transfer, even after the latter's unhelpful performance at Gettysburg.

A measure of Hood's pitiable state of mind after the fall of Atlanta was his telegram to General Bragg on September 10. As he called for reinforcements to his army, Hood wrote, "please ask General Lee to send me my old division." McMurry rightly terms this message "pathetic"; it further confirms one's sad inference that in defeat Hood found some solace by returning in memory to his glory days of two years before.[18]

If Jefferson Davis had held any doubts as to Hood's ability (recall Secretary Seddon's admonition to "be wary no less than bold"), following the loss of Atlanta these concerns could only have grown. They certainly did for Robert E. Lee. In July Lee had recommended that Johnston not be relieved, then had politely suggested Hardee rather than Hood as his successor; after Atlanta Lee boldly recommended that Hood be relieved in favor of General Beauregard. Instead, Davis began to think of appointing General Beauregard to oversee Hood's operations in the kind of super-departmental position he had created for Johnston in late 1862. It seems that Davis asked Lee to talk with Beauregard about such an assignment and get a feel for his interest. Lee met with Beauregard, and so reported to the president in a letter of September 19. But Lee turned the conversation into one about Beauregard assuming command of "*that army*" (emphasis added), meaning the Army of Tennessee—which meant in turn that Lee was trying to persuade the president to relieve Hood of army command. He made his purpose clear when he wrote further, "should you deem, therefore, a change in commander of the army in Georgia advantageous, and select General Beauregard for that position, I think you may feel assured that he understands the general condition of affairs...." Beauregard was interested in returning to the command of a field army, and (probably at Lee's suggestion) affirmed his interest in writing.

Evidently others were also calling for Beauregard to take command of the Army of Tennessee, and word of such a possible switch leaked out to the Southern press. But the president was not about to relieve Hood. To do so would have been a confession of having made a mistake in firing Joe Johnston, and Jefferson Davis had too much pride for that. In-

---

[18] Hood to Bragg, September 10, *OR*, vol. 39, pt. 2, 826; McMurry, *Hood*, 154.

stead, he stuck to his idea of appointing Beauregard in loose authority over Hood's and Richard Taylor's departments, an area extending from Georgia to eastern Louisiana.[19]

President Davis traveled to Georgia in late September to tell Hood personally of his idea for Beauregard's role over him and Dick Taylor and to discuss Hood's plan for future operations. Evidence is spotty and historians are not in agreement, but a strong case can be made that Davis, Hood, and the corps commanders discussed a plan for the army not only to march into north Georgia to disrupt Sherman's communications, but continue marching into Tennessee. Such an offensive would carry the war to the enemy, perhaps reclaim parts of that state and maybe even win new recruits to the cause.

The march of the Army of Tennessee, begun September 29 from Palmetto, was Hood's first strategic offensive. He had not replaced General Shoup, and thus had no chief of staff; that "carelessness off the battlefield" of which General Lee had warned President Davis began to manifest itself in logistical matters. Louisianan Robert Patrick commented on October 1 that "General Hood or somebody else at the head of affairs doesn't seem to understand his business."[20]

Hood's march northward showed promise of good result when the Confederates reached the Western & Atlantic around Big Shanty (today's Kennesaw) and started tearing up railroad track. Hood sent French's Division to Allatoona to obstruct the railroad pass through the tall mountain. Presence of a strong Federal garrison there prevented French from achieving his mission and from capturing or destroying tons of supplies. Hood was not present on October 5, when French's troops attacked the enemy garrison at Allatoona and withdrew after a three hours' battle. More important is that Hood's orders had sent Stewart's farthest division to Allatoona, gave it an arduous task (filling the railroad gap with trees, rocks and dirt), and then sent them five miles farther to burn the Etowah River bridge. And French was to do this in all of two days; Hood was further unaware that a strong enemy garrison held the place, well fortified. No wonder French failed. Such unrealistic instructions, as we have seen before (e.g., to Hardee for the battle of July 22),

---

[19] Seddon to Hood, July 17, *OR*, vol. 38, pt. 5, 885; Lee to Davis, September 19, *OR*, vol. 39, pt. 2, 846; Roman, *Beauregard*, 2:275.

[20] Patrick, *Reluctant Rebel*, 231.

indicate the unrealistic, not-thought-through nature of Hood's style of tactical planning.

As Hood shifted his tactics—marching toward Alabama to lure Sherman into battle, breaking the Western & Atlantic, shirking battle when Sherman's forces approached, eventually raiding into Tennessee—perplexed historians have followed Hood's re-shifting of "my original plan." The impression is one of indecisiveness and inexperience as Hood conducted his first raid into enemy-held territory. When Robert E. Lee had to extemporize in *his* first raid into Maryland, at least he could blame his "Lost Dispatch," Special Orders No. 191.[21]

As he marched farther into north Alabama, Hood's refusal to inform General Beauregard of his progress and plans reflects a stubborn reluctance to acknowledge the latter's authority over him. The mere issuance of orders by the military division commander, observes Roman, produced an "unfortunate effect upon General Hood, who began to chafe under the supervision exercised over him by General Beauregard." This virtual insubordination further damages Hood's record as army commander. Still, Beauregard tried to help Hood with his supply and communications problems, and for good reason. "He began to fear that General Hood was disposed to be oblivious of those details which play an important part in the operations of a campaign," as Roman puts it, "and upon which the question of success or failure often hinges."[22]

Bad weather, Forrest's delay in joining the army, and the need to accumulate supplies stalled the start of Hood's Tennessee Campaign until November 21. The weather had turned cold, and Hood himself predicted that there would be breadless days ahead. But as they marched northward, "the men were cheerful and dutiful," according to Col. Ellison Capers.[23]

With Union Major General Schofield and his two corps falling back toward Nashville, Hood hoped by quick marches to interpose his army between Schofield and Nashville, then engage and defeat the enemy before they could unite with Thomas's forces. Hood's maneuver at Columbia on November 28–29 reminds one of how often he spoke of flanking the enemy or striking him in rear. While S. D. Lee engaged Schofield's attention south of the town, Hood succeeded in crossing his other two

---

[21] Hood, *A & R*, 258; Murfin, *Gleam of Bayonets*, 328.

[22] Roman, *Beauregard*, 2:286, 299.

[23] Capers report, January 4, 1865, *OR*, vol. 45, pt. 1, 735.

infantry corps upriver. Mindful of having sent Hardee out on July 22, this time Hood himself got up at three in the morning of the 29th and rode with the advance of the flanking column—an instance of energetic, sound generalship worth our notice.

But once across the Duck, poor roads slowed the Confederates' march, as did the enemy, when on the morning of the 29th the Southerners had to wheel left to guard against a watchful enemy brigade sent out by Schofield. Hood sent Forrest ahead to seize Spring Hill, a town about ten miles north of Columbia, and block the enemy retreat. But once Schofield got his troops marching, the excellent turnpike used by the Federals meant that the Confederates lost the race to Spring Hill. George Wagner's Union infantry division repelled Forrest's attack, and Federals controlled the town by mid-afternoon.

Absence of written orders from Hood to his corps and division commanders that hectic day prevents historians from establishing Hood's plan at Spring Hill, as well as assigning blame when it fell apart. Believing that the enemy were marching on the pike from Columbia, Hood seems to have ordered Frank Cheatham to get his infantry across the pike to block the enemy's route, or to attack the enemy column in flank, or both. The participants' vague recollections of what happened got vaguer with the passing of years. Cheatham, without a campaign report, had his major paper on the subject published a year after Hood had died. Cleburne was dead, but Bate, Brown, Harris, and others wrote extensively of what they had "seen" and "heard" on the 29th, with the result that the mystery of Spring Hill still enshrouds Hood's Tennessee Campaign. My take is that Hood sensed he had a chance to block Schofield's march to Nashville and gave Cheatham the lead role in doing it. Beginning the next morning, November 30, he began blaming Cheatham for Schofield's escape, without attention to the fact that the Federals might have used other routes west of the turnpike, or that an enemy infantry division already at Spring Hill may have turned on the *Confederates'* rear. Postwar correspondence that has recently come to light seems to justify Hood's criticism of Frank Cheatham as blunderer. William W. Old, who had served on Ed. Johnson's staff, wrote Hood in 1877 he had heard "that Genl. Johnson did go to Gen. Cheatham and beg him to let him attack with his division, stating he did not even require any support, but that Genl. Cheatham refused, stating finally he was opposed to night attacks." Even without the benefit of Old's letter, Wiley Sword asserted two decades ago that "Cheatham ordered a halt in the attacks." Perhaps the mystery of Spring Hill is beginning to be deshrouded.[24]

There is a long view to take as well, and it pertains to Hood's generalship. It must be said that on the afternoon of November 29, when the sun began to set at 5 P.M., and after his troops had marched for miles over poor roads the whole day, John B. Hood, the still-inexperienced army commander, had once more asked his men to do too much.

On the 30th, Schofield's 22,000 infantry got to Franklin, a dozen miles northeast of Spring Hill, and dug in, building on works already constructed two years before. Hood concluded that this was his last chance to damage Schofield before his troops retreated to the safety of Nashville's defenses and Thomas's forces assembled there. Several of his key generals expressed reservations about the attack; Forrest wanted to maneuver around the enemy's left. But Hood wanted to destroy the enemy, not take the town or hasten Schofield's retreat. So with most of Lee's Corps and the army's artillery still on the road, Hood ordered all of his available infantry (18,000 men) to attack the enemy position. He explained that his situation was like Lee's two and a half years before in Virginia: a frontal assault had become "a necessity as imperative" as the one he had led at Gaines' Mill.

But the war had changed much in those two and a half years, even if John Bell Hood's tactical thinking had not. Hood demonstrated this unhappy fact in his memoir, both by his defense of "the Lee and Jackson school" of open-field attacking, and by his belabored argument that constant committal of troops to entrenchments dulled their battle-ardor.[25] Hood had in mind the glorious days of 1862 and 1863 when he ordered the frontal assault at Franklin. The Confederates attacked there with great gallantry and determination, reached the enemy works at some places, and even overran them, but were eventually repulsed. The strength of the Federals' fortifications is the key reason for the disparity in the killed/wounded figures for the two sides at Franklin: 5,550 Confederates to 1,222 Federals, a gruesome 4.5-to-1 ratio. Many Confederates died in the ditches before the enemy parapet, unable to climb out.

Given these heavy casualties, especially among officers such as Pat Cleburne ("the leadership of the army had been shot away," as McMurry puts it), Hood's order for a renewal of assault on December 1 strikes one

---

[24] Hood, *Lost Papers*, 120; Sword, *Embrace an Angry Wind*, 131.

[25] Hood, *A & R*, 296.

as a particularly callous decision. But it was not an anomaly, if one recalls that he briefly considered renewing his attack at Peachtree Creek on July 21. There is also that nebulous mention in Shoup's journal on the night of July 28 of "the proposed night attack" after Lee's repulse at Ezra Church. In each of these three instances Hood's soldiers would have been grateful that their commander did not follow through on such impulses.

The retreat of Schofield's forces the night after the battle allowed Hood to claim Franklin as something of a victory. More, the enemy withdrawal back to Nashville called for pursuit by Forrest's cavalry and the rest of the army. The result was that by December 4 Hood's troops were taking position some six miles south of Nashville. They were too weak (23,000 troops) to attack Thomas's now reinforced army of 55,000 in its fortifications. Having advanced this far into Tennessee, however, Hood also could not retreat. The Confederates, without any other option, therefore dug in and awaited the inevitable Federal attack, which Hood could only hope might lead to a defensive victory. Given the odds, though, Hood was gambling recklessly with his army at stake. Stanley Horn cannot be countered when he opines, "seldom has an inferior military force so obligingly waited near by while the adversary completed preparations to demolish it."

Worsening the odds was Hood's "side bet," as Wiley Sword calls it: sending Forrest and Bate off to capture some blockhouses, wreck some railroad, and threaten Murfreesboro. If Hood had no choice but to advance after Franklin, once at Nashville he definitely had the choice to keep his army together in the face of Thomas's hordes. That he did not do so is a gamble even more reckless than his march to Nashville.[26]

Meanwhile Hood continued to insult General Beauregard with childlike obstinacy. His refusal to send his report of December 11 to his superior officer (instead, sending it to Richmond and only copying Beauregard) caused the military division commander to complain, "General Hood does not seem to understand that he is responsible directly to these Headquarters and *not* to the War Department." Beauregard's tolerance of these slights marks him the far better gentleman than his subordinate officer was.[27]

---

[26] McMurry, *Hood*, 176; Horn, *Army of Tennessee*, 406; Sword, *Embrace an Angry Wind*, 282.

[27] Roman, *Beauregard*, 2:614 (Beauregard endorsement of Hood report).

With his preparations complete, Thomas launched his attack against Hood's army on the morning of December 15. The Southerners were stretched too thin to man their line adequately, and the main Union assault against the Rebel left overwhelmed it, sending the Confederates fleeing. Hood was lucky that darkness broke off the fighting before the enemy could inflict more damage. He ordered his troops to fall back to a new line hurriedly laid out a mile and a quarter to the south.

Thomas attacked again on the 16th. A sortie against the Confederate right was repulsed handily. But on the left, at Shy's Hill, the previous day's trouble was repeated. The Federals in that sector outnumbered Confederates three-to-one, used their artillery effectively, and then attacked with heavy force against a Rebel line that had been imperfectly laid out. In a pitifully short time soldiers of four Confederate divisions were fleeing for the rear. The panic spread, and soon the entire Army of Tennessee was being routed from the battlefield. Efforts to rally the troops failed. Years later Hood recalled that at Nashville, this was the first time he had ever seen a Confederate army fleeing a battlefield.[28]

After a limited pursuit by enemy cavalry, the army was able to retreat to the Tennessee River, get across it, and limp back into Mississippi. Freezing weather and a shortage of shoes, clothing, and tents led to further demoralization among the men. Hood's shortcomings as logistician were on woeful display when, encamped at Tupelo, the army essentially began falling apart. Hood sought to hide the extent of the army's loss and suffering from General Beauregard and the administration. Even before definite news of the Nashville disaster reached Richmond, Hood almost certainly connived to have Tennessee governor Isham G. Harris write a letter to President Davis on December 25. "I am not able to see anything that General Hood has done that he should not," Harris asserted, "or neglected anything that he should have done which it was possible to do." The governor then declared, "I…regret to say that if all had performed their parts as well as he, the results would have been very different." Hood was reverting to his shameful practice of not only hiding details of his failures, but also blaming others for them.[29]

Irritated by the lack of news from Hood as to his actions and plans, Beauregard resolved to travel to Tupelo to see things for himself. Before he left, hearsay from the front and gloating in Yankee newspapers had

---

[28] Hood, *A&R*, 303.

[29] Harris to Davis, December 25, 1864, *OR*, vol. 45, pt. 2, 732.

wafted sufficiently for authorities to assume that Hood had been defeated and had retreated back across the Tennessee. This news, combined with Hood's surly conduct toward him, led Beauregard to conclude that he would relieve Hood from command. President Davis concurred, but Hood beat them to the punch. With no further means of concealing his failure, on January 13 he asked to be relieved from command of the army. The War Department quickly granted his request. The Confederate States Congress added further insult a month later when it formally rescinded his full general's rank. When he applied for parole after the surrender, Hood signed himself as lieutenant general.

Thus one can see how Hood succeeded up to division level and performed at least adequately in his six months of corps command. As army commander for the next six months, judged by the criterion of battlefield victories alone, Hood failed.

The current, catchy consensus among students of the war is to consign Hood as another Civil War victim of the "Peter Principle"—he had risen to a level of authority beyond his capacity to function in it well. When one Peter-principles John Bell Hood, one shoves him into a rather large if not distinguished group of other Civil War officers at all levels of rank. They come from both sides. Ambrose Burnside was promoted from corps to army command, and two and a half months later was demoted after losing at Fredericksburg. Joseph Hooker followed the same course; promoted from corps level, he lost his army command after Chancellorsville. In the Confederate army, one may view Richard Ewell's record similarly. Though health problems were involved in Lee's relieving him from corps command after Spotsylvania, one may say that Dick Ewell had risen to a position beyond his ability (he had been excellent at divisional level). Recently Chris Mackowski has likewise noted A. P. Hill's talent leading the Light Division, but that his "lackluster performance at Gettysburg and near-disastrous performance at Bristoe Station showed that he had risen to a level of command beyond his abilities." Not so Pat Cleburne, whose great reputation as division commander in the Army of Tennessee rests luckily on the fact that he was not Peter-principled into disgrace. "Cleburne was unquestionably a superb division commander," writes Craig Symonds, but "he lacked that spark of independent initiative necessary to be effective in the command of a corps." "Cleburne had reached his proper rank as a major general," the biographer concludes, and thus maintains a great reputation in Civil War history, in part because he was not thrust beyond his level of competence.[30]

Lengthening the list of Civil War officers who rose beyond the level of their ability is to oversimplify the complex interplay of individual personality, soldierly talent, tactical possibilities on the battlefield, and the ever-present consequences of military failure. In Hood's case, a not overly complicated conclusion is that Hood's distinguishing characteristics—bravery and boldness—led him, *nay*, compelled him to commit his army to a campaign into Tennessee. Put another way, as his strategic and tactical options diminished in late 1864, Hood tended to rely on bold initiatives, which, against ever-growing enemy strength, offered fewer and fewer chances of success. As his potential for failure increased, so too did the likelihood of disastrous consequences. With the destruction of his army at Nashville and in the month following, that is exactly what happened.

Instead of finding a formula to explain John Bell Hood's military record, it is wiser, surely more beneficial, to keep one's focus on the man and his deeds. Thus a soldier who, glorying in his own fame, had blazed with ambition, tested his ambition to its utmost limit, found himself wanting, and failed abysmally. Such was John Bell Hood's record as Confederate officer. To use his own phrasing, he had won his *spurs*, but he had not won *greatness*. Like so many tragic mortals, he had aspired but had failed.

Sometimes a novelist is able to express best an individual's sense of loss and tragedy. In Robert Hicks's *A Separate Country*, John Bell Hood living in New Orleans after the war confesses, "I *had* deserved command, and I had commanded in victory and defeat. There's not a general who did not suffer defeat." Then with emphasis he adds, "I suffered defeat. I did. I suffered."

With his memoirs unpublished, Hicks's Hood may be thought of as still suffering at the time of his death. It is the ghost of General Hood that haunts the Louisiana landscape in another novel, James Lee Burke's *In the Electric Mist with Confederate Dead*. Burke's story closes with his protagonist envisioning "old compatriots in butternut brown wending their way in an out of history—gallant, Arthurian, their canister-ripped

[30] Pfanz, "Richard Stoddert Ewell," 2:112; Mackowski and White, *Simply Murder*, 144; Symonds, *Cleburne*, 223.

colors unfurled in the roiling smoke, the fatal light in their faces a reminder that the contest is never quite over, the field never quite ours."[31]

The contest never quite over, the field never quite won: such is the measure of John Bell Hood's generalship in 1864.

[31] Hicks, *A Separate Country*, 33; Burke, *In the Electric Mist with Confederate Dead*, 346.

Works Cited

Adams, John Q. "Hold the Fort!" In Kerksis, *The Atlanta Papers*, 789–99.

Alexander, E. P. *Fighting for the Confederacy: The Personal Recollections of General Edward Porter Alexander*. Edited by Gary W. Gallagher. Chapel Hill: University of North Carolina Press, 1989.

Ambrose, Stephen E. *Halleck: Lincoln's Chief of Staff.* Baton Rouge: Louisiana State University Press, 1962.

Andrews, J. Cutler. *The South Reports the Civil War*. Princeton: Princeton University Press, 1970.

Atchison, Ray M. "*Scott's Monthly Magazine*: A Georgia Post-Bellum Periodical of Literature and Military History." *Georgia Historical Quarterly* 49/3 (September 1965): 294–305.

Axelrod, Alan. *America's Wars*. New York: John Wiley & Sons, 2002.

Bailey, Anne. "Edmund Kirby Smith." In W. C. Davis, *The Confederate General*, 5:162–71.

———. *The Chessboard of War: Sherman and Hood in the Autumn Campaigns of 1864*. Lincoln: University of Nebraska Press, 2000.

———. "The USCT in the Confederate Heartland." In *Black Soldiers in Blue: African American Troops in the Civil War Era*, edited by John David Smith, 227–48. Chapel Hill: University of North Carolina Press, 2002.

Ballard, Michael B. "Breakdown in Macon." *Civil War Times Illustrated* 19/6 (October 1980): 31–33.

Banks, R. W. *The Battle of Franklin November 30, 1864: The Bloodiest Engagement of the War Between the States*. New York: Neale Publishing Company, 1908.

Barefoot, Daniel W. *Let Us Die Like Brave Men: Behind the Dying Words of Confederate Warriors*. Winston-Salem, NC: John F. Blair, Publisher, 2005.

Barnwell, Valentine T. *Barnwell's Atlanta City Directory*. Atlanta: Intelligencer and Job Office, 1867.

Basso, Hamilton. *Beauregard: The Great Creole*. New York: C. Scribner's Sons, 1933.

Beach, J. N. "From the Union Side at Franklin." *Confederate Veteran* 2/8 (August 1894): 239.

Bearss, Edwin C. *Forrest at Brice's Cross Road and in North Mississippi on 1864*. Dayton OH: Press of Morningside Bookshop, 1974.

———. "Francis Asbury Shoup." In W. C. Davis, *The Confederate General*, 5:150–51.

———. "Grierson's Winter Raid on the Mobile and Ohio Railroad." *Military Affairs* 24/1 (Spring 1960): 20–37.

Beasley, Paul H., and C. Buford Gotto. "Fortress Nashville." *Civil War Times Illustrated* 3/8 (December 1964): 25–26.

Bennett, Stewart. "'The Storm Broke in All Its Fury': The Struggle for Allatoona Pass." In Woodworth and Grear, *The Tennessee Campaign of 1864*, 23–45.

Bergeron, Arthur W., Jr. *Confederate Mobile.* Jackson: University Press of Mississippi, 1991.

———. "Randall Lee Gibson." In W. C. Davis, *The Confederate General*, 2:185–88.

Bishop, W. P. "Twenty-ninth Tennessee Infantry." In Lindsley, *Military Annals of Tennessee*, 1:433–39. Nashville: J. M. Lindsley & Co., 1886.

Black, Robert C. III. *The Railroads of the Confederacy.* Chapel Hill: University of North Carolina Press, 1952.

Bledsoe, Andrew S. "The Destruction of the Army of Tennessee's Officer Corps at the Battle of Franklin." In Woodworth and Grear, *The Tennessee Campaign of 1864*, 66–80.

Bobrick, Benson. *The Battle of Nashville: General George H. Thomas & the Most Decisive Battle of the Civil War.* New York: Alfred A. Knopf, 2010.

Bohannon, Keith S. "'A Bold Fighter' Promoted beyond His Abilities: Gen. John Bell Hood." In *Leaders of the Lost Cause: New Perspectives on the Confederate High Command*, edited by Gary W. Gallagher and Joseph T. Glatthaar, 249–87. Mechanicsburg, PA: Stackpole Books, 2004.

———. "'He exhibited the highest soldierly qualities': Edward C. Walthall." In Hewitt and Schott, *Confederate Generals in the Western Theater*, 4:217–249.

Bonner, James C. "Bushrod Washington Frobel." In *Dictionary of Georgia Biography*, edited by Kenneth Coleman and Charles Stephen Gurr, 1:328–29. Athens: University of Georgia Press, 1983.

Boynton, H. V. *Sherman's Historical Raid. The Memoirs in the Light of the Record: A Review Based upon Compilations from the Files of the War Office.* Cincinnati: Wilstach, Baldwin & Co., 1875.

Brennan, Patrick. "The Battle of Franklin." *North & South* 8/1 (January 2005): 26–46.

———. "Last Stand in the Heartland: The Fight for Nashville, December 1864." *North & South* 8/3 (May 2005): 20–45.

Brewer, George E. "How Lee's Corps Crossed the Duck River." *Confederate Veteran* 18/7 (July 1910): 326.

Bright, John M. "The States in the Confederate War." *Confederate Veteran* 17/8 (August 1909): 393–99.

Brinsfield, John W., William C. Davis, Benedict Maryniak, and James I. Robertson, Jr. *Faith in the Fight: Civil War Chaplains.* Mechanicsburg, PA: Stackpole Books, 2003.

Brooksher, William R., and David K. Snider. "Devil on the River." *Civil War Times Illustrated* 15/5 (August 1976): 12–19.

Brown, Campbell H. "Forrest's Johnsonville Raid." *Civil War Times Illustrated* 4/3 (June 1965): 48–57.

———. "The Myth of the 5 Dead Rebel Generals." *Civil War Times Illustrated* 8/5 (August 1969): 14–15.

———. "To Rescue the Confederacy: Why and How Hood Invaded Tennessee." *Civil War Times Illustrated* 3/8 (December 1964): 13–15, 44–48.

Brown, Fred E. "The Battle of Allatoona." *Civil War History* 6/3 (September 1960): 277–97.

Brown, Joseph M. *The Battle of Allatoona. October 5, 1864.* Atlanta: Record Publishing Co., 1890.

Brown, Russell K. *Our Connection with Savannah: A History of the 1st Battalion Georgia Sharpshooters.* Macon: Mercer University Press, 2004.
Brown, Thomas J. *John Bell Hood: Extracting Truth from History.* Privately printed, 2012.
———. "John Bell Hood and the Lost Cause." Foreword to Hood, *John Bell Hood: The Rise, Fall, and Resurrection of a Confederate General*, xii–xxix. El Dorado Hills, CA: Savas Beatie, 2013.
Bryan, Charles F., Jr. "The Siege of Yorktown Part II." *Civil War Times Illustrated* 21/5 (September 1982): 28.
Buck, Irving A. "Military View of the Battle of Franklin." *Confederate Veteran* 17/8 (August 1909): 383.
Burke, James Lee. *In the Electric Mist with Confederate Dead.* New York: Avon Books, 1994.
Butkovich, Brad. *The Battle of Allatoona Pass: Civil War Skirmish in Bartow County, Georgia.* Charleston: History Press, 2014.
Capers, Walter B. *The Soldier-Bishop Ellison Capers.* New York: Neale Publishing Company, 1912.
Carbone, John S. *The Civil War in Coastal North Carolina.* Raleigh: North Carolina Division of Archives and History, 2001.
Carmichael, Peter S. "'Every Map of the Field Cries Out about It': The Failure of Confederate Artillery at Pickett's Charge." In Gallagher, *Three Days at Gettysburg*, 270–83.
Carpenter, Noel. *A Slight Demonstration—Decatur, October 1864: Clumsy Beginning of Gen. John B. Hood's Tennessee Campaign.* Austin: Legacy Books & Letters, Publisher, 2007.
Cartwright, Thomas Y. "Franklin: The Valley of Death." In Joslyn, *A Meteor Shining Brightly*, 265–84.
Castel, Albert. "The Atlanta Campaign and the Presidential Election of 1864: How the South Almost Won by Not Losing." In *Winning and Losing in the Civil War: Essays and Stories*, 15–32. Columbia: University of South Carolina Press, 1996.
———. *Decision in the West: The Atlanta Campaign of 1864.* Lawrence: University Press of Kansas, 1992.
———. *General Sterling Price and the Civil War in the West.* Baton Rouge: Louisiana State University Press, 1968.
———. "Prevaricating through Georgia: Sherman's *Memoirs* as a Source on the Atlanta Campaign." In *Winning and Losing in the Civil War: Essays and Stories*, 89–116. Columbia: University of South Carolina Press, 1996.
Cate, Wirt Armistead, ed. *Two Soldiers: The Campaign Diaries of Thomas J. Key, C.S.A. December 7, 1863–May 17, 1865 and Robert J. Campbell, U.S.A. January 1, 1864-July 21, 1864.* Chapel Hill: University of North Carolina Press, 1938.
Catton, Bruce. *The American Heritage Picture History of the Civil War.* New York: American Heritage Publishing Co., 1960.
Chalaron, J. A. "Hood's Campaign at Murfreesboro." *Confederate Veteran* 11/10 (October 1903): 438–40.
Chalmers, J. R. "Forrest and His Campaigns." *Southern Historical Society Papers* 7/10 (October 1879): 451–86.
Chambers, William Pitt. *Blood & Sacrifice: The Civil War Journal of a Confederate Soldier.* Edited by Richard A. Baumgartner. Huntington, WV: Blue Acorn Press, 1994.

Cheatham, B. F. "The Lost Opportunity at Spring Hill, Tenn.—General Cheatham's Reply to General Hood." *Southern Historical Society Papers* 9/10–12 (October–December 1881): 524–41.

Chesnut, Mary Boykin. *A Diary from Dixie.* Edited by Ben Ames Williams. Boston: Houghton Mifflin Company, 1949.

Clampitt, Bradley R. *The Confederate Heartland: Military and Civilian Morale in the Western Confederacy.* Baton Rouge: Louisiana State University Press, 2011.

Clark, Thomas D. *The Rampaging Frontier: Manners and Humors of Pioneer Days in the South and Middle West.* Indianapolis: Bobbs-Merrill Company, 1939.

Clauss, Errol MacGregor. "The Atlanta Campaign, 18 July–2 September 1864." Ph.D. dissertation, Emory University, 1965.

———. "The Battle of Jonesborough and the Fall of Atlanta." Master's thesis, Emory University, 1961.

———. "Sherman's Failure at Atlanta." *Georgia Historical Quarterly* 53/3 (September 1969): 321–29.

Coddington, Edwin B. *The Gettysburg Campaign: A Study in Command.* New York: Charles Scribner's Sons, 1968.

"Confederate Monument at Franklin." *Confederate Veteran* 8/1 (January 1900): 5–11.

"Confederates Buried at Ashwood Church." *Confederate Veteran* 7/10 (October 1899): 460.

Connelly, Donald B. *John M. Schofield and the Politics of Generalship.* Chapel Hill: University of North Carolina Press, 2006.

Connelly, Thomas Lawrence. *Autumn of Glory: The Army of Tennessee, 1862–1865.* Baton Rouge: Louisiana State University Press, 1971.

Connelly, Thomas Lawrence, and Archer Jones. *The Politics of Command: Factions and Ideas in Confederate Strategy.* Baton Rouge: Louisiana State University Press, 1973.

Cooke, James J. "Feeding Sherman's Army: Union Logistics in the Campaign for Atlanta." In Savas and Woodbury, *The Campaign for Atlanta & Sherman's March to the Sea*, 97–114.

Cooling, Benjamin Franklin. *To the Battles of Franklin and Nashville and Beyond: Stabilization and Reconstruction in Tennessee and Kentucky, 1864–1866.* Knoxville: University of Tennessee Press, 2011.

Cooper, James L. "Service with the Twentieth Tennessee." Diary edited by Deering J. Roberts. *Confederate Veteran* 33/5 (May 1925): 180–81.

Cooper, William J., Jr. *Jefferson Davis, American.* New York: Alfred A. Knopf, 2000.

———. *Jefferson Davis and the Civil War Era.* Baton Rouge: Louisiana State University Press, 2008.

———. "A Reassessment of Jefferson Davis as War Leader: The Case from Atlanta to Nashville." *Journal of Southern History* 36/2 (May 1970): 189–204.

Cox, Jacob D. *The Battle of Franklin Tennessee November 30, 1864: A Monograph.* Dayton, OH: Kessinger Publishing, 1983 [1897].

———. *The March to the Sea: Franklin and Nashville.* New York: C. Scribner's Sons, 1882.

Crawford, W. T. "The Mystery of Spring Hill." *Civil War History* 1/2 (June 1955): 101–26.

Craven, John J. *Prison Life of Jefferson Davis Embracing Details and Incidents of his Captivity, Particulars concerning his Health and Habits, Together with Many Conversations on Topics of Great Public Interest.* New York: G. W. Dillingham Company, 1905 [1866].

Crist, Lynda Lasswell, Kenneth H. Williams, and Peggy L. Dillard, eds. *The Papers of Jefferson Davis.* 14 vols. Baton Rouge: Louisiana State University Press, 1971–2018.

Cross, Harold A. *They Sleep Beneath the Mockingbird: Mississippi Burial Sites and Biographies of Confederate Generals.* Murfreesboro, TN: Southern Heritage Press, 1994.

Crownover, Sims. "The Battle of Franklin." *Tennessee Historical Quarterly* 14/4 (December 1955): 291–322.

Cubbison, Douglas R. "John Bell Hood and the Campaign in North Alabama and Middle Tennessee, October–November 1864." *Military History of the West* 34 (2004): 51–74.

Cunningham. S. A. "Battle of Franklin." *Confederate Veteran* 1/4 (April 1893): 101–102.

———. "Disastrous Campaign in Tennessee." *Confederate Veteran* 12/7 (July 1904): 338–41. (Reprint of "Battle of Franklin".)

———. "Events Leading to the Battle [of Franklin]." *Confederate Veteran* 18/1 (January 1910): 17–20.

Dabney, T. G. "Gen. A. P. Stewart on Strong Topics." *Confederate Veteran* 17/1 (January 1909): 31–32.

———. "When Hood Superseded Johnston." *Confederate Veteran* 22/9 (September 1914): 406–407.

Daniel, John M. *The* Richmond Examiner *during the War.* New York: printed for the author, 1868.

Daniel, Larry J. *Cannoneers in Gray: The Field Artillery of the Army of Tennessee.* Tuscaloosa: University of Alabama Press, 1989.

———. *Soldiering in the Army of Tennessee: A Portrait of Life in a Confederate Army.* Chapel Hill: University of North Carolina Press, 1991.

———. "The South Almost Won by Not Losing: A Rebuttal." *North & South* 3 (February 1998): 44–48, 50–51.

———, and Lynn N. Bock. *Island No. 10: Struggle for the Mississippi Valley.* Tuscaloosa: University of Alabama Press, 1996.

———. *Sherman's March.* New York: Random House, 1980.

Davis, Jefferson. *The Rise and Fall of the Confederate Government.* 2 vols. New York: Thomas Yoseloff, 1958 [1881].

Davis, Stephen. "The Death and Burials of General Richard Brooke Garnett." *Gettysburg Magazine* 5 (July 1991): 107–16.

———. "Full-throated Defender of Hood." *Civil War News* 42/4 (May 2016): 16–17.

———. *What the Yankees Did to Us: Sherman's Bombardment and Wrecking of Atlanta.* Macon: Mercer University Press, 2012.

———. "Would P. G. T. Lead the A. o. T?" *Civil War Times Illustrated* 56/2 (April 2017): 36–41.

Davis, Steve. "Citizenship for General Lee." *Lincoln Herald* 81/2 (Summer 1979): 94–102.

———. "Franklin: Where Did Those Generals Go?" *Morningside Bookshop Catalog #25* (September 1988): 15–16.

Davis, Varina. *Jefferson Davis, Ex-President of the Confederate States of America: A Memoir by his Wife.* 2 vols. Baltimore: Nautical & Aviation Publishing Company of America, 1990 [1890].

Davis, William C. "Arnold Elzey." In *The Confederate General*, 2:98–103.
———. *The Cause Lost: Myths and Realities of the Confederacy*. Lawrence: University Press of Kansas, 1996.
———, ed. *The Confederate General*. 6 vols. Harrisburg, PA: National Historical Society, 1991.
———. *An Honorable Defeat: The Last Days of the Confederate Government*. New York: Harcourt, 2001.
———. *Jefferson Davis—The Man and His Hour: A Biography*. New York: HarperCollins Publishers, 1991.
Davison, Eddy W., and Daniel Foxx. *Nathan Bedford Forrest: In Search of an Enigma*. Gretna, LA: Pelican Publishing Company, 2007.
"The Death of General John B. Hood." *Southern Historical Society Papers* 7/10 (October 1879): 496.
"Death of Gen. Strahl." *Confederate Veteran* 1/1 (January 1893): 31.
Derden, John K. *The World's Largest Prison: The Story of Camp Lawton*. Macon: Mercer University Press, 2012.
Dew, Charles B. *Ironmaker to the Confederacy: Joseph R. Anderson and the Tredegar Iron Works*. Wilmington, NC: Broadfoot Publishing Company, 1987 [1966].
Dinkins, James. *1861 to 1865 by an Old Johnnie: Personal Recollections and Experiences in the Confederate Army*. Dayton, OH: Press of Morningside Bookshop, 1975 [1897].
Dixon, William D. *The Blues in Gray: The Civil War Journal of William Daniel Dixon and the Republican Blues Daybook*. Edited by Roger S. Durham. Knoxville: University of Tennessee Press, 2000.
Dodd, W. O. "Reminiscences of Hood's Tennessee Campaign." *Southern Historical Society Papers* 9/10–12 (October–December 1881): 518–24.
Douglas, James P. "Losses in the Atlanta Campaign." *Confederate Veteran* 8/6 (June 1900): 273.
Douglas, Lucia Rutherford, comp. *Douglas's Texas Battery, CSA*. Waco, TX: Texian Press, 1966.
Dowdey, Clifford, and Louis H. Manarin, eds. *The Wartime Papers of R. E. Lee*. Boston: Little, Brown and Company 1961.
Downey, Fairfax. *Storming of the Gateway: Chattanooga, 1863*. New York: David McKay Company, 1960.
Dozier, J. A. "Concerning Hood's Tennessee Campaign." *Confederate Veteran* 16/4 (April 1908): 192.
Durham, Roger S. *Guardian of Savannah: Fort McAllister, Georgia, in the Civil War and Beyond*. Columbia: University of South Carolina Press, 2008.
Dyer, John P. *"Fightin' Joe" Wheeler*. Baton Rouge: Louisiana State University Press, 1941.
———. *The Gallant Hood*. Indianapolis: Bobbs-Merrill Company, 1950.
Eckenrode, H. J., and Bryan Conrad. *James Longstreet: Lee's War Horse*. Chapel Hill: University of North Carolina Press, 1936.
Eckert, Edward J. *"Fiction Distorting Fact": Prison Life, Annotated by Jefferson Davis*. Macon: Mercer University Press, 1987.
Eicher, David J. *The Longest Night: A Military History of the Civil War*. New York: Simon & Schuster, 2001.

Einolf, Christopher J. *George Thomas: Virginian for the Union.* Norman: University of Oklahoma Press, 2007.
Eisenschiml, Otto. "Sherman: Hero or War Criminal?" *Civil War Times Illustrated* 2/9 (January 1964): 7–9, 29–31, 34–36.
Elliott, Sam Davis. *Isham G. Harris of Tennessee: Confederate Governor and United States Senator.* Baton Rouge: Louisiana State University Press, 2010.
Emery, Tom. *Hold the Fort: The Battle of Allatoona Pass.* Carlinville, IL: History in Print, 2001.
Escott, Paul D. *After Secession: Jefferson Davis and the Failure of Confederate Nationalism.* Baton Rouge: Louisiana State University Press, 1978.
Esposito, Vincent J., ed. *The West Point Atlas of American Wars.* 2 vols. New York: Frederick A. Praeger, Publishers, 1959.
Evans, Clement A., ed., *Confederate Military History.* Twelve volumes. New York: Thomas Yoseloff, 1962 [1899].
Evans, David. "Hold the Fort!" *America's Civil War* 1/5 (January 1989): 18–25.
Field, Henry M. *Bright Skies and Dark Shadows.* New York: C. Scribner's Sons, 1890.
Figuers, H. P. "A Boy's Impressions of the Battle of Franklin." *Confederate Veteran* 23/1 (January 1915): 4–7, 44.
Finlay, Luke W. "Another Report on Hood's Campaign." *Confederate Veteran* 15/9 (September 1907): 404–407.
Flint, Mortimer R. "The Battle of Allatoona." In Kerksis, *The Atlanta Papers*, 721–44.
Flood, Charles Bracelen. *Grant and Sherman: The Friendship That Won the Civil War.* New York: Harper Perennial, 2006 [2005].
Foote, Shelby. *The Civil War: A Narrative.* 3 vols. New York: Random House, 1958–1974.
Foster, Gaines M. *Ghosts of the Confederacy: Defeat, the Lost Cause, and the Emergence of the New South.* New York: Oxford University Press, 1987.
Foster, Samuel T. *One of Cleburne's Command: The Civil War Reminiscences and Diary of Capt. Samuel T. Foster, Granbury's Texas Brigade, CSA.* Edited by Norman D. Brown. Austin: University of Texas Press, 1980.
Fowler, John D. "The Finishing Stoke to the Independence of the Southern Confederacy: Hood's Tennessee Campaign." In *Tennessee in the Civil War*, edited by Carroll Van West, 140–59. Nashville: Tennessee Historical Society, 2011.
———. *Mountaineers in Gray: The Nineteenth Tennessee Volunteer Infantry Regiment, C.S.A.* Knoxville: University of Tennessee Press, 2004.
Freeman, Douglas Southall. *Lee's Lieutenants: A Study in Command.* 3 vols. New York: Charles Scribner's Sons, 1942–1944.
———. *R. E. Lee.* Abridged by Richard Harwell. New York: Chares Scribner's Sons, 1961.
French, S. G. "Errors in Reports of Battle of Franklin." *Confederate Veteran* 1/8 (August 1893): 230–31.
———. "Movement against Allatoona—Letter from General S. G. French." *Southern Historical Society Papers* 10/8–9 (August–September 1882): 402–406.
———. *Two Wars: An Autobiography of Gen. Samuel G. French.* Huntington, WV: Blue Acorn Press, 1999 [1901].
Frey, Donald J. *Longstreet's Assault—Pickett's Charge: The Lost Record of Pickett's Wounded.* Shippensburg, PA: Burd Street Press, 2000.

[Frobel, Bushrod W.] "The Georgia Campaign; or, a South-side View of Sherman's March to the Sea. By the author of 'Field and Camp.'" *Scott's Monthly Magazine* 6/4 (October 1868): 694–701; 6/5 (November 1868): 756–61; 6/6 (December 1868): 840–46.

Frye, Dennis E. "Drama between the Rivers: Harpers Ferry in the 1862 Maryland Campaign." In *Antietam: Essays on the 1862 Maryland Campaign*, edited by Gary W. Gallagher, 14–34. Kent: Kent State University Press, 1989.

"Fund for the Children of Gen. J. B. Hood." *Confederate Veteran* 10/1 (January 1902): 15.

Furgurson, Ernest B. *Ashes of Glory: Richmond at War.* New York: Alfred A. Knopf, 1996.

Gaines, John J. "Civilian Participants and Observers during the Franklin-Nashville Campaign." In Woodworth and Grear, *The Tennessee Campaign of 1864*, 187–200.

Gale, W. D. "Confederate Disaster at Nashville." *Confederate Veteran* 2/2 (February 1894): 46–47.

[Gale, William D.] "Hood's Campaign in Tennessee. *Confederate Veteran* 2/1 (January 1894): 4–5.

Gallagher, Gary W. *Lee and His Army in Confederate History.* Chapel Hill: University of North Carolina Press, 2001.

———. "Pierre Gustave Toutant Beauregard." In W. C. Davis, *The Confederate General*, 1:85–90.

Gallagher, Gary W., Stephen D. Engle, Robert K. Krick, and Joseph T. Glatthaar. *The American Civil War: This Mighty Scourge of War.* Oxford, England: Osprey Publishing, 2003.

Garrard, Louis F. "Gen. S. D. Lee's Part in Checking the Rout." *Confederate Veteran* 12/7 (July 1904): 350–53.

Gates, Robert. "Sixth Tennessee Infantry." In Lindsley, ed., *Military Annals of Tennessee*, 1:204–25.

Gayarre, Charles. "Military Operations of General Beauregard." *Southern Historical Society Papers* 12/10–12 (October–December 1884): 433–47.

"Gen. John Adams at Franklin." *Confederate Veteran* 5/6 (June 1897): 299–302.

Gibson, Randall L. "Report of Brigadier-General R. L. Gibson of Operations in Vicinity of Nashville." *Southern Historical Society Papers* 4/3 (September 1877): 140–43.

Gillum, Jamie. *Twenty-five Hours to Tragedy—The Battle of Spring Hill and Operations on November 29, 1864: Precursor to the Battle of Franklin.* Springfield, TN: Jamie Gillum, 2014.

Glatthaar, Joseph T. "Battlefield Tactics." In *Writing the Civil War: The Quest of Understanding*, edited by James M. McPherson and William J. Cooper, Jr., 60–80. Columbia: University of South Carolina Press, 1998.

Goff, Richard D. *Confederate Supply.* Durham, NC: Duke University Press, 1969.

Goodwin, Doris Kearns. *Team of Rivals: The Political Genius of Abraham Lincoln.* New York: Simon & Schuster, 2005.

Gorgas, Josiah. *The Civil War Diary of General Josiah Gorgas.* Edited by Frank E. Vandiver. Tuscaloosa: University of Alabama Press, 1947.

Gott, Kendall D. *Where the South Lost the War: An Analysis of the Fort Henry-Fort Donelson Campaign, February 1862.* Mechanicsburg, PA: Stackpole Books, 2003.

Gottschalk, Phil. *In Deadly Earnest: The Missouri Brigade CSA.* Columbia: Missouri River Press, 1991.

Govan, Gilbert E., and James W. Livingood. *A Different Valor: The Story of General Joseph E. Johnston, C. S. A.* Westport, CT: Greenwood Press, 1973 [1956].

"Graves of Gen. John B. Hood and Wife." *Confederate Veteran* 20/3 (March 1912): 123.

Grear, Charles D. "'Our Peoples Are Depressed in Spirit': Texans' Reactions to Hood's Tennessee Campaign." In Woodworth and Grear, *The Tennessee Campaign of 1864*, 201–16.

Grimsley, Mark. "Learning to Say 'Enough': Southern Generals and the Final Weeks of the Confederacy." In *The Collapse of the Confederacy*, edited by Mark Grimsley and Brooks D. Simpson, 40–79. Lincoln: University of Nebraska Press, 2001.

Groom, Winston. *Shrouds of Glory—From Atlanta to Nashville: The Last Great Campaign of the Civil War.* New York: Atlantic Monthly Press, 1995.

Hagerman, Edward. *The American Civil War and the Origins of Modern Warfare: Ideas, Organization and Field Command.* Bloomington: Indiana University Press, 1988.

Hallock, Judith Lee. *Braxton Bragg and Confederate Defeat.* Vol. 2. Tuscaloosa: University of Alabama Press, 1991.

Hanna, A. J. *Flight into Oblivion.* Richmond: Johnson Publishing Company, 1938.

Hardin, David. *After the War: The Lives and Images of Major Civil War Figures after the Shooting Stopped.* Chicago: Ivan R. Dee, 2010.

Hattaway, Herman. *General Stephen D. Lee.* Jackson: University Press of Mississippi, 1976.

——— and Archer Jones. *How the North Won: A Military History of the Civil War.* Urbana: University of Illinois Press, 1983.

Hay, Thomas Robson. *Hood's Tennessee Campaign.* Dayton, OH: Press of Morningside Bookshop, 1976 [1929].

Hay, Thomas Robson, ed. *Cleburne and His Command by Capt. Irving A. Buck and Pat Cleburne Stonewall Jackson of the West by Thomas Robson Hay.* Jackson, TN: McCowat-Mercer Press, 1959.

Hearn, Chester G. *The Capture of New Orleans 1862.* Baton Rouge: Louisiana State University Press, 1995.

Hennessy, John J. *Return to Bull Run: The Campaign and Battle of Second Manassas.* New York: Simon & Schuster, 1993.

Henry, Robert Selph. *"First With the Most" Forrest.* Jackson, TN: McCowat-Mercer Press, 1969 [1944].

Hess, Earl J. *Braxton Bragg: The Most Hated Man in the Confederacy.* Chapel Hill: University of North Carolina Press, 2016.

———. *Kennesaw Mountain: Sherman, Johnston and the Atlanta Campaign.* Chapel Hill: University of North Carolina Press, 2013.

———. *Pickett's Charge—The Last Attack at Gettysburg.* Chapel Hill: University of North Carolina Press, 2001.

Hettle, Wallace, ed. *The Confederate Homefront: A History in Documents.* Baton Rouge: Louisiana State University Press, 2017.

Hewett, James B., Noah Andre Trudeau, and Bryce Suderow, eds. *Supplement to the Official Records of the Union and Confederate Armies.* 100 volumes. Wilmington, NC: Broadfoot Publishing Company, 1994–2001.

———. *Port Hudson, Confederate Bastion on the Mississippi.* Baton Rouge: Louisiana State University Press, 1987.

Hewitt, Lawrence Lee and Arthur W. Bergeron, Jr., ed., *Confederates Generals in the Western Theater: Essays on America's Civil War.* 4 vols. Knoxville: University of Tennessee Press, 2010–2018. (Volume 4 ed. Hewitt and Thomas E. Schott.)

Hicken, Victor. "'Hold the Fort.'" *Civil War Times Illustrated* 7/3 (June 1968): 18–23, 26–27.

Hicks, Robert. *A Separate Country*. New York, 2011 [2009].

Hirshson, Stanley P. *The White Tecumseh: A Biography of William T. Sherman.* New York: John Wiley & Sons, 1997.

Hood, J. B. *Advance and Retreat: Personal Experiences in the United States and Confederate States Armies.* New Orleans: Pub. for the Hood Orphan Memorial Fund, 1880. Cited as *A&R.*

Hood, John B. Book IV Field Dispatches & Orders Sept. 11, 1864–Jan 14, 1865. John Bell Hood Papers, entry 122, Record Group 109, War Department Collection of Confederate Papers, National Archives and Records Administration, Washington, DC.

Hood, John Bell. Compiled Military Service File, NATF 86. National Archives and Records Administration, Washington, DC. Cited as Hood Service Record. No pagination.

Hood, O. C. *The Army of Tennessee in Retreat: From Defeat at Nashville through "the Sternest Trials of the War."* Jefferson NC: McFarland, 2018.

Hood, Stephen M. *John Bell Hood: The Rise, Fall, and Resurrection of a Confederate General.* El Dorado Hills, CA: Savas Beatie, 2013.

———. *The Lost Papers of Confederate General John Bell Hood.* El Dorado Hills, CA: Savas Beatie, 2015.

[———]. "Once a Confederate, Twice an American: Gen. Hood Requests Amnesty." *John Bell Hood Historical Society Newsletter.* 3/1 (Winter 2006): 1–2.

Horn, Stanley F. *The Army of Tennessee: A Military History.* Norman: University of Oklahoma Press, 1953 [1941].

———. *The Decisive Battle of Nashville.* Baton Rouge: Louisiana State University Press, 1956.

———. "Most Decisive Battle of the War. *Civil War Times* 3/8 (December 1961): 4–6, 18–19.

———. "Nashville during the Civil War." *Tennessee Historical Quarterly* 4/1 (March 1945): 3–22.

———. "Nashville: The Most Decisive Battle of the War." *Civil War Times Illustrated* 3/8 (December 1964): 5–11, 31, 36.

———. "The Spring Hill Legend." *Civil War Times Illustrated* 8/1 (April 1969): 20–32.

———, ed. *Tennessee's War 1861–1865 Described by Participants.* Nashville: Tennessee Civil War Centennial Commission, 1965.

Houchens, Harry W. "The Making of General John B. Hood: A Study of Command." Carlisle Barracks, PA: U.S. Army War College, 1993.

Hudson, Leonne M. *The Odyssey of a Southerner: The Life and Times of Gustavus Woodson Smith.* Macon: Mercer University Press, 1998.

Huff, Frederick Ware, comp. *Four Families: Winn, Thomas, Ware, Garrett of the Southern United States from 1600s to 1993.* Kennesaw, GA: F. W. Huff, 1993.

Hughes, Nathaniel Cheairs, Jr. *Brigadier General Tyree H. Bell, C. S. A.: Forrest's Fighting Lieutenant.* Knoxville: University of Tennessee Press, 2004.

———. *William J. Hardee: Old Reliable.* Baton Rouge: Louisiana State University Press, 1965.

Hurst, Jack. *Nathan Bedford Forrest: A Biography.* New York: Alfred A. Knopf, 1993.

Irby, H. C. "Ninth Tennessee Infantry." In Lindsley, *Military Annals of Tennessee*, 1:267–80.

Jacobson, Eric A. *For Cause & For Country: A Study of the Affair at Spring Hill and the Battle of Franklin.* Franklin, TN: O'More Publishing, 2008.

———. "Hood's Tennessee Campaign: From the Fall of Atlanta to the Battle of Franklin Sept. 2 to Nov. 30, 1864." *Blue & Gray* 30/4 (2014): 6–9, 19–26, 42–50.

———. *The McGavock Confederate Cemetery.* Nashville: Vaughan Printing, 2007.

Jarman, Robert A. "A Mississippian at Nashville." Edited by Richard M. McMurry. *Civil War Times Illustrated* 12/2 (May 1973): 8–11, 14–15.

Johnson, Robert Underwood and Clarence Clough Buel, eds., *Battles and Leaders of the Civil War*. 4 vols. New York: Thomas Yoseloff, 1956 [1888].

Johnson, W. C. "The March to the Sea." In Kerksis, *The Atlanta Papers*, 801–30.

———. *Narrative of Military Operations, Directed, during the Late War Between the States.* New York: D. Appleton and Company, 1874.

Johnston, William Preston. *The Life of Gen. Albert Sidney Johnston, Embracing His Services in the Armies of the United States, the Republic of Texas, and the Confederate States.* New York: Da Capo Press, 1997 [1879].

Jones, A. C. "Criticism of General Hood." *Confederate Veteran* 22/11 (November 1914): 507–508.

Jones, J. B. *A Rebel War Clerk's Diary at the Confederate States Capital.* 2 vols. Philadelphia: J. B. Lippincott & Co., 1866.

———. *A Rebel War Clerk's Diary At the Confederate States Capital.* Edited by James I. Robertson, Jr. 2 vols. Lawrence: University Press of Kansas, 2015.

Jones, John B. *A Rebel War Clerk's Diary.* Edited by Earl Schenck Miers. New York: Sagamore Press, 1958 [1866].

Jones, John Beauchamp. *A Rebel War Clerk's Diary at the Confederate States Capital.* Edited by Howard Swiggett. 2 vols. New York: Old Hickory Bookshop, 1935 [1866].

Jones, Terry L. "Harry Thompson Hays." In W. C. Davis, *The Confederate General*, 3:78–80.

Jordan, Thomas, and J. P. Pryor. *The Campaigns of Lieut.-Gen. N. B. Forrest and of Forrest's Cavalry.* Dayton, OH: Press of Morningside Bookshop, 1977 [1868].

Joslyn, Mauriel Phillips, ed. *Charlotte's Boys: Civil War Letters of the Branch Family of Savannah.* Gretna, LA: Pelican Publishing Company, 2010 [1996].

———. *A Meteor Shining Brightly: Essays on Maj. Gen. Patrick R. Cleburne.* Milledgeville, GA: Terrell House Publishing, 1997.

Kean, Robert G. H. *Inside the Confederate Government: The Diary of Robert Garlick Hill Kean.* Edited by Edward Younger. New York: Oxford University Press, 1957.

Keenan, Jerry. *Wilson's Cavalry Corps: Union Campaigns in the Western Theater, October 1864 through Spring 1865.* Jefferson, NC: McFarland, 1998.

Kennett, Lee. *Sherman: A Soldier's Life.* New York: HarperCollins, 2001.

Kerksis, Sydney C., comp. *The Atlanta Papers.* Dayton, OH: Press of Morningside Bookshop, 1980.

King, Alvy L. *Louis T. Wigfall: Southern Fire-Eater.* Baton Rouge: Louisiana State University Press, 1970.

Klinger, Michael J. "Gallant Charge Repulsed." *America's Civil War* 1/5 (January 1989): 26–33.
Knight, James R. *The Battle of Franklin: When the Devil Had Full Possession of the Earth.* Charleston: History Press, 2009.
———. *Hood's Tennessee Campaign: The Desperate Venture of a Desperate Man.* Charleston: History Press, 2014.
Kundahl, George G. *Alexandria Goes to War: Beyond Robert E. Lee.* Knoxville: University of Tennessee Press, 2004.
Kurtz, Wilbur G. "General Joseph E. Johnston's Review of Sherman's Memoirs, published in Drake's Annals of the Army of Tennessee, Nashville, April 1878." Unpublished manuscript (July 20, 1960), box 81, folder 3. Kurtz Collection. Atlanta History Center.
Lane, Bryan. "The Familiar Road: The Life of Confederate Brigadier General John Adams." *Civil War Times Illustrated* 35/5 (October 1996): 40–46.
Lane, Bryan W. *Tennessee Hero: Confederate Brigadier General John Adams.* Charleston: History Press, 2017.
———. *Where No Sorrows Come: The Life and Death of Confederate Brigadier General John Adams.* Spring Hill, TN: self-published, 2014.
Lee, S. D. "From Palmetto, Ga., to Defeat at Nashville." *Confederate Veteran* 16/6 (June 1908): 257–59.
Lee, Stephen D. Papers. #2440. Southern Historical Collection, Wilson Library, University of North Carolina at Chapel Hill.
Lewis, Lloyd. *Sherman Fighting Prophet.* New York: Harcourt, Brace & Co., 1932.
Liddell Hart, B. H. *Sherman Soldier Realist American.* New York: Dodd, Mead & Company, 1929.
Lindsay, R. H. "Capture of Florence, Ala., Under Hood." *Confederate Veteran* 4/12 (December 1896): 423.
———. "The Retreat from Nashville." *Confederate Veteran* 7/7 (July 1899): 311.
Lindsley, John Berrien, ed. *Military Annals of Tennessee.* 2 vols. Nashville: J. M. Lindsley & Co., 1886.
Livermore, Thomas L. *Numbers and Losses in the Civil War in America, 1861–1865.* New York: Kraus Reprint Co., 1969 [1900].
Logsdon, David R., ed. *Eyewitnesses at the Battle of Nashville.* Nashville: Kettle Mills Press, 2000.
Long, David E. *The Jewel of Liberty: Abraham Lincoln's Re-Election and the End of Slavery.* Mechanicsburg, PA: Stackpole Books, 1994.
Longacre, Edward G. *Cavalry of the Heartland: The Mounted Forces of the Army of Tennessee.* Yardley, PA: Westholme Publishing, 2009.
———. *From Union Stars to Top Hat: A Biography of the Extraordinary General James Harrison Wilson.* Harrisburg, PA: Stackpole Books, 1972.
Longstreet, James. *From Manassas to Appomattox: Memoirs of the Civil War in America.* Bloomington: Indiana University Press, 1960 [1896].
Losson, Christopher. *Tennessee's Forgotten Warriors: Frank Cheatham and His Confederate Division.* Knoxville: University of Tennessee Press, 1989.
"A Louisiana Branch of the Southern Historical Society." *Southern Historical Society Papers* 7/5 (May 1879): 253.

Lovett, Bobby. "Nashville's Fort Negley: A Symbol of Blacks' Involvement with the Union Army." In *Tennessee in the Civil War*, edited by Carroll Van West, 123–39. Nashville: Tennessee Historical Society, 2011.

Lowrey, M. P. "An Autobiography." *Southern Historical Society Papers* 16 (January–December 1888): 365–76.

Lowry, Don. *Dark and Cruel War: The Decisive Months of the Civil War September–December 1864.* New York: Hippocrene Books, 1993.

Lundberg, John R. "Errant Moves on the Chessboard of War: The Battle of Spring Hill, November 29, 1863." In Woodworth and Grear, *The Tennessee Campaign of 1864*, 46–65.

———. *The Finishing Stroke: Texans in the 1864 Tennessee Campaign.* Abilene: McWhiney Foundation Press, 2002.

———. *Granbury's Texas Brigade: Diehard Western Confederates.* Baton Rouge: Louisiana State University Press, 2012.

Mackowski, Chris, and Kristopher D. White. *Simply Murder: The Battle of Fredericksburg, December 13, 1862.* El Dorado Hills, CA: Savas Beatie, 2012.

Madaus, Howard Michael. *The Battle Flags of the Confederate Army of Tennessee.* Milwaukee: Milwaukee Public Museum, 1976.

Mahr, Theodore C. *The Battle of Cedar Creek: Showdown in the Shenandoah October 1–30, 1864.* Lynchburg, VA: H. E. Howard, 1992.

Maier, Larry B. *Gateway to Gettysburg: The Second Battle of Winchester.* Shippensburg, PA: Burd Street Press, 2002.

Manigault, Arthur M. *A Carolinian Goes to War: The Civil War Narrative of Arthur Middleton Manigault, Brigadier General, C. S. A.* Edited by R. Lockwood Tower. Columbia: University of South Carolina Press, 1983.

Marshall, Park. "Artillery in the Battle of Franklin." *Confederate Veteran* 23/3 (March 1915): 100–102.

———. "Shy's Hill—Whence the Name?" *Confederate Veteran* 20/11 (November 1912): 522–23.

Martin, Charles B. "Jackson's Brigade in Battle of Nashville." *Confederate Veteran* 17/1 (January 1909): 11–13.

Martin, Samuel J. *The Road to Glory: Confederate General Richard S. Ewell.* Indianapolis: Guild Press of Indiana, 1991.

Marvel, William. *Andersonville: The Last Depot.* Chapel Hill: University of North Carolina Press, 1994.

Mathes, J. Harvey. *General Forrest.* New York: D. Appleton and Company, 1902.

Matter, William D. *If It Takes All Summer: The Battle of Spotsylvania.* Chapel Hill: University of North Carolina Press, 1988.

Maury, Dabney H. "The Southern Historical Society: Its Origin and History." *Southern Historical Society Papers* 18 (January –December 1890): 349–65.

Maury, Dabney Herndon. *Recollections of a Virginian in the Mexican, Indian and Civil Wars.* New York: Charles Scribner's Sons, 1894.

McAlister, Bryan. "Forrest on Fort Pillow." In vol. 8 of *The Rebellion Record: A Diary of American Events*, edited by Frank Moore, 56. New York: Arno Press, 1977 [1861–1868].

McBride, Mary Gorton. *Randall Lee Gibson of Louisiana: Confederate General and New South Reformer.* Baton Rouge: Louisiana State University Press, 2007.

McCardle, Linda S., ed. *A Just and Holy Cause: The Civil War Letters of Marcus Bethune and Martha Frances Ely 1862–1865.* Macon: Mercer University Press, 2016.

McCarley, J. Britt. "'He Was So Well Provided for that He Could Sweep the World for Gain': The Supply of Sherman's Armies during the Atlanta Campaign, 1864." Master's thesis, Georgia State University, 1982.

McDonough, James L. "John McAllister Schofield." *Civil War Times Illustrated* 13/5 (August 1974): 10–17.

———. *Nashville: The Western Confederacy's Final Gamble.* Knoxville: University of Tennessee Press, 2004.

———. *Schofield: Union General in the Civil War and Reconstruction.* Tallahassee: Florida State University Press, 1972.

McDonough, James Lee. "The Battle of Franklin, Tennessee, November 30, 1864." *Blue & Gray* 2/1 (August-September 1984): 18–28, 33.

———. *Chattanooga: A Death Grip on the Confederacy.* Knoxville: University of Tennessee Press, 1984.

———. *Stones River: Bloody Winter in Tennessee.* Knoxville: University of Tennessee Press, 1980.

———, and Thomas L. Connelly. *Five Tragic Hours: The Battle of Franklin.* Knoxville: University of Tennessee Press, 1983.

McElroy, Robert. *Jefferson Davis: The Unreal and the Real.* New York: Kraus Reprint Co., 1969 [1937].

McFeely, William S. *Grant: A Biography.* New York: W. W. Norton & Company, 1981.

McInvale, Morton R. "Foredoomed to Oblivion." *America's Civil War* 61 (March 1977): 28–33.

McKinney, Francis F. *Education in Violence: The Life of George H. Thomas and the History of the Army of the Cumberland.* Chicago: Americana House, 1991 [1961].

McMahon. T. W. *Cause and Contrast: An Essay on the American Crisis.* Richmond: West & Johnston, 1862.

McMurray, W. J. "Causes of Failure at Spring Hill." *Confederate Veteran* 12/8 (August 1904): 395–96.

McMurry, Richard M. *Atlanta 1864: Last Chance for the Confederacy.* Lincoln: University of Nebraska Press, 2000.

———. "Confederate Morale in the Atlanta Campaign of 1864." *Georgia Historical Quarterly* 54/2 (Summer 1970): 226–43.

———. "Disappointment in History: The Papers of John Bell Hood." *Prologue: The Journal of the National Archives* 4 (1974): 161–64.

———. *John Bell Hood and the War for Southern Independence.* Lexington: University Press of Kentucky, 1982.

———. "The *Official Records* of the Rebellion." In *American Historians, 1866–1912*, edited by Clyde N. Wilson, 367–72. Detroit: Gale Research Company, 1986.

———. *Two Great Rebel Armies: An Essay in Confederate Military History.* Chapel Hill: University of North Carolina Press, 1989.

M'Neilly, J. H. "Bloody Franklin." *Confederate Veteran* 29/1 (January 1921): 5–6.

McNeilly, James H. "A Great Game of Strategy." *Confederate Veteran* 27/10 (October 1919): 377–84.

McPherson, James M. "American Victory, American Defeat." In *Why the Confederacy Lost*, edited by Gabor S. Boritt, 17–42, 166–68. New York: Oxford University Press, 1992.

———. *Battle Cry of Freedom: The Civil War Era.* New York: Oxford University Press, 1988.

———. *Embattled Rebel: Jefferson Davis as Commander in Chief.* New York: Penguin Press, 2014.

McWhirter, George I. "Soldier in the Western Army." *Confederate Veteran* 5/12 (December 1897): 623.

Melton, Brian C. *Sherman's Forgotten General: Henry W. Slocum.* Columbia: University of Missouri Press, 2007.

Message from the Executive of the Commonwealth, Transmitting a Report of Wm. H. Peters, Commissioner, Appointed by the Governor to Make an Inventory of Property Taken from the United States Government, at the Navy Yard, Gosport, and in and Near Portsmouth, Virginia (November 30, 1861, Doc. No. XL). Richmond, VA: Office of the Governor, 1861. Copy courtesy of Greg Biggs, Clarksville, TN.

Miller, Brian Craig. *John Bell Hood and the Fight for Civil War Memory.* Knoxville: University of Tennessee Press, 2010.

Mohr, Clarence L. *On the Threshold of Freedom: Masters and Slaves in Civil War Georgia.* Athens: University of Georgia Press, 1986.

Moody, Wesley. *Demon of the Lost Cause: Sherman and Civil War History.* Columbia: University of Missouri Press, 2011.

Morton, John Watson. *The Artillery of Nathan Bedford Forrest's Cavalry, "The Wizard of the Saddle."* Kennesaw, GA: Continental Book Company, 1962 [1909].

Moore, Christopher C. *Apostle of the Lost Cause: J. William Jones, Baptists, and the Development of Confederate Memory.* Knoxville: University of Tennessee Press, 2019.

Morton, M. B. "Battle of Nashville." *Confederate Veteran* 17/ 1 (January 1909): 17–21.

Muhlenfield, Elisabeth. *Mary Boykin Chesnut: A Biography.* Baton Rouge: Louisiana State University Press, 1981.

Murfin, James V. *The Gleam of Bayonets: The Battle of Antietam and the Maryland Campaign of 1862.* New York: Thomas Yoseloff, 1965.

Naisawald, L. VanLoan. *Grape and Canister: The Story of the Field Artillery of the Army of the Potomac, 1861–1865.* Mechanicsburg, PA: Stackpole Books, 1999 [1960].

Neff, Robert O. *Tennessee's Battered Brigadier (The Life of General Joseph B. Palmer).* Edited by Martha O. Hewes. Nashville, TN: Historic Travelers' Rest, 1988.

Nelson, Larry E. *Bullets, Ballots, and Rhetoric: Confederate Policy for the United States Presidential Contest of 1864.* Tuscaloosa: University of Alabama Press, 1980.

Newton, Steven H. *Lost for the Cause: The Confederate Army in 1864.* Mason City, IA: Savas Publishing Company, 2000.

Norris, Herbert S., and James R. Long. "The Road to Redemption." *Civil War Times Illustrated* 36/4 (August 1997): 32–38, 55–57.

O'Connor, Richard. *Hood: Cavalier General.* New York: Prentice-Hall, 1949.

"The Opposing Forces in the Operations at New Orleans, La." In Johnson and Buel, *Battles and Leaders of the Civil War*, 2:73–75.

"Origin and History of our Society." *Southern Historical Society Papers* 6/6 (December 1878): 244–45.

Osborne, Charles C. *Jubal: The Life and Times of General Jubal A. Early, C S A, Defender of the Cause.* Chapel Hill: Algonquin Books of Chapel Hill, 1992.
Owen, Richard and James Owen. *Generals at Rest: The Grave Sites of the 425 Official Confederate Generals.* Shippensburg, PA: White Mane Publishing Co., 1997.
Parker, William H. "The Gold and Silver in the Confederate States Treasury." *Southern Historical Society Papers* 21 (January–December 1893): 304–13.
Parrish, T. Michael. "Richard Taylor." In W. C. Davis, *The Confederate General*, 6:29–31.
———. *Richard Taylor: Soldier Prince of Dixie.* Chapel Hill: University of North Carolina Press, 1992.
Patrick, Rembert W. *The Reconstruction of the Nation.* New York: Oxford University Press, 1967.
Patrick, Robert. *Reluctant Rebel: The Secret Diary of Robert Patrick 1861–1865.* Edited by F. Jay Taylor. Baton Rouge: Louisiana State University Press, 1959.
Patterson, D. H. "Battle of Franklin." *Confederate Veteran* 9/3 (March 1901): 116–117.
Peddy, George W. *Saddle Bag and Spinning Wheel: Being the Civil War Letters of George W. Peddy, MD Surgeon, 56th Georgia Volunteer Regiment, C. S. A.* Edited by George W. Cuttino. Macon: Mercer University Press, 1981.
Pennington, Estill Curtis. *The Last Meeting's Lost Cause.* Spartanburg, SC: Robert M. Hicklin, Jr., Inc., 1988.
Perman, Michael. *Reunion without Compromise: The South and Reconstruction 1865–1868.* New York: Cambridge University Press, 1973.
Perret, Geoffrey. *Lincoln's War: The Untold Story of America's Greatest President as Commander in Chief.* New York: Random House, 2004.
Pfanz, Donald C. "Richard Stoddert Ewell." In W. C. Davis, *The Confederate General*, 2:111–12.
Phillips, Brenda D., ed. *Personal Reminiscences of a Confederate Soldier Boy: Robert M. Magill, Co. F, 39th GA Regt. Inf.* Milledgeville, GA: Boyd Publishing Co., 1993.
Phillips, G. C. "Witness to the Battle of Franklin." *Confederate Veteran* 14/6 (June 1906): 261–62.
Phillips, Jason. *Diehard Rebels: The Confederate Culture of Invincibility.* Athens: University of Georgia Press, 2007.
Piston, William Garrett. *Lee's Tarnished Lieutenant: James Longstreet and His Place in Southern History.* Athens: University of Georgia Press, 1987.
Pollard, E. A. *Southern History of the War.* 2 vols. New York: Charles B. Richardson, 1866.
Pollard, Edward A. *The Lost Cause; A New Southern History of the War of the Confederates.* New York: E. B. Treat & Co., 1867.
Porter, Horace. *Campaigning with Grant.* Bloomington: Indiana University Press, 1961 [1897].
Porter, James D. "Spring Hill and Battle of Franklin." *Confederate Veteran* 12/7 (July 1904): 341–48.
———. *Tennessee.* Volume 8 of Evans, ed., *Confederate Military History.* New York: Thomas Yoseloff, 1962 [1899].
Povall, Allie Stuart. *Rebels in Repose: Confederate Commanders after the War.* Charleston: History Press, 2019.
Purdue, Howell and Elizabeth Purdue. *Pat Cleburne Confederate General.* Gaithersburg, MD: Olde Soldier Books, 1987 [1973].

Quintard, Charles Todd. *Doctor Quintard, Chaplain C. S. A. and Second Bishop of Tennessee: The Memoir and Civil War Diary of Charles Todd Quintard.* Edited by Sam Davis Elliott. Baton Rouge: Louisiana State University Press, 2003.

Rable, George C. *God's Almost Chosen Peoples: A Religious History of the American Civil War.* Chapel Hill: University of North Carolina Press, 2010.

Rees, W. H. "Cleburne's Men at Franklin." *Confederate Veteran* 9/2 (February 1901): 54.

Remington, J. D. "Cause of Hood's Failure at Spring Hill." *Confederate Veteran* 21/12 (December 1913): 569–71.

Richardson, James D., comp. *A Compilation of the Messages and Papers of the Confederacy Including the Diplomatic Correspondence 1861–1865.* 2 vols. Nashville: United States Publishing Company, 1906.

Ridley, B. L. "The Last Night of Sixty-Four." *Confederate Veteran* 8/12 (December 1900): 539.

Riley, Harris D., Jr. "A Gallant Adopted Son of Tennessee—General John C. Carter, C. S. A." *Tennessee Historical Quarterly* 48/4 (Winter 1989): 195–208.

Ritter, William L. "Sketch of the Third Maryland Artillery: Retreat from Nashville." *Southern Historical Society Papers* 12/4 (April 1884): 170–72.

Robertson, James I., Jr. *Civil War Virginia: Battleground for a Nation.* Charlottesville: University of Press of Virginia, 1991.

———. *Stonewall Jackson: The Man, the Soldier, the Legend.* New York: Macmillan Publishing USA, 1997.

Robison, Dan M. "The Carter House: Focus of the Battle of Franklin." *Tennessee Historical Quarterly* 22/1 (March 1963): 3–21.

Roland, Charles P. *An American Iliad: The Story of the Civil War.* Lexington: University Press of Kentucky, 2004 [1991].

———. *Albert Sidney Johnston: Soldier of Three Republics.* Austin: University of Texas Press, 1964.

Rollins, Richard. "The Failure of Confederate Artillery at Gettysburg: Ordnance and Logistics." *North & South* 3/2 (January 2000): 44–54.

———. "The Failure of Confederate Artillery in Pickett's Charge." *North & South* 3/4 (April 2000): 26–42.

———, ed. *The Returned Battle Flags.* Redondo Beach, CA: Rank and File Publications, 1995 [1905].

Roman, Alfred. *The Military Operations of General Beauregard.* 2 vols. New York: Da Capo Press, 1994 [1884].

Rosen, Robert N. *The Jewish Confederates.* Columbia: University of South Carolina Press, 2000.

Roth, David E. "That Cruel Hole in His Brow: The Death of Col. William Shy." *Blue & Gray* 11/2 (December 1993): 49.

———. "The Mysteries of Spring Hill, Tennessee." *Blue & Gray* 2/2 (October–November 1984): 12–28, 33.

Rottman, Gordon L. *The Great Locomotive Chase: The Andrews Raid 1862.* New York: Osprey Publishing, 2009.

Rowland, Dunbar, ed. *Jefferson Davis Constitutionalist: His Letters, Papers and Speeches.* 10 vols. Jackson: Mississippi Department of Archives and History, 1923.

Roy, T. B. "General Hardee and the Military Operations around Atlanta." *Southern Historical Society Papers* 8/8–9 (August–September 1880): 337–87.

Rubin, Anne Sarah. *Through the Heart of Dixie: Sherman's March and American Memory.* Chapel Hill: University of North Carolina Press, 2014.

Ruffin, Edmund. *The Diary of Edmund Ruffin.* Edited by William Kauffman Scarborough. 3 vols. Baton Rouge: Louisiana State University Press, 1972–1989.

Sanders, D. W. "Autobiography of Maj. D. W. Sanders." *Confederate Veteran* 18/8 (August 1910): 370–72.

———. "Hood's Tennessee Campaign." *Confederate Veteran* 15/9 (September 1907): 401–404.

———. "Hood's Tennessee Campaign. Chapter II." *Southern Bivouac* 3/5 (January 1885): 193–203.

———. "Hood's Tennessee Campaign. Chapter IV." *Southern Bivouac* 3/6 (February 1885): 241–52.

———. "Hood's Tennessee Campaign. Chapter V." *Southern Bivouac* 3/7 (March 1885): 289–94.

———. "Hood's Tennessee Campaign. Chapter VI." *Southern Bivouac* 3/8 (April 1885): 350–66.

———. "Strange Order of Hood at Allatoona." *Confederate Veteran* 9/11 (November 1901): 507.

Sanger, Donald Bridgman, and Thomas Robson Hay. *James Longstreet: Soldier, Politician, Officeholder, and Writer.* Baton Rouge: Louisiana State University Press, 1952.

Sayers, Alethea D. "The Last Campaign." In Joslyn, *A Meteor Shining Brightly*, 233–64.

Scaife, William R. *Allatoona Pass: A Needless Effusion of Blood.* Cartersville, GA: Etowah Valley Historical Society, 1995.

———. *Hood's Campaign for Tennessee: March to Certain Doom.* Kennesaw, GA: Kennesaw Mountain Historical Association, 1986.

——— and William Harris Bragg. *Joe Brown's Pets: The Georgia Militia 1861–1865.* Macon: Mercer University Press, 2004.

Scharf, J. Thomas. *History of the Confederate States Navy.* New York: Fairfax Press, 1977 [1887].

Schiller, Herbert M. *Sumter is Avenged! The Siege and Reduction of Fort Pulaski.* Shippensburg, PA: White Mane Publishing Company, 1995.

Schloemer, Christopher N. "General John Bell Hood: His Leadership during the 1864 Tennessee Campaign." *Saber and Scroll* 5/2 (September 2016): 23–36.

Schofield, John M. *Forty-Six Years in the Army.* New York: Century Co., 1897.

Scofield, Levi T. *The Retreat from Pulaski to Nashville, Tenn.* Cleveland, OH: Press of the Caxton Co., 1909.

Seabrook, Lochlainn. *Encyclopedia of the Battle of Franklin: A Comprehensive Guide to the Conflict that Changed the Civil War.* Franklin, TN: Sea Raven Press, 2012.

Sears, Stephen W. *Gettysburg.* Boston: Houghton Mifflin Company, 2003.

———. *Landscape Turned Red: The Battle of Antietam.* New Haven: Ticknor & Fields, 1983.

———. "McClellan and the Peace Plank of 1864: A Reappraisal." *Civil War History* 36/1 (March 1990): 57–64.

———. *To the Gates of Richmond: The Peninsula Campaign.* New York: Ticknor & Fields, 1992.

Seitz, Don C. *Braxton Bragg: General of the Confederacy.* Columbia, SC: The State Company, 1924.

Shands, G. D. "Dr. Lea Williamson." *Confederate Veteran* 8/8. (August 1900): 370.
Shapard, E. "At Spring Hill and Franklin Again." *Confederate Veteran* 24/3 (March 1916): 138–39.
Shellenbarger, John K. "The Fighting at Spring Hill, Tennessee." *Confederate Veteran* 36/3 (March 1928): 100–103.
Sherman, William T. "The Grand Strategy of the Last Year of the War." In Johnson and Buel, *Battles and Leaders of the Civil War*, 4:247–59.
———. *Memoirs of General William T. Sherman. By Himself.* 2 vols. New York: D. Appleton & Co., 1875.
Shotwell, R. H. "Hood's Campaign through North Georgia." *Confederate Veteran* 1/10 (October 1893): 311–12.
Shoup, F. A. "Dalton Campaign—Works at Chattahoochee River—Interesting History." *Confederate Veteran* 3/9 (September 1895): 262–65.
Silver, James W. *Confederate Morale and Church Propaganda.* New York: W. W. Norton & Company, 1967 [1957].
Simmons, J. W. "Pathetic Burial at Midnight." *Confederate Veteran* 7/8 (August 1899): 358–59.
Simpson, Brooks D. "A Failure to Communicate: Grant, Thomas, and the Nashville Campaign." In Woodworth and Grear, *The Tennessee Campaign of 1864*, 105–22.
———, and Jean D. Berlin, eds. *Sherman's Civil War: Selected Correspondence of William T. Sherman, 1860–1865.* Chapel Hill: University of North Carolina Press, 1999.
Simpson, Harold B. *Hood's Texas Brigade in Reunion and Memory.* Hillsboro, TX: Hill College Press, 1974.
Simpson, John A. *S. A. Cunningham and the Confederate Heritage.* Athens: University of Georgia Press, 1994.
Smith, Derek. *In the Lion's Mouth: Hood's Tragic Retreat from Nashville, 1864.* Mechanicsburg, PA: Stackpole books, 2011.
Smith, Ed W. "Douglas's Battery in the Battle of Franklin." *Confederate Veteran* 12/11 (November 1904): 531–32.
Smith, Michael Thomas. *The 1864 Franklin-Nashville Campaign.* Santa Barbara, CA: Praeger, 2014.
Smith, R. D. "Record of Walthall's Division at Franklin and Nashville Late in 1864." *Confederate Veteran* 7/6 (June 1899): 265.
Smith, Tress T. Recollections. Typescript courtesy of Dr. Dent Purcell, Savannah, GA.
Snow, William P. *Lee and His Generals.* New York: Fairfax Press, 1982 [1867].
"The Southern Historical Society." *Southern Historical Society Papers* 1/1 (January 1876): 39–45.
Speer, Lonnie R. *Portals to Hell: Military Prisons of the Civil War.* Mechanicsburg, PA: Stackpole Books, 1997.
Stephenson, Philip D. *The Civil War Memoir of Philip Daingerfield Stephenson, D. D.* Edited by Nathaniel Cheairs Hughes, Jr. Conway: University of Central Arkansas Press, 1995.
Steplyk, Jonathan M. "Killing at Franklin: Anatomy of Slaughter." In Woodworth and Grear, *The Tennessee Campaign of 1864*, 81–104.
Stevenson, Carter L. "Report of Major-General Carter L. Stevenson of the Tennessee Campaign." *Southern Historical Society Papers* 3/4 (April 1877): 161–67.

Stewart, Alex. P. "The Army of Tennessee: A Sketch." In Lindsley, *Military Annals of Tennessee*, 1:55–111.

———. "'A Critical Narrative.'" *Confederate Veteran* 16/9 (September 1908): 462–64.

Stewart, Bruce H., Jr. *Invisible Hero: Patrick R. Cleburne.* Macon: Mercer University Press, 2009.

Stewart, George R. *Pickett's Charge: A Microhistory of the Final Attack at Gettysburg, July 3, 1863.* Dayton, OH: Press of Morningside Bookshop, 1980 [1959].

Stickles, Arndt M. *Simon Bolivar Buckner: Borderland Knight.* Chapel Hill: University of North Carolina Press, 1940.

Stockdale, Paul H. *The Death of an Army: The Battle of Nashville and Hood's Retreat.* Murfreesboro, TN: Southern Heritage Press, 1992.

Stoker, Donald. *The Grand Design: Strategy in the Civil War.* New York: Oxford University Press, 2010.

Stone, Henry. "Repelling Hood's Invasion of Tennessee." In Johnson and Buel, *Battles and Leaders of the Civil War*, 4:440–64.

Stout, Harry S. *Upon the Altar of the Nation: A Moral History of the American Civil War.* New York: Viking, 2006.

Strode, Hudson. *Jefferson Davis Tragic Hero 1864–1889.* New York: Harcourt, Brace & World,

Sutherland, Daniel E. "No Better Officer in the Confederacy: The Wartime Career of Daniel C. Govan." In Hewitt and Bergeron, *Confederate Generals in the Western Theater*, 1:229–55.

Swanberg, W. A. *First Blood: The Story of Fort Sumter.* New York: Charles Scribner's Sons, 1957.

Sword, Wiley. "The Battle of Nashville." *Blue & Gray* 11/2 (December 1993): 12–16, 18–20, 22–23, 38–39, 42–48, 50.

———. *Embrace an Angry Wind—The Confederacy's Last Hurrah: Spring Hill, Franklin and Nashville.* New York: HarperCollins, 1992.

Symonds, Craig L. *Joseph E. Johnston: A Civil War Biography.* New York: W. W. Norton & Company, 1992.

———. *Stonewall of the West: Patrick Cleburne and the Civil War.* Lawrence: University Press of Kansas, 1997.

Taylor, John M. "Twenty-seventh Tennessee Infantry." In Lindsley, *Military Annals of Tennessee*, 1:415–428.

Taylor, Richard. *Destruction and Reconstruction: Personal Experiences in the Late War.* Edited by Richard B. Harwell. New York: Longmans, Green and Co., 1955 [1879].

———. "The Last Confederate Surrender." *Southern Historical Society Papers* 3/3 (March 1877): 155–58.

Thomas, Emory M. *The Confederate Nation: 1861–1865.* New York: Harper & Row, 1979.

Thomas, Wilbur. *General George H. Thomas: The Indomitable Warrior.* New York: Exposition Press, 1964.

Thompson, William C. "A Confederate's Recollection of the Battle of Franklin." *Civil War Times Illustrated* 3/8 (December 1964): 24.

Trask, W. L. War Journal. Typescript at Kennesaw Mountain National Battlefield Park (courtesy of Dennis Kelly): Microfilm copy of handwritten original at Emory University, MARBL.

Trudeau, Noah Andre. *Gettysburg: A Testing of Courage*. New York: HarperCollins, 2002.
———. *Like Men of War: Black Troops in the Civil War 1862–1865*. New York: Little, Brown and Company, 1998.
Tsouras, Peter G., ed. *The Greenhill Dictionary of Military Quotations*. London: Greenhill Books, 2000.
Tucker, Leslie R. *Brigadier General John Adams CSA: A Biography*. Jefferson, NC: McFarland, 2014.
Tucker, Phillip Thomas. *The Forgotten "Stonewall Jackson of the West": Major General John Stevens Bowen*. Macon: Mercer University Press, 1997.
———. *Irish Confederates: The Civil War's Forgotten Soldiers*. Abilene, TX: McWhiney Foundation Press, 2006.
———. *Pickett's Charge: A New Look at Gettysburg's Final Attack*. New York: Skyhorse Publishing, 2016.
Tyler, H. A. "Forrest Covers Hood's Retreat." *Confederate Veteran* 12/9 (September 1904): 436.
———. Untitled reminiscence. *Confederate Veteran* 22/1 (January 1914): 15.
U.S. War Department. *The War of the Rebellion: A Compilation of the Official Records of the Union and Confederate Armies*. 128 volumes. Washington: Government Printing Office, 1880–1901. Cited as *OR*. Please note that citations refer to Series 1 unless otherwise indicated.
Walker, Hugh F. "Bloody Franklin." *Civil War Times Illustrated* 3/8 (December 1964): 16–25.
Ward, Clyde. "Twisting Confederate Iron." *Civil War Times Illustrated* 25/8 (December 1986): 24–27.
Warner, Ezra J. *Generals in Gray: Lives of the Confederate Commanders*. Baton Rouge: Louisiana State University Press, 1959.
———, and W. Buck Yearns. *Biographical Register of the Confederate Congress*. Baton Rouge: University of Louisiana State University Press, 1975.
Warren, Daniel C. "John Bell Hood: A Bid for Fame." Master's thesis, U.S. Army Command and General Staff College, 1974.
Washburn, W. A. "Cleburne's Division at Franklin." *Confederate Veteran* 13/1 (January 1905): 27–28.
Watkins, S. R. "Reminiscences of Hood's Tennessee Campaign." *Southern Bivouac* 2/9 (May 1884): 399–402.
Watkins, Sam R. *Co. "Aytch" Maury Grays, First Tennessee Regiment or, A Side Show of the Big Show*. Edited by Ruth Hill Fulton McAllister. Franklin, TN: Providence House Publishers. 2007 [1882].
Waugh, John C. *Reelecting Lincoln: The Battle for the 1864 Presidency*. New York: Crown Publishers, 1997.
Webb, Ashley. "The Confederate Dead at Franklin." Appendix B in *The Aftermath of Battle: The Burial of the Civil War Dead*, by Meg Groeling, 155–157. El Dorado Hills, CA: Savas Beatie, 2015.
Weigley, Russell F. *A Great Civil War: A Military and Political History, 1861–1865*. Bloomington: Indiana University Press, 2000.
Welcher, Frank J. *The Union Army, 1861–1865: Organization and Operations*. 2 vols. Bloomington: Indiana University Press, 1989.

Welsh, Jack D. *Medical Histories of Confederate Generals.* Kent, OH: Kent State University Press, 1995.

Wert, Jeffry D. *Gettysburg: Day Three.* New York: Simon & Schuster, 2001.

Wesley, Charles H., and Patricia W. Romero. *Negro Americans in the Civil War: From Slavery to Citizenship.* New York: Publishers Company, 1967.

West, Richard S. *Mr. Lincoln's Navy.* New York: Longmans, Green and Company, 1957.

"What Will Become of the Children [?:] General Hood's Family Faces an Uncertain Future." *Civil War Times Illustrated* 25/9 (January 1987): 36–37.

White, William Lee. *Let Us Die Like Free Men: The Battle of Franklin, November 30, 1864.* El Dorado Hills CA: Savas Beatie, 2019.

_____, ed. "The Long Lost Diary of Major General Patrick R. Cleburne." In Woodworth and Grear, *The Tennessee Campaign of 1864*, 7–22.

Wigginton, Thomas A. "Cavalry Operations." *Civil War Times Illustrated* 3/8 (December 1964): 40–43.

Wiley, Bell I. "Jefferson Davis, an Appraisal." *Civil War Times Illustrated* 6/1 (April 1967): 4–11, 44–49.

Williams, Hiram S. *This War So Horrible: The Civil War Diary of Hiram Smith Williams.* Edited by Lewis N. Wynne and Robert A. Taylor. Tuscaloosa: University of Alabama Press, 1993.

Williams, T. Harry. *P. G. T. Beauregard: Napoleon in Gray.* Baton Rouge: Louisiana State University Press, 1955.

Wills, Brian Steel. *A Battle from the Start: The Life of Nathan Bedford Forrest.* New York: HarperCollins, 1992.

———. *George Henry Thomas: As True as Steel.* Lawrence: University Press of Kansas, 2012.

Wilson, Charles Reagan. *Baptized in Blood: The Religion of the Lost Cause, 1865–1920.* Athens: University of Georgia Press, 1980.

Winschel, Terrence J. *Triumph & Defeat: The Vicksburg Campaign.* Mason City, IA: Savas Publishing, 1999.

Wood, W. J. *Civil War Generalship: The Art of Command.* Westport, CT: Praeger, 1997.

Woodward, C. Vann, ed. *Mary Chesnut's Civil War.* New Haven, CT: Yale University Press, 1981.

Woodworth, Steven E. *Jefferson Davis and His Generals: The Failure of Confederate Command in the West.* Lawrence: University Press of Kansas, 1990.

———, and Charles D. Grear, eds. *The Tennessee Campaign of 1864.* Carbondale: Southern Illinois University Press, 2016.

Wooster, Ralph A. *Texas and Texans in the Civil War.* Austin: Eakin Press, 1995.

Wright, Mrs. D. Giraud [Louise Wigfall]. *A Southern Girl in '61: The War-Time Memories of a Confederate Senator's Daughter.* New York: Doubleday, Page & Company, 1905.

Wright, J. W. A. "Hood's Nashville Campaign." In *The New Annals of the Civil War*, edited by Peter Cozzens and Robert I. Girardi, 468–87. Mechanicsburg, PA: Stackpole Books, 2004.

Wyeth, John A. *Life of General Nathan Bedford Forrest.* Dayton, OH: Press of Morningside Bookshop, 1975 [1899].

Young, J. P. "Hood's Failure at Spring Hill." *Confederate Veteran* 16/1 (January 1908): 25–41.

Zimmerman, Mark. *Guide to Civil War Nashville.* Nashville: Battle of Nashville Preservation Society, 2004.

Index